Lecture Notes in Physics

Edited by J. Ehlers, München, K. Hepp, Zürich, and
H. A. Weidenmüller, Heidelberg
Managing Editor: W. Beiglböck, Heidelberg

22

Proceedings of the Europhysics Study Conference on Intermediate Processes in Nuclear Reactions

August 31 – September 5, 1972
Plitvice Lakes, Yugoslavia

Edited by Nikola Cindro, Petar Kulišić, and Theo Mayer-Kuckuk

Springer-Verlag Berlin Heidelberg GmbH

The Conference was sponsored by
    The Council of Scientific Research of the
    S.R. of Croatia
    The Nuclear Physics Division of the Euro-
    pean Physical Society
    The Stiftung Volkswagenwerk

The following institutions have participated in
organizing the Conference:
    The Union of Societies of Mathemeticians,
    Physicists and Astronomers of Yugoslavia
    The Institute "Rudjer Bošković", Zagreb
    The Physics Department, Faculty of
    Electrical Engineering, Zagreb
    The Institute of Nuclear Physics, Bonn.

Secretary of the Conference
    Petar Kulišić, Institute "Rudjer Bošković"
    and the University, Zagreb

ISBN 978-3-540-06526-5   ISBN 978-3-540-37824-2 (eBook)
DOI 10.1007/978-3-540-37824-2

© Springer-Verlag Berlin Heidelberg 1973. Library of Congress Catalog Card Number 73-16618.
Originally published by Springer-Verlag Berlin Heidelberg New York in 1973

Offsetprinting and bookbinding: Julius Beltz, Hemsbach/Bergstr.

PREFACE

This Conference was devoted to the study of intermediate processes in nuclear reactions. Included in this somewhat vague term today
are reactions that can not be interpreted in terms of either the direct
reaction model or the statistical model of nuclear reactions. They give
rise to a particular energy dependence of the average cross section;
a dependence characterized by an intermediate width of typically several hundred keV. In this sense "intermediate structure" in the cross
sections can be understood as a deviation from the statistical model,
localized in energy.

Progress in understanding the dynamics of reaction processes has
led to the idea that simple model of excitation may be able to reproduce this characteristic energy dependence (i.e. increased particle
widths) just as the dipole state produces a localized enhancement of
the radiative width. Intermediate reactions would be then one of several line broadenings observed in nuclear physics.

The next step in understanding these processes was to identify
the simple configurations associated with intermediate structure in nuclear reactions. The discovery of intermediate analogue resonances
showed that nuclear models can be useful even at the highest excitation
energies. This discovery prompted in some way the marriage of simple
nuclear structure models, in particular the shell model, to nuclear reaction theory. The first successful result of this marriage was the
concept of doorway states, i.e. simple states strongly coupled to the
entrance channel. However, the concept of simplicity of a nuclear configuration is very much model dependent. A single particle state is
simple in a shell model representation but fairly complicated in a pho-
non representation and vice versa. This implies that the nature and the
number of doorway states will be model dependent. In spite of this difficulty the doorway state approach has the great advantage that it enables the intermediate resonances to be treated as any other simple resonances, provided that we add to the total width a term called the
spreading width $\Gamma^{\downarrow}$. The spreading width accounts for the fact that the
doorway state is not an eigenvalue of the nuclear hamiltonian and that
it may dissolve into more complex configurations. Thus the probability
of finding a nucleus in a doorway state decreases with time.

The following questions then arise: Is the physics of intermediate processes in nuclear reactions interesting? Is it a general phenomenon or does it occur only in very special cases? If the former is
true, why it is not seen more often? In fact, only three classes of nuclear reactions can be clearly and unambiguosly classified as intermediate resonances:
- the isobaric analogue resonances
- the giant dipole resonance and
- the resonances in neutron induced fission.
What about the large variety of other nuclear processes? These problems
find - or do not find - answers in the ten papers that comprise the
Proceedings, summarized in the contribution by H. Feshbach.

*

The Proceedings start, appropriately, with the survey of C. Mahaux on the present state of intermediate reaction theories and experiments. Mahaux's approach is model dependent, and, for many and obvious reasons, is based on the shell model, which has been so success-
fully employed in nuclear structure calculations.

The two following papers on the preequilibrium emission of particles (by M. Blann and by E. Gadioli and L. Milazzo-Colli) are in a
sense the application of these concepts. They answer the question of
what happened on the way to equilibrium? While the exciton model (Griffin, 1966) used by the Milano group has the physical transparency of
the early statistical models, the hybrid model of Blann allows also the
calculation of absolute cross sections by introducing a microscopic
description of the collision process.

A large group of papers (D. Sperber, W. Scheid et al.,R. Stokstadt et al.and M. Petrascu) review the occurrence of intermediate states in different nuclear processes.

D. Sperber calculates the neutron evaporation and prompt fission
in a rigorous way with no apparent adjustable parameters. The intermediate structure in these cross sections is reproduced by modifying the
statistical treatment.

Two papers (W. Scheid et al., R. Stokstadt et al.) relate the hot
topic of intermediate structure in heavy ion reactions. The concept of
simple configuration in this case is related to the spatial distribution of two large, well separated groups of nucleons. The quasi-molecular
and/or alpha particle configurations are introduced as possible doorway states. Paradoxically, the problem in using such concepts to treat
intermediate structure in heavy ion reactions is not to explain how and
why resonances stick out in a region where the density of compound nu-

cleus levels reaches 10^4/ MeV, but why it does not happen in all or most
of the cases.

Although isobaric analogue resonances are considered intermediate
resonances par excellence, this subject has not been treated explicitly
in the Conference. The article by M. Petrascu relates the topics relevant, at the present stage of knowledge, to the subject of intermediate
analogue resonances as intermediate structure. Similarly, F. Cvelbar
introduces fast neutron radiative capture as intermediate processes
governed by a direct-semidirect mechanism; the gross structure and
the energy spectra arise from the coupling of the incident particle to
the collective states of the target nucleus. The contribution of P.
Brentano deals with doorway states as poles of the average S-matrix.
The treatment is slightly more restrictive then the usual one, since it
requires additional limitations on the spreading width, $\Gamma^{\downarrow}$.

Finally, the subject of simple structure in the exit channels is
treated by L. Papineau. Here the simple structure is produced by selection rules other those related to isospin conservation.

*

A conference is always both a scientific and organizational en-
deavour. The Europhysics Study Conferences were modelled having in mind
the Gordon Research Conferences, so popular in the U.S. This has meant
devoting more time to invited review papers than to short communica-
tions and also providing both a place and time for the participants to
interact informally. We felt, thus, that the location of the first
Europhysics Study Conference in the field of nuclear physics in the
marvelous setting of the Plitvice Lakes was quite appropriate. In another sense, this Conference continued the heritage of the Adriatic
Summer Meetings in Physics, one of the earliest regular international
physics meetings in Europe.

The Conference was made possible by the financial support of the
Council for Scientific Research of the Socialist Republic of Croatia,
the Volkswagen Foundation and the Institute "Rudjer Bošković"; we
acknowledge here our indebtedness to these organizations. Thanks are
also due to the Institutions which contributed organizational support
to the Conference: the Union of Physicists, Mathematicians and Astro-
nomers of Yugoslavia, the Physics Department, Faculty of Electrical
Engineering, Zagreb and the Institute of Nuclear Physics of the University, Bonn.

The manuscript was typed and prepared by Miss Božena Zubić, whose
efforts and skill are kindly acknowledged.

Zagreb and Bonn, June 1973   The Editors

Contents

List of Participants

1. ALBRECHT, R. — Max Planck Institut für Kernphysik, Heidelberg, Germany
2. BISPLINGHOFF, J. — Institut für Strahlen und Kernphysik der Universität Bonn, Bonn, Germany
3. BLANN, M. — Nuclear Structure Laboratory, Univ. of Rochester, Rochester, USA
4. BLEULER, K. — Institut für Theoretische Kernphysik der Universität Bonn, Bonn, Germany
5. BLINOWSKA, K. — Institute of Experimental Physics, Warsaw, Poland
6. BOHLEN, H.G. — Max Planck Institut für Kernphysik, Heidelberg, Germany
7. BOHNE, W. — Hahn-Meitner Institut, Berlin, Germany
8. BONDORF, J. — Niels Bohr Institute, Copenhagen, Denmark
9. BORMANN, M. — I. Institut für Experimentalphysik, Hamburg, Germany
10. BRENTANO, P. — Institut für Kernphysik, Universität zu Köln, Köln, Germany
11. BRZOSKO, J. — Joint Insitutute of Nuclear Research, Dubna, USSR
12. BUCK, W. — Physikalisches Institut der Universität Tübingen, Germany
13. CHARLES, P. — C E N Saclay, France
14. CHEVARIER, A. — Institut de Physique Nucléaire, Villeurbanne, France
15. CHEVARIER, N. — Institut de Physique Nucléaire, Villeurbanne, France
16. CINDRO, N. — Institute "R. Bošković", Zagreb, Yugoslavia
17. CVELBAR, F. — Institute "J. Stefan", Ljubljana, Yugoslavia
18. CUNHA, J.D. — Laboratorio de Fisica e Engenharia Nucleares, Sacavém, Portugal

19. ČAPLAR, R.	Institute "R. Bošković", Zagreb, Yugoslavia
20. DAVIDSON, W.F.	Kernforschungsanlage Jülich, Germany
21. DEMEYER, A.	Institut de Physique Nucléaire, Villeurbanne, France
22. DERRIEN, H.	C.E.N. de Saclay, France
23. DIEHL, H.	Institut für Theoretische Physik der Universität, Frankfurt/M, Germany
24. DRENTJE, A.G.	Kernphysisch Versneller Instituut, Groningen, The Netherlands
25. ERNST, J.	Institut für Strahlen und Kernphysik der Universität, Bonn, Germany
26. FESHBACH, H.	Massachusetts Institute of Technology, Cambridge, USA
27. GADIOLI, E.	Istituto di Fisica dell'Universita, Milano, Italy
28. GRUHLE, W.	Institut für Kernphysik der Universität Köln, Germany
29. HATEGAN, C.	Institute for Atomic Physics Bucharest, Romania
30. HOLUB, E.	Institute "R. Bošković", Zagreb, Yugoslavia
31. IORI, I.	Istituto di Fisica dell'Universita, Milano, Italy
32. JEAN, M.	Institut de Physique Nucléaire Orsay, France
33. KRETSCHMER, W.	Physikalisches Institut der Universität Erlangen, Germany
34. KULIŠIĆ, P.	Institute "R. Bošković", Zagreb, Yugoslavia
35. KUZMINSKI, J.	Institute of Physics, Silesian University, Katowice, Poland
36. LALOVIĆ, B.	Institute "B. Kidrič",Belgrade, Yugoslavia
37. LIU, Q.	Hahn Meitner Institut Berlin, Germany
38. LOVAS, I.	Central Research Institute for Physics, Budapest, Hungary
39. MAHAUX, C.	Université de Liege, Belgium
40. MARIĆ, Z.	Institute of Physics, University of Belgrade, Yugoslavia

41. MAYER-BÖRICKE, C.   Institut für Kernphysik der KFA,
Jülich, Germany

42. MAYER-KUCKUK, T.   Institut für Strahlen und Kernphysik der Universität Bonn, Germany

43. MITTIG, W.   C.E.N. de Saclay, France

44. MORGENSTERN, H.   Hahn-Meitner Institut Berlin, Germany

45. MYSLEK, B.   Institute for Nuclear Research,
Swierk near Warsaw, Poland

46. NASH, G.F.   Institute "R. Bošković", Zagreb,
Yugoslavia

47. NEWSTEAD, C.   Kernforschungszentrum Karlsruhe,
Germany

48. OBLOŽINSKÝ, P.   Institute of Physics, Slovak Academy
of Science, Bratislava, Czechoslo-
vakia

49. PAAR, V.   Institute "R. Bošković", Zagreb,
Yugoslavia

50. PAPINEAU, L.   C.E.N. Saclay, France

51. PATIN, Y.   C.E.A. Paris, France

52. PETRASCU, M.   Institute of Atomic Physics,
Bucharest, Romania

53. PISK, K.   Institute "R. Bošković", Zagreb,
Yugoslavia

54. RADJA, L.   Faculty of Electrical Engineering
Split, Yugoslavia

55. RAUCH, F.   Institut für Kernphysik der Universität Frankfurt/M, Germany

56. ROHWER, T.   Physikalisches Institut der Universität Tübingen, Germany

57. RONSIN, G.   Institut de Physique Nucléaire,
Orsay, France

58. SAETTA-MENICHELLA, E.   C.I.S.E. Milano, Italy

59. SCHEID, W.   Institut für Theoretische Physik
der Universität Frankfurt/M, Germany

60. SPERBER, D.   Rensselaer Polytechnic Institute,
Troy, USA

61. STOKSTAD, R.G.   Yale University, USA

62. STÄBLER, A.   Physikalisches Institut der Universität Tübingen, Germany

63. STRZALDOWSKI, A.   Institute of Nuclear Physics Cracow,
Poland

64. TOKE, J.   Institute of Experimental Physics
                               Warsaw, Poland
65. TURKIEWICZ, I.   Institute of Nuclear Research
                               Warsaw, Poland
66. TURKIEWICZ, J.   Institute of Nuclear Research
                               Warsaw, Poland
67. VOSS, F.   Institut für Angewandte Kernphysik
                               Kernforschungszentrum, Karlsruhe,
                               Germany
68. VULETIN, J.   Faculty of Electrical Engineering
                               Split, Yugoslavia
69. ZORAN, V.   Max Planck Institut für Kernphysik
                               Heidelberg, Germany

PRESENT STATUS OF INTERMEDIATE REACTION THEORIES

C. MAHAUX,

University of Liege, Belgium

1. Introduction

Until recently, the fields of nuclear structure and of nuclear
reactions had very little overlap. The analysis and interpretation of
compound nuclear reactions, in particular, were almost entirely dis-
connected from the rest of nuclear physics. The reason was that practically nothing was known about the usefulness of simple models for
the compound nuclear states. However, the giant dipole resonance
(cf [10]) shows that simple modes of excitation may exist at high
excitation energy. Until about ten years ago, this phenomenon was
considered as a splendid exception. Its interpretation (cf [27], [58],
[120]) required practically no use of reaction theory except for some
details,like for instance the interpretation of the total width of the
giant dipole resonance (cf [36]). Progress in the dynamical understanding of reaction processes (cf [47], [48], [134]) led to the suggestion (cf [75]) that simple modes of excitation may also be able to
produce a characteristic energy dependence of the average partial
widths in particle channels just like the dipole state produces a
localized enhancement of the radiative widths. The expression "intermediate structure" was coined for this phenomenon. The discovery of
isobaric analogue resonances (cf [105]) provided an ideal guinea pig
for the improvement and extension of intermediate structure theory
(cf [49], [78], [84], [90], [108], [127]). It showed that nuclear models
can be useful at high excitation energy and that, conversely, valuable
dynamical information can sometimes be obtained from resonance reactions. This is in remarkable contrast with the statistical model of
nuclear reactions (cf [19]), which is essentially based on the assumption that the compound nucleus does not retain any simple dynamical
feature.

The list of headings of the sessions of the present Conference
gives a fair idea of the content of the expression "intermediate processes in nuclear reactions". Roughly speaking, an intermediate reaction is one which cannot be interpreted in terms of the direct reaction model or of the statistical model and of their corollary, the
standard optical-model potential (cf [50]). Here, "standard" is

underlined because a suitable optical-model potential can always fit elastic scattering data; this remark should be kept in mind when discussing intermediate processes seen in elastic scattering or in total cross sections, particularly in the elastic scattering between two heavy ions. Since the statistical model and the direct reaction model are based on different statistical assumptions (cf [68]), it is natural that intermediate processes involve yet another type of statistical assumptions. This provides an approach to a model independent definition of intermediate structure (cf [85], [86]). We shall see that intermediate structure can be interpreted by retaining the standard statistical assumptions provided one first singles out one privileged configuration, called the doorway state (cf [24]). More generally, intermediate reactions are interpreted in terms of a model where one can select one or several privileged configurations and then apply the standard statistical assumptions on the quantities (energies, matrix elements) pertaining to the remaining configurations. By extension, one may call intermediate reaction any reaction, direct reaction excluded, for the interpretation of which one can use more or less explicitly some nuclear configurations. Thus, intermediate reactions can provide dynamical information. This justifies their detailed theoretical and experimental investigation.

The definition just given is sufficiently flexible to include many types of phenomena, but is so loose that it renders an unified and exhaustive description of the theory of intermediate reactions somewhat difficult. We have therefore limited our aim in the present review to the following main points. We first briefly describe (in section 2) a theoretical framework which is particularly convenient for the description of intermediate reaction processes. In section 3, we introduce the concept of doorway states, which is useful for the theoretical understanding of the strength function and of the imaginary part of the optical-model potential. Sometimes, as discussed in section 4, it is possible to find the nature of the configuration which is effective as a doorway in a given energy domain. When statistical assumptions are valid for the remaining configurations, an isolated doorway state gives rise to intermediate structure (section 5). The lengthy section 6 is devoted to a discussion of various examples, or tentative examples, of intermediate structure. In section 7, we show that direct reactions can, like common doorway states, lead to correlations between partial widths of different channels.

2. Theoretical Framework

We emphasized in the introduction that the interpretation of intermediate reactions involves a separate treatment of some privileged
configurations. This separation is most conveniently achieved in the
frame of the projection operator technique developed by the MIT group
(cf [7], [47], [49], [69], [127]). This formalism is quite general and
allows the use of practically any nuclear modes. In particular, it
can be applied to the shell-modes (cf [22], [83], [87]), in which case
the explicit construction of the projection operators is especially
simple. Below, we use the conjunction of the notations of refs.[7],
[49] and [87]. In a scattering problem, the basis set of configurations must include potential scattering states, let us call them χ_E^C.
Here, E refers to the energy and c to the channel quantum numbers.
If resonances must be described, the basis should also contain
bound configurations, ϕ_j (j = 1,..., M + D). We assume that the
states ϕ_j and χ_E^C are orthonormalized and antisymmetrised.

$$<\chi_E^C | \chi_{E'}^{c'}> = \delta_{cc'} \, \delta \, (E - E') \qquad , \qquad (2.1)$$

$$< \phi_j | \chi_E^C > = 0 \quad , \quad < \phi_j | \phi_k > = \delta_{jk} \qquad . \qquad (2.2)$$

For nucleon scattering, for instance, χ_E^C is essentially the antisymmetrised product, with appropriate angular momentum coupling, of
a single-particle wave function in a suitable potential well and of
a target state. The latter can be approximated by a collective state
(either rotational or vibrational) or a shell-model configuration. In
the first case, it may be difficult to take the Pauli principle into
account.

If a model energy ε_j is associated with each state ϕ_j, the model
space $\{\phi_j \, , \, \chi_E^C\}$ defines the model Hamiltonian

$$H_o = \sum_j | \phi_j > \varepsilon_j < \phi_j | + \sum_c \int d E | \chi_E^C > E < \chi_E^C | \, . \qquad (2.3)$$

Let V denote some appropriately chosen "residual" interaction. The
cross sections are calculated by diagonalizing the full Hamiltonian

$$H = H_o + V \qquad (2.4)$$

in the space $\{\phi_j \, , \, \chi_E^C\}$, taking the boundary conditions into account.

In other words, one looks for a wave function of the form

$$\Psi_E^{c(+)} = \sum_j b_E^c (j) \, \phi_j + \sum_{c'} \int d E' \, a_E^c (E' ; c') \, \chi_{E'}^{c'} \quad , \quad (2.5)$$

which has an incoming wave in channel c only and is such that

$$< \Psi_E^{c(+)} \mid H \mid \Psi_{E'}^{c'(+)} > \; = E \, \delta_{cc'} \quad \delta (E - E') \quad , \quad (2.6)$$

The asymptotic behaviour of $\Psi_E^{c(+)}$ yields the scattering matrix.

The model just defined is fairly general, but technically very complicated to solve when channels with two composite particles or, even more so, channels with more than two fragments are included. The latter case is confronted with the basic difficulties inherent to the three-body scattering problem. It has recently been studied in the frame of the shell-model approach to nuclear reactions [1], [56] and of an extended R - matrix theory [126] but practical applications appear still remote. A method to construct the states ϕ_j and χ_E^c for channels with composite particles has been developed by de Toledo Piza [128] . It is very closely related to the generator coordinate method (cf [54], [55], [123], [124], [137], [139]) which also includes the resonating group method [14], [136] as a particular case. The generator coordinate method appears quite promising [124]. It has mainly been applied to collisions between light nuclei (cf [31], [66], [109], [124]) but was recently used for the elastic scattering between two heavy ions [51], [138]. In view of the close analogy (cf [14], [44], [138]) between the equations obtained from the generator coordinate method and those derived in the frame of eqs. (2.1) - (2.6), the results described below hold, at least semiquantitatively, for channels with two composite fragments. Henceforth, we assume that eqs. (2.1) and (2.2) are fulfilled. An useful method for achieving this, starting from an arbitrary basis, is described in appendix 1 of ref. [7].

The configurations ϕ_j and χ_E^c are coupled together by the residual interaction:

$$< \phi_k \mid V \mid \phi_j > \; = V_{kj} \quad , \quad (2.7)$$

$$< \phi_k \mid V \mid \chi_E^c > \; = V_k^c \; (E) \quad , \quad (2.8)$$

$$< \chi_E^c \mid V \mid \chi_{E'}^{c'} > \; = V_{EE'}^{cc'} \quad . \quad (2.9)$$

The matrix elements V_{kj} play the same role as in the standard bound
state problem. The bound-continuum coupling V_k^C (E) is responsible for
the occurrence of resonances while the continuum-continuum coupling
$V_{EE'}^{CC'}$ gives rise to direct reactions [68],[87]. Here, we assume that
single-particle resonances have been removed from χ_E^C (cf [7],[35],
[130],[131]. The different physical roles of the matrix elements
(2.7) - (2.9) make it convenient to distinguish between scattering and
bound configurations, as we did.

We emphasized in the introduction that a further subdivision must
be made in the case of an intermediate reaction, since privileged configurations will be singled out. It is thus convenient to divide the
configuration space into three orthogonal parts associated with the
projection operators [49]

$$P = \sum_{p=1}^{P} \int d E \mid \chi_E^p > < \chi_E^p \mid , \qquad (2.10)$$

$$D = \sum_{d=M+1}^{M+D} \mid \phi_d > < \phi_d \mid , \qquad (2.11)$$

$$Q = \sum_{m=1}^{M} \mid \phi_m > < \phi_m \mid + \sum_{q>P} \int d E \mid \chi_E^q > < \chi_E^q \mid . \qquad (2.12)$$

For simplicity, we shall speak about the P, D and Q spaces. The (P + D)
space contains the configurations that we wish to treat explicitly:
the scattering ones are put in P, the bound ones in D. Space P should
include at least the entrance and outgoing channels and possibly a few
channels to which either of them is strongly coupled. The D space
contains the privileged bound configurations, the choice of which
depends upon the nature of the intermediate process. The configurations
contained in the Q space will usually be treated in a statistical way.

In order to exhibit later the effect of the privileged configurations, it is useful to give the expressions which are obtained for
the collision matrix when the space D is omitted. We call $_0\psi_E^{C(+)}$ the
corresponding wave function, eq. (2.5), T_{pot} the transition amplitude,
S_{pot} the scattering matrix. Following Feshbach [47],[48], it is convenient to introduce the notation

$$PHP = H_{PP} \; , \; PHD = H_{PD} \; , \; PH(P + Q) = H_{P,P+Q} \; , \; \ldots \; . \qquad (2.13)$$

We have [7],[49], with abbreviated but standard notations,

$$S_{pot,cc'} = \langle {}_o\psi^{c\,(-)}_E \mid {}_o\psi^{c'\,(+)}_E \rangle = \delta_{cc'} - 2i\pi\, T_{pot,cc'}\,(E), \quad (2.14)$$

$$T_{pot,cc'}\,(E) = \langle\, \chi^{c\,(-)}_E \mid H_{PQ}\left[E^+ - H_{QQ} - H_{QP}(E^+ - H_{PP})^{-1}H_{PQ}\right]^{-1}$$

$$H_{PQ} \mid \chi^{c'\,(+)}_E \rangle \,, \qquad\qquad\qquad (2.15)$$

This transition amplitude displays resonances due to the coupling of the bound states contained in the Q space to the scattering states.

When the D - space is included, the transition matrix is given by the right-hand side of eq. (2.15), with Q replaced by Q + D everywhere:

$$T_{cc'}\,(E) = \langle\, \chi^{c'\,(-)}_E \mid H_{P,Q+D}\left[E^+ - H_{Q+D,Q+D} - H_{Q+D,P}(E^+ - H_{PP})^{-1}\right.$$

$$\left. H_{Q+D,P}\right]^{-1}H_{Q+D,P} \mid \chi^{c\,(+)}_E \rangle \qquad\qquad . \qquad (2.16)$$

An equivalent expression can be obtained by introducing the effective Hamiltonian [7]

$$\mathcal{H}\,(E) = H - H\,Q\left[E^+ - H_{QQ}\right]^{-1}Q\,H \qquad\qquad . \qquad (2.17)$$

One has [7]

$$T_{cc'}\,(E) = T_{pot,cc'}\,(E) + \langle\, {}_o\psi^{c'\,(-)}_E \mid \mathcal{H}_{PD}(E)\left[E^+ - \mathcal{H}_{DD}(E) - \right.$$

$$\mathcal{H}_{DP}(E)\,(E^+ - \mathcal{H}_{PP}(E))^{-1}\,\mathcal{H}_{PD}(E)\right]^{-1}\mathcal{H}_{DP}(E) \mid {}_o\psi^{c\,(+)}_E \rangle \qquad (2.18)$$

This expression presents the advantage that the value of $T_{cc'}$, in the absence of the privileged configurations appears as a separate "background term". At high excitation energy, it is often convenient to introduce the energy average, $\langle T \rangle$, of the transition matrix. Denoting by I the averaging interval, we have

$$\langle T_{cc'}\,(E) \rangle = T_{cc'}\,(E + i\,I) \qquad\qquad . \qquad (2.19)$$

3. Doorway States

Espressions (2.16) and (2.18) for the transition matrix are valid

for any separation of the full space into P , D and Q. The precise
definition of these subspaces can be adapted to the particular reaction or to the specific aspect of a given reaction that one wants to
analyse, In the present section, we discuss one possible choice for
the D - space, which was used by Block and Feshbach [24] and Shakin
[114] in their study of the neutron strength function.

In the scattering of low energy neutrons from even-even targets,
only one channel is open, since only s - wave neutrons need to be
considered. Let us include only that channel in P, and neglect the
other channels. The operator whose inverse appears on the r.h.s. of
eq. (2.16) is represented by a (M + D) x (M + D) matrix in the Q + D
space. The zeros of the determinant of that matrix give the complex
resonance energies

$$E_j = \zeta_j - \frac{1}{2} i \; \Gamma_j \; , \quad (j = 1, \; ..., \; M + D) \qquad . \qquad (3.1)$$

The scattering function reads [87]

$$S_{cc} = \exp \; (2 \; i \; \delta_c) \; [1 \; - \; i \; \sum_{j=1}^{M+D} \frac{\gamma_{jc}^2}{E \; - \; \zeta_j \; + \; \frac{1}{2} \; i \; \Gamma_j}] \quad , \qquad (3.2)$$

where δ_c is the potential scattering phase shift associated with χ_E^c.
The following sum rules hold

$$\sum_{j=1}^{M+D} \Gamma_j = \sum_{j=1}^{M+D} \gamma_{jc}^2 = 2 \; \pi \; \sum_{j=1}^{M+D} < \phi_j \; | \; V \; | \; \chi_E^c >^2 \qquad . \qquad (3.3)$$

In practice, the states ϕ_j and χ_E^c are chosen as eigenstates of
some nuclear model Hamiltonian H_o, and the residual interaction V is
assumed to be simple, for instance a sum of two-body interactions.
Because of the simplicity of ϕ_j, χ_E^c and V, it usually happens that
only a few states ϕ_j are coupled to χ_E^c by the residual interaction.
These configurations are called <u>doorway states</u> and could be considered
as <u>privileged</u>. It should be noted that the number and the nature of
these doorway configurations depend upon the choice of the model
Hamiltonian. Let us call ϕ_d , d= M + 1, ..., M + D, the doorway states which span the privileged D space. We denote by a_d and a_r the average distance between neighbouring doorway states and resonances, respectively, and by ΔE the size of the energy
interval where the model is believed to apply. We have, <u>assuming</u> that
all states of interest lie in the interval ΔE ,

$$\Delta E \ = \ D \ a_d \ = \ (M + D) \ a_r \qquad\qquad (3.4)$$

From eqs. (3.3) and (3.4), we obtain

$$\frac{< \Gamma >}{a_r} \ = \ \frac{< \gamma_c^2 >}{a_r} \ = \ \frac{< \Gamma_c^\dagger >}{a_d} \qquad\qquad , \qquad (3.5)$$

where $< \Gamma >$ is the average of Γ_j in ΔE, $< \gamma_c^2 >$ that of γ_{jc}^2 and $\Gamma_c^\dagger$ the average of the **"escape widths"**

$$\Gamma_{dc}^\dagger \ = \ 2 \ \pi \ < \phi_d \ | \ V \ | \ \chi_E^c >^2 \qquad\qquad . \qquad (3.6)$$

We conclude from (3.5) that the strength function can be computed from the doorway states alone, _if_ our assumptions hold. It is, however, not possible to include in a calculation all the doorway states which influence the cross section in ΔE, because some of them lie outside the interval ΔE. Conversely, some doorway states contained in ΔE may loose part of their strength by coupling with configurations which lie outside of ΔE. It is therefore necessary, in practice, to modify eq. (3.5) in order to take these effects into account [114]. The values of the strength function in the tin isotopes and in ^{209}Pb are well reproduced in this way [114]. Another approach consists in focusing one's interest on the _deviation_ of the strength function from the value prodicted from the standard optical-model [24]. More microscopic calculations of this type should be performed.

Hay [64] studied the compound nucleus ^{41}Ca. He diagonalized the full Hamiltonian in the space of two particle - one hole (2p - 1h) states, assuming that V is an effective zero-range interaction. Hay computed the values of the corresponding doorway energies $< \phi_d \ | \ H \ | \ \phi_d >$ and widths $\Gamma_{dc}^\dagger$. Fig. 1, taken from ref. [65], shows that the observed density of resonances is, as expected, larger than that of doorway states (fig. 1 (a)), but that the sum rule (3.3) is roughly fulfilled locally, i.e. in a range $E \simeq 0.5$ MeV (fig. 1 (b)). Fig. 1 (c) shows that the strength function at low energy is fairly well given by eq. (3.5). Fig. 2 (from ref. [99]) shows that the same kind of agreement is obtained [38] in ^{88}Sr + n, when one takes as doorway configurations either 2p - 1h states, or a neutron bound single-particle orbital coupled to vibrational states in ^{88}Sr. The fair

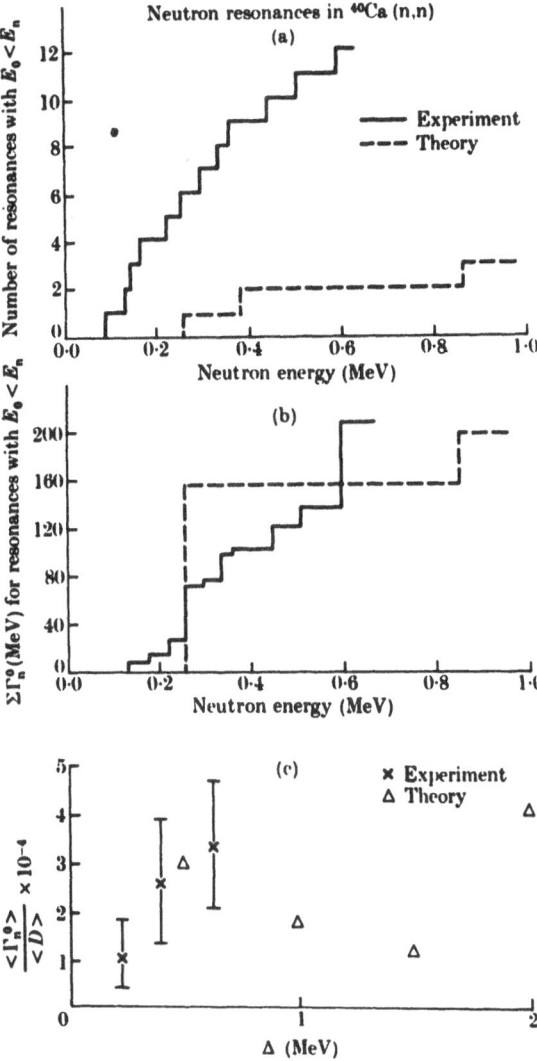

Fig. 1. From ref. [65]. Cumulative plots of
the numbers (a) and renormalized
widths (b) of the resonances in
^{40}Ca + n, for the experimental data
and for the calculations of Hay [64].
In (c), one gives the experimental
and theoretical [64] values of the
strength function

10

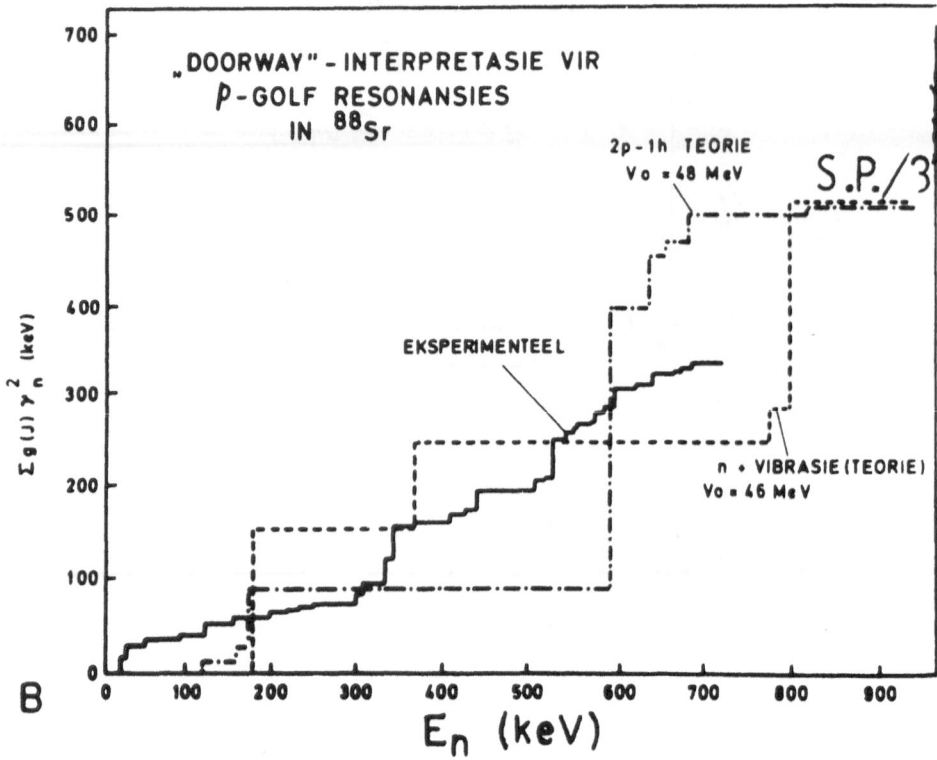

Fig. 2. From ref. [99]. Cumulative plot of
the resonance widths in ^{88}Sr + n .
The full histogram gives the experimental data, the other curves show
the result of a theoretical calculation [38] including as doorway configurations either 2p - 1h states
(dash-and dot) or neutron + vibrational states (dots)

agreement between experiment and theory obtained in refs. [24], [38],
[64], [114] shows that the few doorway configurations considered by
these authors suffice to explain the local value of the sum of the
resonance widths. In a more complete calculation one would therefore
need to add only bound configurations which are not coupled to the
elastic channel.

A "model" is characterized by both a Hamiltonian and the selection of a truncated basis. We noted that the number and the nature of
the doorway states are model dependent. For instance, it is always
possible to make an unitary transformation of the basis in the D -
space in such a way that, for a given channel, only a few of the resulting configurations are still doorway states. In particular, it is
always possible to reduce the space of doorways to only one configuration, namely

$$\phi_d^c = N \sum_{j=M+1}^{M+D} | \phi_j > < \phi_j | V | \chi_E^c > \quad , \tag{3.7}$$

where N is a normalisation coefficient. The configuration (3.7),
however, usually cannot be associated with a physical state, because
its life-time is too short. There exists at least one counterexample
to the latter statement. Indeed, the configuration

$$\phi_{dip} = N \sum_{j=1}^{\infty} | \phi_j > < \phi_j | E1 | \Psi_0 > \quad , \tag{3.8}$$

where E 1 is the dipole operator and Ψ_0 the ground state wave function, has a physical meaning as a state: it is the giant dipole state.

4. Isolated Doorway State

We emphasized above that it is always possible to perform an unitary transformation of the basis in such a way that only one doorway
state exists, eq. (3.7). Hence, it is not surprizing that any elastic
scattering data can always be interpreted in terms of only one doorway
state, and a number of more complicated states included in the Q -
space (cf [73] , [85] , [86]). In order to see this more clearly, let us
use eqs. (2.15), (2.18). Since the Q - space is not connected to the
P - space, H_{QP} = 0 and, neglecting for simplicity all channels but c,

$$S_{pot} = \exp (2 i \delta_c) \quad , \quad {}_o\Psi_E^{c(+)} = \chi_E^{c(+)} \quad . \tag{4.1}$$

Assuming for simplicity that

$$Q V Q = 0 \quad , \tag{4.2}$$

eq. (2.18) gives

$$S_{cc} = \exp (2 i \delta_c) \frac{c.\ c.}{E - \varepsilon_d + i \Gamma_d^{\uparrow} - \sum\limits_{j=1}^{M} \frac{v_j^2}{E - e_j}} ,$$ (4.3)

where

$$\varepsilon_d = < \phi_d \mid H \mid \phi_d > + P \int d E' (E - E')^{-1} < \chi_{E'}^c \mid H \mid \phi_d >^2 ,$$ (4.4)

$$\Gamma_d^{\uparrow} = 2 \pi < \phi_d \mid H \mid \chi_E^c >^2 ,$$ (4.5)

$$v_j = < \phi_d \mid V \mid \phi_j > , \quad < \phi_j \mid H \mid \phi_m > = \delta_{jm} e_j .$$ (4.6)

Eq. (4.3) can always be written in the form

$$S_{cc} = \exp (2 i \delta_c) \frac{1 - i \sum\limits_{\lambda=1}^{M+1} \frac{\alpha_\lambda^2}{E - E_\lambda}}{1 + i \sum\limits_{\lambda=1}^{M+1} \frac{\alpha_\lambda^2}{E - E_\lambda}} .$$ (4.7)

This parametric form is identical to the M + 1 level approximation of R - matrix theory [79] and is known to provide good fits to experimental data. Hence, the one doorway state model is always in agreement with elastic scattering data provided that the parameters are suitably adjusted.

Starting from an analysis based on eqs. (4.3) or (4.7), it is sometimes possible to find the structure of the doorway configuration ϕ_d. The comparison between the experimental and theoretical values of $\Gamma_d^{\uparrow}$ and ε_d provides a check for the correctness of the assumed nature of ϕ_d. The crudest test consists in comparing $\Gamma_d^{\uparrow}$ with the sum of the widths of the resonances. Eq. (3.3) gives, in the case of only one doorway state,

$$\Gamma_d^{\uparrow} = 2 \pi < \phi_d \mid V \mid \chi_E^c >^2 = \sum\limits_{j=1}^{M+1} \Gamma_j .$$ (4.8)

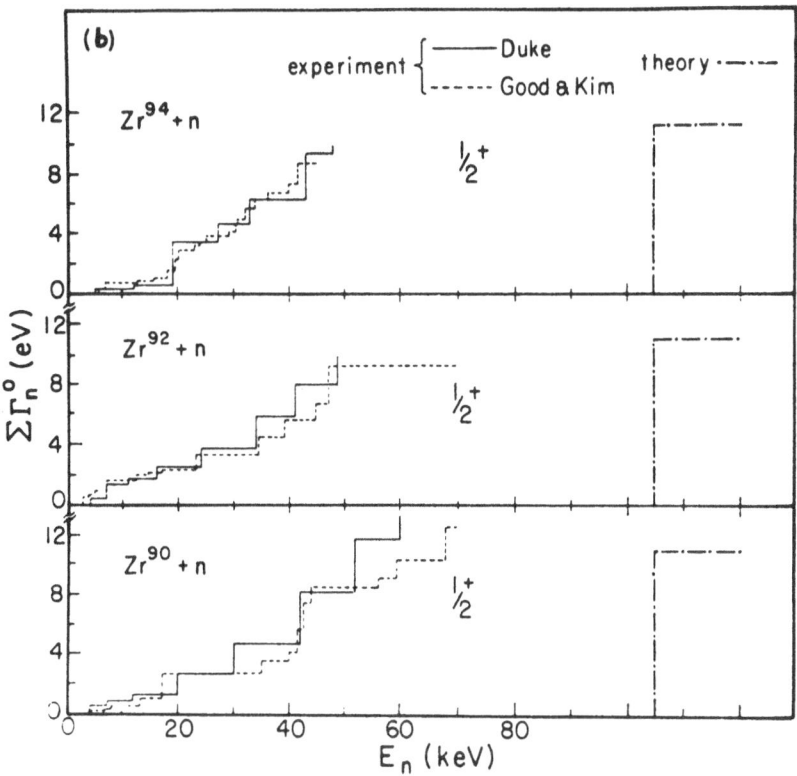

Fig. 3. From. ref. [4o]. Comparison between the theoretical (dash-and-dot)
value of $\Gamma_d^{\uparrow}$ and the experimental
cumulative plot of the widths for
the targets ^{90}Zr, ^{92}Zr and ^{94}Zr

Fig. 3 shows the cumulative plot of the renormalized widths of
the observed resonances in ^{94}Zr + n , ^{92}Zr + n and ^{90}Zr + n . The data are obtained from refs. [100] (full line) and [59] (dotted line).
The dash-and-dot line indicates the value of $\Gamma_d^{\uparrow}$ and of ε_d for a
doorway state resulting from diagonalization among 2p - 1h states
[39] . From fig. 3 , one may conclude that this doorway state is sufficient to explain the sum of the widths of the observed resonances,
so that, in a more detailed interpretation, one needs only to add to
the basis a number of complicated (i.e. non-doorway) model states.
Similar results are obtained for ^{58}Ni + n and ^{54}Fe + n [40] .

Fig. 4 shows [41] the sum of the reduced widths of s - wave resonances in ^{208}Pb + n ((a), thick curve) and ^{206}Pb + n ((b) thick
curve). The theoretical values of $\Gamma_d^{\uparrow}$ and ε_d are shown for a door-

14

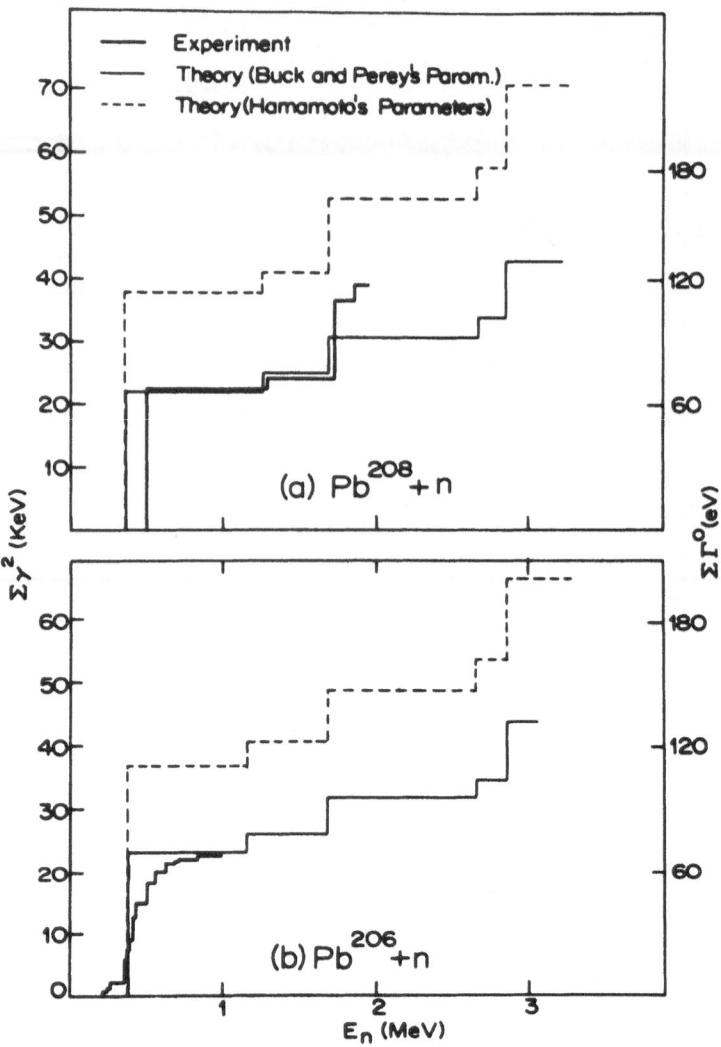

Fig. 4. From ref. [41]. Comparison between experimental (thick line) and
theoretical values of the sum of
the widths, for ^{208}Pb + n (a) and
^{206}Pb + n (b)

way configuration where a g 9/2 neutron is coupled to a vibrational
(4$^+$) excited state of the target. The two theoretical results correspond to two choices for the potential well used for calculating χ_E^c:
the dotted line is based on a potential taken from ref. [63], and the
thin line, on a potential from ref. [32]. The agrement found in the

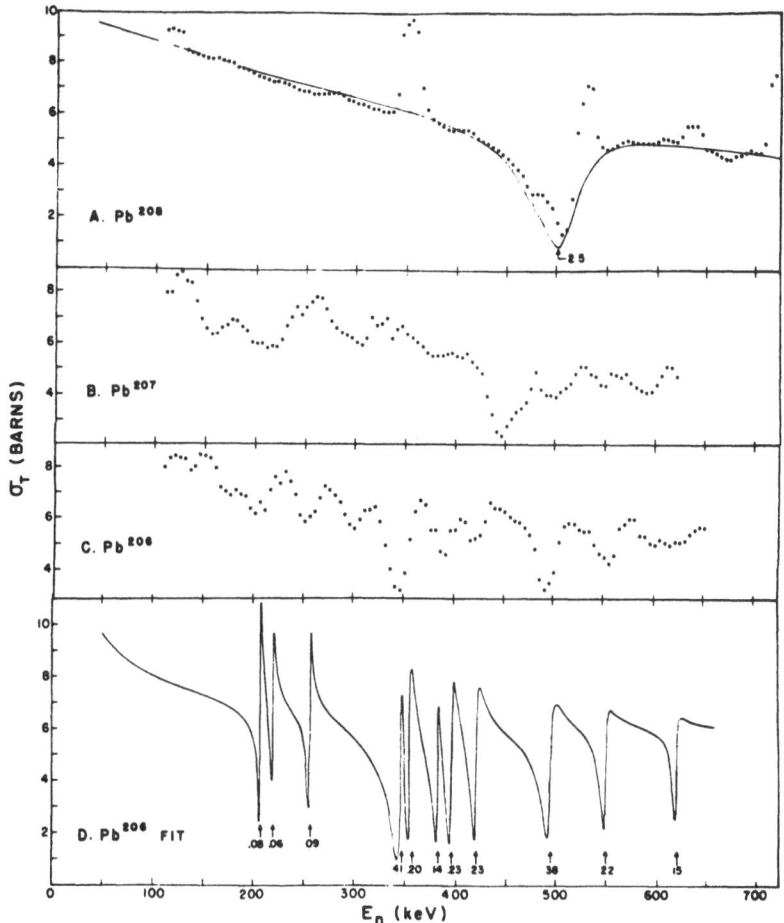

Fig. 5. From ref. [46]. Experimental total
cross sections for n on the targets ^{208}Pb (A) , ^{207}Pb (B) , and
^{206}Pb (C). Part D shows the fit
to the s - wave part of the ^{206}Pb +
n cross section

case of ^{208}Pb + n shows that no other bound configuration need to
be added to the model space. In the case ^{206}Pb + n , many resonances
are observed near 500 keV , but the sum of their widths is remarkably
close to the escape width of the nearby theoretical doorway configuration ϕ_d. The doorway configuration is a g 9/2 bound neutron coupled to
the 4^+ vibrational state of ^{206}Pb. Fig. 5 shows [46] the experimental
cross sections for the targets ^{208}Pb (A) , ^{207}Pb (B) , ^{206}Pb (C).
In part D of the figure, the fit to the s - wave part of the cross

section is shown. We conclude from fig. 4 (b) that the curve in D can
be interpreted in terms of the doorway configuration described above
plus ten "complicated" configurations, whose nature is much harder to
guess [15]. In such a case, where one doorway configuration can be
found which explains the sum of the widths of the resonances in a whole energy range, one speaks of an isolated doorway state. We note that
this theoretical interpretation of the data does not imply that the
data display any specific property. This is clear from figs. 1 and 3,
where one doorway configuration also explains the sum of the widths of
many resonances, but the data do not show anything particular. The total cross section of ^{206}Pb + n displays, however, the special feature
that an enhancement exists for the neutron widths in the range 100 -
900 keV . Such an enhancement is in contradiction with the statistical
model and is characteristic of the so-called "intermediate structure"
phenomenon, which is discussed in the following section.

5. Intermediate Structure

"Intermediate structure" is a deviation from the statistical model, which is localized in energy [86]. Let us briefly recall the basic
assumptions of the statistical model. We write the scattering matrix in
the form

$$S_{c'c} = \exp (i \delta_c + i \delta_{c'}) \left[S_{c'c}^{BG} - i \sum_{\lambda} \frac{\gamma_{\lambda c} \, \gamma_{\lambda c'}}{E - \zeta_\lambda + \frac{1}{2} i \Gamma_\lambda} \right] . \qquad (5.1)$$

The three main assumptions of the statistical model are

(i) $S_{c'c}^{BG}$ is diagonal:

$$S_{c'c}^{BG} = S_c^{BG} \delta_{cc'} \qquad \qquad . \qquad (5.2a)$$

(ii) $I^{-1} \sum_{\lambda \text{ in } I} \gamma_{\lambda c} \, \gamma_{\lambda c'}^* = s_c \delta_{cc'} \qquad \qquad . \qquad (5.2b)$

(iii) The strength function s_c (E) is independent of energy.
$$(5.2c)$$

In assumption (iii), E is the energy on which the averaging interval
I in eq. (5.3) is centered. We neglect the monotonic, and therefore

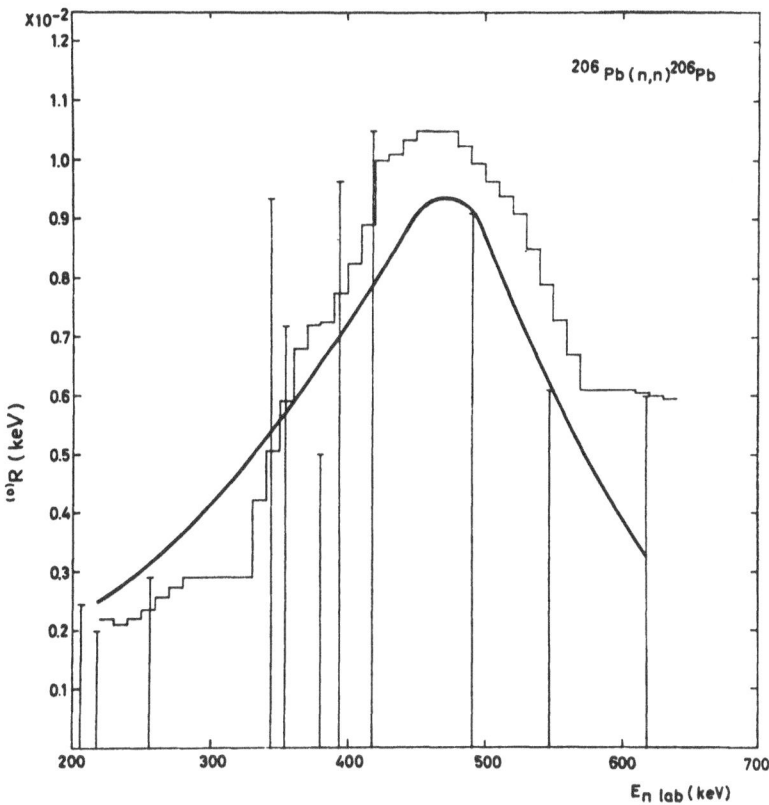

Fig. 6. From ref. [81]. Energy dependence
of s_c x a_r , for ^{206}Pb + n . The
vertical lines show the quantities
$|\gamma_{\lambda c}|^2$, the histogram gives their
average over ± 20 keV and the full
curve is a Lorentzian fitted to the
histogram

trivial, energy dependence of s_c due to penetration effects.

In the case ^{206}Pb + n , assumption (iii) is apparently violated.
This is displayed in fig. 6, which shows the energy dependence of s_c
x a_r , where a_r = 40 keV is the average separation between resonances [81] . We see that the shape of the strength function s_c is approximately Lorentzian. It must be checked that this grouping of large
widths is not accidental. This was studied by Baglan, Berman and Bowman [9] who concluded from Monte-Carlo calculations that such an
accidental grouping would occur only once per 4 MeV .

The physical interpretation of the enhancement is fairly clear:

the doorway state is mixed with the non-doorway configurations by the
residual interaction. If the latter is not too strong and if all the
matrix elements have about the same magnitude, the mixing occurs locally and the resulting resonance states acquire a sizable width only
in the vicinity of the doorway state. The width of the intermediate
structure phenomenon is related to the life-time of the doorway configuration (for decay into other bound states by nuclear collisions)
by the uncertainty relation.

These qualitative considerations can be put in a more quantitative
form. Returning to the general idea described in section 1, we introduce statistical assumptions on the matrix elements v_j (eq. (4.6))
involving the "complicated" states ϕ_j of the Q- space: we assume
that the function

$$R (E) = \sum_{j=1}^{M} \frac{v_j^2}{e_j - E}$$

is a statistical R - function [79] . This essentially amounts to
assume that the relations

$$- i R (E + i I) = \Gamma^{\downarrow} = 2 \pi \frac{v^2}{a_r} \tag{5.3a}$$

hold, where v^2 is a real constant equal to the average of v_j^2 in
the energy interval $\Delta = (E - \frac{I}{2} , E + \frac{I}{2})$. The following three state-
ments are equivalent, and any one of them imply the two other ones
[73]

(a) $\Gamma^{\downarrow}$ is independent of energy ,

(b) $<S_{cc} (E)> = \exp (2i\delta_c) \left[1 - i \dfrac{\Gamma_d^{\uparrow}}{E - \varepsilon_d + \frac{1}{2} i (\Gamma_d^{\uparrow} + \Gamma^{\downarrow} + 2I)}\right]$,

$$\tag{5.3b}$$

(c) $\alpha^2 = \dfrac{a_r}{4\pi} \dfrac{\Gamma_d^{\uparrow} \Gamma^{\downarrow}}{(E - \varepsilon_o) + \frac{1}{4} (\Gamma^{\downarrow} + 2I)^2}$, $\tag{5.3c}$

where α^2 is the average of α_λ^2 (eq. (4.7)) in the energy interval
Δ . Eq. (5.3b) shows that the partial widths are enhanced in the vi-

cinity of the doorway configuration. This confirms the results of the
qualitative discussion given above. According to eq. (5.3b), the average total cross section should display a resonance. One might therefore believe that the measurement of the average total cross section
suffices to determine the interesting doorway parameters ε_d, $\Gamma_d^{\uparrow}$,
$\Gamma^{\downarrow}$. Unfortunately, this is not always true. Indeed, intermediate
structure is associated with a given angular momentum, while the experimental average cross section usually contains contributions from
several angular momentum states. A measurement of the fine structure
is often necessary in order to disentangle them. This is exhibited in
fig. 7 (from [99]), which shows the energy averaged total neutron
cross sections for neutrons for the targets ^{209}Bi, ^{208}Pb , ^{207}Pb ,
^{206}Pb , ^{204}Pb and Tl . We note, in particular, that the intermediate
structure in ^{206}Pb + n is not visible in the energy average total
cross section. Even the average of the s - wave part does not exhibit
a nice Breit-Wigner shape [81]. This could be expected from fig. 6,
since the statistical assumption (5.2a) is only approximately fulfilled.
It is only for very strong enhancements, i.e. for (eq. (5.3))

$$\frac{\Gamma_d^{\uparrow}}{\Gamma^{\downarrow}} \gtrless 1 \quad \text{and} \quad \Gamma^{\uparrow} + \Gamma^{\downarrow} \overset{<}{\sim} a_d \qquad , \qquad (5.4)$$

that intermediate structure is visible in the average cross section.
This is the case for the giant dipole resonances, the isobaric analogue resonances and the intermediate structure in neutron induced
fission. In the last two cases, selection rules render $\Gamma^{\downarrow}$ particularly small. In the giant dipole resonance, the background due to other
multipolarities is very small. We return to this point in section
6.f.

Eqs. (5.3b) and (5.3c) should be generalized to take into account
the following two facts:

(1) The enhancement due to a given doorway configuration is superposed on a smooth background due to far-away other doorway configurations.

(2) Several channels may be open.

Eq. (2.18) provides a good starting point for these generalizati-
ons [7]. The statistical assumption (5.2a) is replaced by the assumption that the optical-model Hamiltonian describing the background is
given by (eq. (2.17))

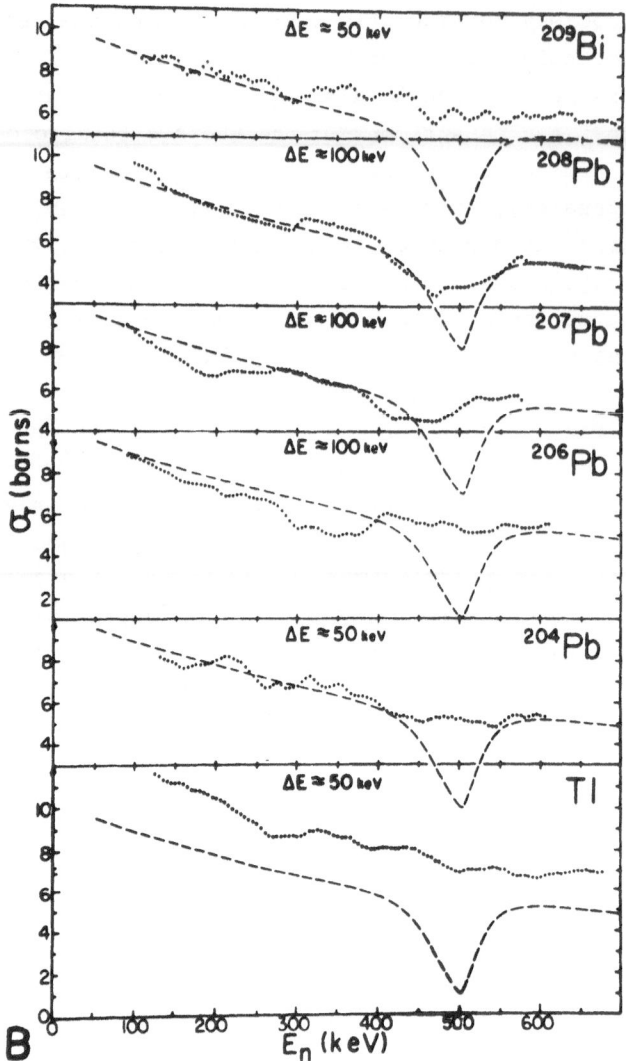

Fig. 7. From ref. [99]. Average neutron total cross sections for the indicated targets, with an averaging interval ΔE = 50 or 100 keV

$$H_{opt} = H_{PP} (E + i I) \qquad . \qquad (5.5)$$

The validity of this assumption depends on the definition of the D -
space and is usually hard to appreciate in detail. Introducing opti-

cal-model wave functions through

$$(H_{opt} - E) \mid \omega_E^{c(+)} > = 0 \qquad , \qquad (5.6a)$$

$$< \tilde{\omega}_E^{c(-)} \mid (H_{opt}^+ - E) = 0 \qquad , \qquad (5.6b)$$

one finds that (cf [7])

$$<T_{cc'} > = T_{opt,cc'} + (2 \pi)^{-1} \frac{\tilde{\alpha}_{dc'} \, \alpha_{dc}}{E - E_d + \frac{1}{2} i \Gamma} \qquad , \qquad (5.7)$$

where

$$\tilde{\alpha}_{dc'} = < \omega_E^{c'(-)} \mid H_{PD} \mid \phi_d > \qquad , \qquad (5.8a)$$

$$\alpha_{dc} = < \phi_d \mid H_{DP} \mid \omega_E^{c(+)} > \qquad , \qquad (5.8b)$$

and with suitable definitions for E_d and [7]. We recall that in
this formalism, the essential **assumption** is that the experimental
background or, equivalently, the standard (i.e. smoothly energy dependent) optical-model Hamiltonian can be identified with (5.5).

Let us for simplicity assume that direct reactions are negligible.
Then, the optical-model potential (5.5) and $T_{opt,cc'}$ are diagonal
in the channel indices. The fact that $< T_{c'c} >$ ($c' \neq c$) does not va-
nish in the energy range ($E - \Gamma$, $E + \Gamma$) implies, according to eq.
(5.1), that the partial widths amplitudes $\gamma_{\lambda c}$ and $\gamma_{\lambda c'}$ are correlated, i.e. that assumption (5.2b) is violated. This can be understood
qualitatively in the following way. The bound-bound coupling (2.7)
gives rise to configuration mixing between the model states. The true
compound states are, in first approximation, obtained by diagonalizing
H in the bound configurations of the (D + Q) space. We call the
resulting states :

$$\Omega_\lambda = \sum_{j=1}^{M+D} 0_{\lambda j} \, \phi_j \qquad , \qquad (5.9)$$

and their energies ζ_λ . The partial width amplitudes are approximately
given by

$$\gamma_{\lambda c} = (2 \pi)^{\frac{1}{2}} < \chi_E^c \mid V \mid \Omega_\lambda > \qquad . \qquad (5.10)$$

If statistical assumptions hold for the matrix elements v_j (eq. (5.2))
we have [73]

$$0_{\lambda d}^2 = \frac{a_r}{2\pi} \frac{\Gamma^\downarrow}{(\zeta_\lambda - \varepsilon_d)^2 + \frac{1}{4} (\Gamma^\downarrow)^2} \qquad (5.11)$$

When only one doorway state exists for <u>both</u> channels c and c' , the
following relations are obtained:

$$\gamma_{\lambda c} = (2 \pi)^{\frac{1}{2}} < \chi_E^c \mid V \mid \phi_d > 0_{\lambda d} = \gamma_{dc}^\uparrow 0_{\lambda d} \qquad , \qquad (5.12)$$

$$\gamma_{\lambda c'} = (2 \pi)^{\frac{1}{2}} < \chi_E^{c'} \mid V \mid \phi_d > 0_{\lambda d} = \gamma_{dc'}^\uparrow 0_{\lambda d} \qquad , \qquad (5.13)$$

$$\frac{\gamma_{\lambda c}}{\gamma_{\lambda c'}} = \frac{\gamma_{dc}^\uparrow}{\gamma_{dc'}^\uparrow} \qquad \text{(independent of } \lambda \text{)} \qquad . \qquad (5.14)$$

The channel-channel correlation implied by eq. (5.14) is a violation of
the statistical assumption (5.2b). It is very hard to find experimenta-
ly, because it is difficult to resolve the resonances above inelastic
threshold. Moreover, it is exceptional that two channels c and c'
have a <u>common</u> strong doorway state. Fig. 8 shows (cf [97]) the photon,
elastic and inelastic proton widths of the fine structure peaks of an
isobaric analogue resonance in ^{55}Mn , the parent state being the
ground state of ^{55}Cr . The elastic and inelastic proton widths are
clearly correlated, also with the radiative widths to the ground state.
    Fig. 9, taken from ref. [9], indicates that the neutron and pho-
ton widths are apparently correlated in the case of the fine structure
in ^{207}Pb , shown in figs. 4-7. If one assumes that this correlation is
due to a common doorway state in the neutron and photon channels, the
escape radiative width ($\approx$36.5 eV) of the doorway state can be obtained
from the data [9]. In the case of photonuclear reactions, however, an

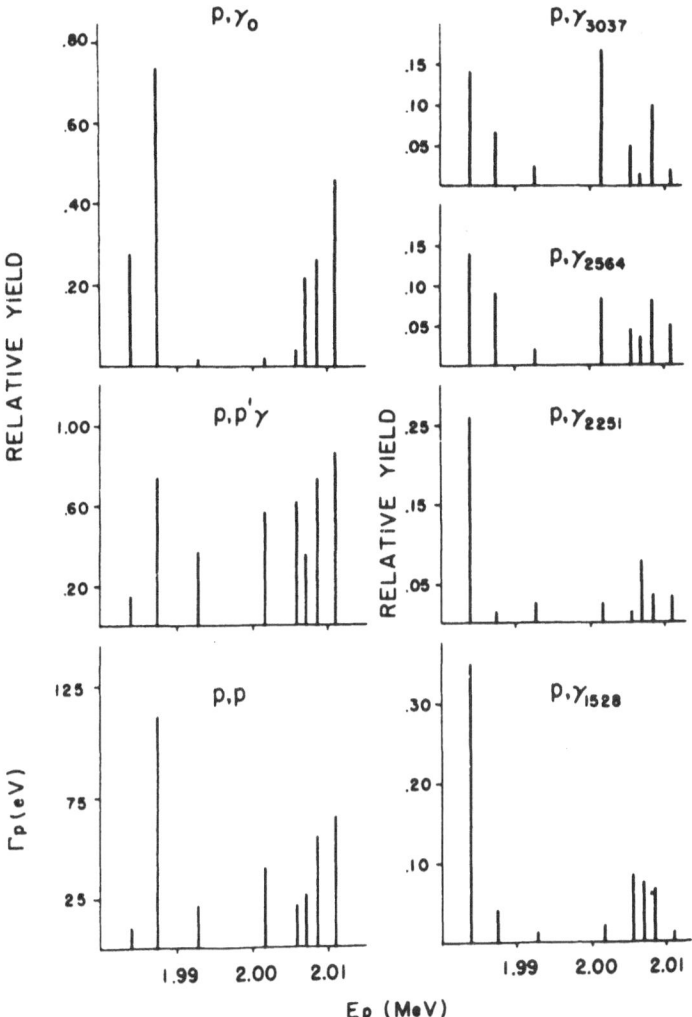

Fig. 8. From ref. [97]. Relative strengths
for gamma, elastic and inelastic
proton widths of the fine structure
of an isobaric analogue resonance
in ^{55}Mn

additional and sometimes important correlation may arise from the direct
capture process [86] , as described in section 7. Beres and Divadeenam
[16] find a theoretical radiative width of 23 eV for the doorway state
in ^{207}Pb , by including the coupling of the neutron + vibration doorway
state with the giant dipole resonance, without using any free parameter.
The agreement with the experimental value is impressive but could also

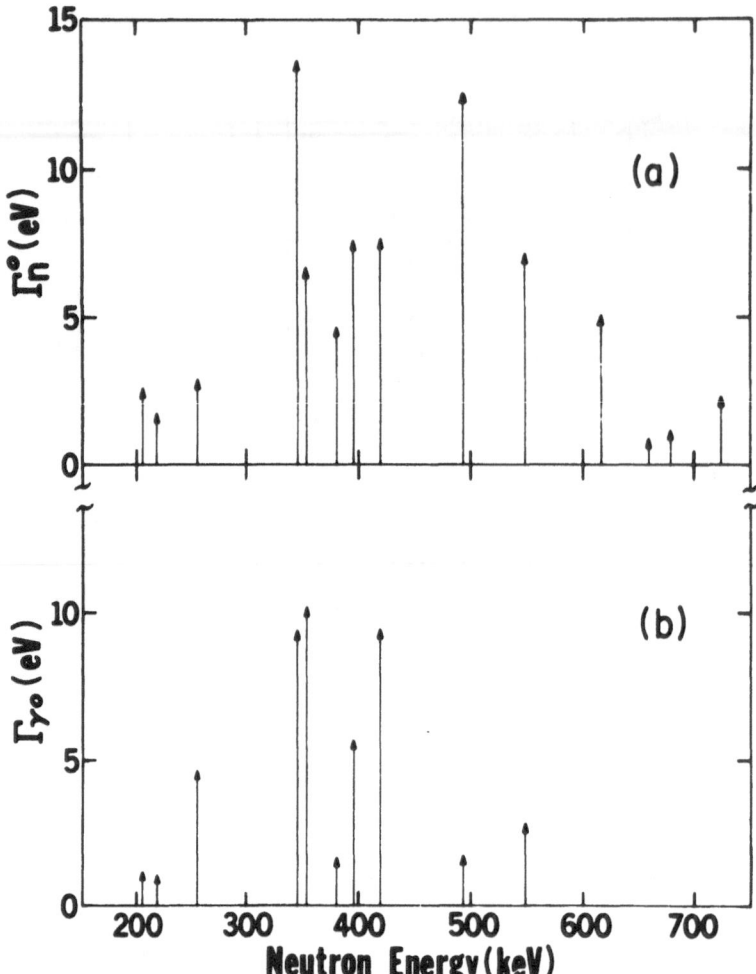

Fig. 9. From ref. [9]. Neutron and photon
widths of the resonances shown in
fig. 5, for the compound nucleus
^{207}Pb

be quite fortuitious. Indeed, recent measurements [3] indicate that
the photon widths in fig. 9 do not belong to the same resonances as
the neutron widths, and that the photon widths of the $\frac{1}{2}^+$ states is not
enhanced. This would imply that the doorway state of the neutron channel is not a strong doorway for the photon channel.

6. Examples of Intermediate Structure
6.a. Introduction
The analysis of an intermediate structure phenomenon involves the
following questions.

(a) Do the data show a **significant** deviation from the statistical model? The answer usually implies a careful use of the methods of
statistical analysis.

(b) Is the phenomenon characterized by a given angular momentum
and parity? This problem often requires the measurement of angular distributions and/or the resolution of the fine structure peaks.

(c) Can a doorway configuration be found which reproduces the
energy and escape widths obtained from the analysis of the data? This
step is delicate because many possibilities and free parameters are
at one's disposal and the identification may therefore be ambiguous.

(d) Is the intermediate structure seen in various channels? This
may help answering question (a) by discarding an interpretation in
terms of statistical fluctuations and question (c) by providing a
guide to the physical intuition.

6. b. Giant Dipole Resonance

In the case of the **giant dipole resonance**, the data yield the
energy of the doorway state E_{dip}, its photon width $\Gamma_{dp}^{\uparrow}$ and its
total width Γ. The latter quantity is the sum of the spreading width
$\Gamma^{\downarrow}$ and of the escape widths in the particle channels. Recent data [18]
show a systematic dependence of Γ on mass number, with minima at magic numbers. These results are not yet quantitatively understood. The
value of $\Gamma^{\downarrow}$ has been computed for ^{208}Pb (cf 37 , 42), with moderate success. The value of E_{dip} in the case of ^{208}Pb is not well
understood [76] . The giant dipole resonance can show some substructure. In medium weight and heavy nuclei, this can be interpreted in
terms of the coupling of the dipole state with the surface vibrations
[67] , [129] ,with fair success [119] . For light nuclei, a more microscopic approach to the fine structure is necessary. This is a difficult
task in view of the unknown nature of the states lying in the vicinity
of the dipole state. In the case of ^{16}O , it has been successively
proposed that the fine structure is due to 2p - 2h and 4p - 4h
states [53] , to 3p - 3h states [115] , [131] or to particle-hole excitations of the deformed 0^{+} excited states at 6.05 MeV [60] . All
these suggestions together provide more fine structure than needed. It
is likely that none of these interpretations is complete by itself,
and that each fine structure peak is a combination of all the proposed structures. It is tempting to guess the nature of a peak from the
way it shows up in another channel, for instance by proposing that

a strong peak seen in ^{12}C (α,γ) is a 4p - 4h state [53]. This type
of conclusion is, however, dangerous because α - particle widths
are difficult to interpret microscopically, and one can moreover not
be sure that a peak seen in ^{15}N (p,γ) is the same as the one seen
in ^{12}C (α,γ). The calculation reported in ref. [131] is also a good
example of the delicacy of the problem of the identification of the fine structure peaks. In an earlier work [115] and in ref. [131], the
authors achieve an impressive agreement with experiment with only a
small adjustement of the energy of the fine structure peaks and of
their coupling to the dipole state. However, they later realized that
their calculation contains an error[131] which renders necessary the
introduction of larger adjustements. In fact, any kind of configurations will always be able to fit the data if their energies and coupling
strengths are suitably adjusted. Care must always be taken not to adjust too many free parameters, otherwise the agreement with experiment
is meaningless.

6. c. Isobaric Analogue Resonances

The theory of isobaric analogue resonances (IAR) was first developed by Robson [108]. Alternative formulations can be found in refs.
[87],[90] and a detailed theory in the frame of Feshbach's projection
operator formalism has been published recently [7]. We refer to the
latter reference and to another session of the present Conference for
a discussion of the present experimental and theoretical status. In
short, one can say that the proton escape widths of the IAR are well
reproduced by the theory. Recent experimental data [13],[106] suggest
that some IAR may acquire an appreciable escape width in isospin for-
bidden channels (neutron, deuteron,or alpha channels), probably by
indirect coupling [106],[128]. In the frame of the formalism presented
in section 5, this means that ω_E^C (c=neutron, for instance) is obtained from a coupled channel calculation involving proton and neutron
channels.

The calculated value of the spreading width of IAR is too large.
It has been proposed, but not yet demonstrated, that taking into account the monopole state may resolve this difficulty and may also give
better agreement between the computed and experimental Coulomb displa-
cement energies [7]. We finally remark that the excitation energy of
IAR is a very sensitive test for the value of the residual interaction
[104], but that it must be applied with caution since spurious isospin
components are usually included in the calculation.

27

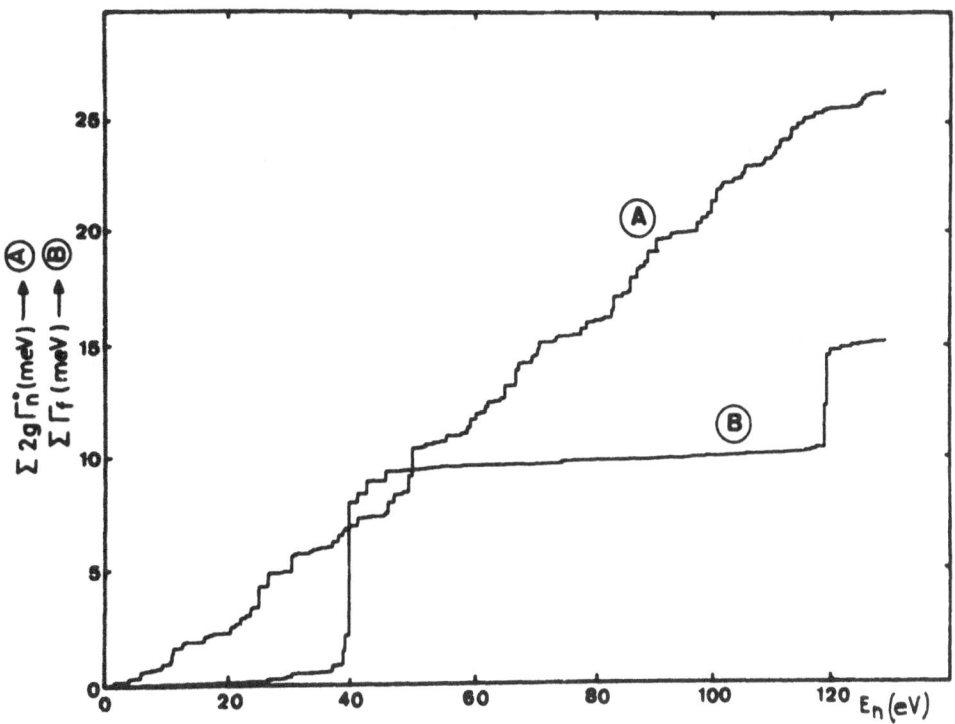

Fig. 10. From ref.[52]. Cumulative plot of
the neutron widths (A) and fission
widths (B) for ^{237}Np(n,f)

6. d. Intermediate Structure in Fission

Intermediate structure has been observed in subthreshold neutron-
induced fission [52],[96]. Curve A in fig. 10. (cf [52]) shows a cumulative plot of the neutron widths of ^{237}Np , while curve B gives the
fission widths. It is clear that the enhancement is due to a strong
doorway state in the fission channel. The enhancement is very strong,
of order 10^3 . Fig. 11 (cf [92]) shows that there exists a very dense
background of very narrow resonances, presumably of spin or parity
different from those which are enhanced. The interpretation of this
type of phenomenon in terms of a double-humped fission barrier [122]
has been given by Lynn [82] and Weigmann [132] . A very complete picture is available (cf 20) for ^{240}Pu , showing the rotational band
[117] (0^+ , 2^+ , 4^+ , 6^+ , 8^+) of a shape isomer, some β - vibrational states, seen from (d, pf) stripping [118], and, higher up,
doorway states seen in neutron-induced fission [96] . The moment of

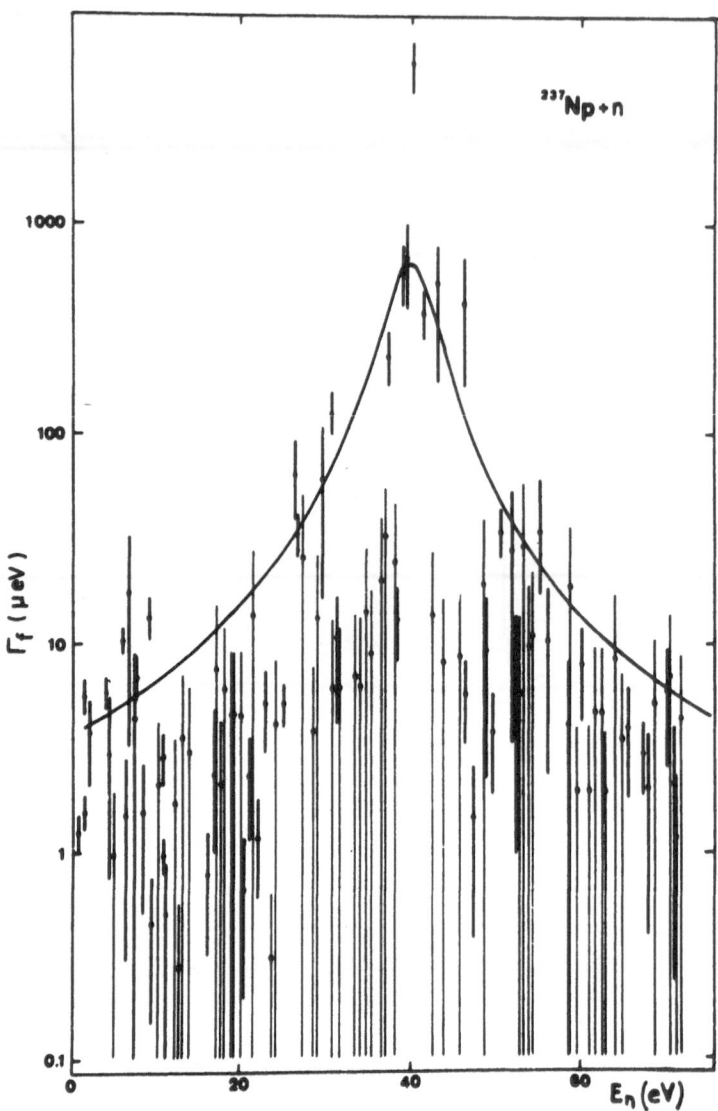

Fig. 11. From ref. [92]. Fission widths for
the ^{237}Np + n resonances. The solid curve is a Lorentzian line
drawn through the largest fission
widths

inertia obtained from the rotational band based on the shape isomer is
twice smaller than that of the ground state rotational band [117], thus
confirming the larger deformation of the shape isomer. This moment of
inertia is well reproduced by dynamical calculations [26], [116]. We

note that intermediate structure in fission can present the two extre-
me situations $\Gamma^\uparrow \gg \Gamma^\downarrow$ and $\Gamma^\uparrow \ll \Gamma^\downarrow$, depending on the relative
sizes of the internal and external barriers [133].

6. e. Miscellaneous Examples

The three types of intermediate structure briefly described
above are convincing because of two reasons. Firstly, the large value
of the enhancement of the partial widths implies a deviation from the
statistical model. Secondly, the nature of the doorway state is well
understood. Besides these examples, only very few convincing examples
of intermediate structure have been found. In isolated cases, the
interpretation is doubtful or not existing, because no systematic trend
is observed. Growing evidence from heavy ion reactions points, however, to the existence of quasimolecular states. This will be discussed briefly in the next sections. Here, we briefly list a number of
other candidates for intermediate structure.

(1) ^{19}F + n (cf [45]). Six bumps in the average total cross
section between 0.3 and 2.0 MeV . Their statistical significance
has not been quantitatively studied and does not appear very convincing. Angular distributions appear to associate most bumps with a given angular momentum. The identification of doorway configurations is
only partly successful [2].

(2) ^{56}Fe + n (cf [98]). One apparent enhancement of neutron
widths between 200 and 600 keV . Significance level not investigated. Energy dependence of $\Gamma^\downarrow$ (eq. (5.2)) is fairly strong. Doorway
state not identified. No correlation between neutron and photon widths
[9].

(3) ^{206}Pb + n [46] . Discussed in section 4. Good significance
level [9], given angular momentum and parity ($\frac{1}{2}^+$) . Doorway configuration presumably identified [41], [17]. Disputed [9], [16], [3] correlation between neutron and photon widths.

(4) ^{28}Si + n [99], [111], [71]. Significance level of enhancement between 0.4 and 1.2 MeV not yet calculated. Correlation between neutron and photon widths, probably due to direct capture [71].
Spin and parity $\frac{3}{2}$. Tentative doorway configuration: 2 p 3/2 neutron
coupled to excited state of ^{28}Si at 750 keV [71].

(5) ^{207}Pb + n [112]. Weak inhancement near 400 keV (?).
Mainly 1^- levels but a few 0^-. Sum of widths close to that of
^{208}Pb + n and ^{206}Pb + n (see section 4) for the same kinetic energy.
Estimated $\Gamma^\downarrow \simeq 50$ keV.

(6) ^{209}Bi + n [99] . Very weak enhancement near 400 keV (?).

Sum of the widths about the same as in ^{206}Pb , ^{207}Pb and ^{208}Pb + n.
Estimated $\Gamma^\downarrow \simeq$ 250 keV .

(7) ^{90}Zr + γ [8]. Enhancement of radiative transition strength between 11 and 11.5 MeV . Significance level not known, nor angular
momentum and parity or doorway configuration. Doubtful.

(8) ^{57}Fe + γ [9]. Enhancement near 250 keV believed to be due to
$\frac{1}{2}^+$ resonances. Disproved in ref. [72] .

(9) ^{57}Fe + γ [72]. Three $\frac{3}{2}^-$ resonances near $E_n \neq$ 230 keV with
large radiative width. One $\frac{1}{2}$ resonance with very large radiative
width at 606 keV . Extreme cases with only two ($\frac{3}{2}$) or zero ($\frac{1}{2}$)
complicated states. Very speculative proposals for the doorway configurations given in ref. [52] .

(10) ^{116}In + γ [34] . Enhancement of radiative widths for high
energy photons seen in ^{115}In (n,γ) . Width of enhancement only 100
eV . Claimed significance level 99% but spins and parity not measured.
Correlation observed between averaged radiative strength to low-lying
states and averaged neutron strength function. Doorway configuration
not identified. Spreading width amazingly small. Spins should be measured.

(11) ^{187}Re + γ [121] . Enhancement similar to case (10), near
E_n = 100 eV , but only 30 eV wide. Claimed significance level 99,5% ,
but spins and parity not identified.

(12) ^{70}Ge + p [125]. Narrow substructures observed within an iso-
lated $\frac{1}{2}^+$ isobaric analogue resonance in $^{71}_{33}$As . High significance level
if isolated resonances. Spin and parity $\frac{1}{2}^+$. Doorway configurations unknown. An experiment is in progress [91] to refine the measurement of
ref. [125] .

6. f. Discussion

We now briefly discuss the reason why so few intermediate structure phenomena have been found, besides the standard three cases recalled
in sections 6. b - 6. d. The observation of intermediate structure is
possible only if it is sufficiently marked to stick out from the background. Eq. (5.3b) shows that a necessary condition is

$$\frac{\Gamma^\uparrow_{dc}}{\Gamma^\downarrow} \gtrsim s_\alpha + 1 \qquad\qquad , \qquad\qquad (6.1)$$

where the background s_α is the sum of all background strength functions pertaining to the same partition fragments as channel c . If the
angular momentum and parity of each resonance is known, condition (6.1)

reduces to

$$\frac{\Gamma_{dc}^{\uparrow}}{\Gamma^{\downarrow}} \gtrsim s_c + 1 \qquad . \qquad (6.2)$$

Moreover, one should have

$$\Gamma^{\downarrow} < a_d \qquad ; \qquad \Gamma_{d'}^{\downarrow} < a_d \qquad , \qquad (6.3)$$

where a_d is the distance to the nearest strong doorway state, and $\Gamma_{d'}^{\downarrow}$ the spreading width of that doorway state.

Estimates by Payne [103] show that condition (6.2) is usually not fulfilled for 2p - 1h doorway states, except perhaps near doubly magic nuclei. This appears confirmed by experiment [99]. Condition (6. 1) is even less likely to be realized, mainly at higher energy where many channels are open. Hence, a conjunction of favourable circumstances is needed, which render $\Gamma^{\downarrow}$ very small while leaving $\Gamma_{dc}^{\uparrow}$ large. In the case of the giant dipole resonance, the value of s_{α} is small and that of $\Gamma_{dc}^{\uparrow}$ very large. The value of $\Gamma^{\downarrow}$ is exceptionally small for IAR (isospin selection rule) and for the fission isomers (difference in shape between doorway and complicated states). The value of $\Gamma^{\downarrow}$ may also become small when the density of complicated states of same spin and parity is very small. This may occur for the first few states of a given high angular momentum. These can for instance be excited in heavy ion reactions (sections 6. g and 6. h).

6. g. Alpha-nucleus Scattering

The existence of low-lying deformed excited states is well established in alpha-particle like nuclei. They can be interpreted in terms of alpha-clusters [30] or of quartet states [5]. The latter interpretation also applies to heavier nuclei. Alpha-particle transfer experiments appear to lead to well-defined groups of states at higher excitation energy [57], [94], [95], although their interpretation is somewhat ambiguous [101]. It has been proposed that these highly excited states have the character of an α - particle coupled to a core. If they lie above threshold, these states should be visible as compound nuclear resonances [135]. Compound nuclear reactions present the advantage of being less influenced by kinematical factors than direct transfer reactions.

Recently, a systematic experimental investigation of back-angle enhancements in elastic α - scattering cross sections was performed

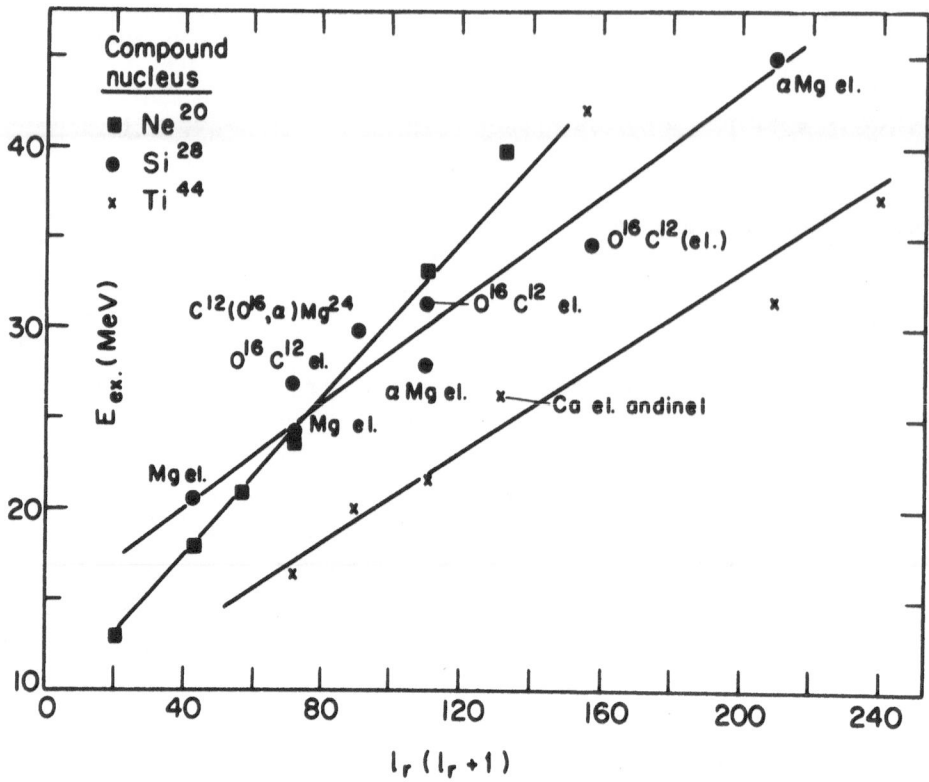

Fig. 12. From ref. ⌊107⌋ . Resonance energies
associated with each l_r as a function of $l_r(l_r + 1)$, for the compound nuclei ^{20}Ne , ^{28}Si and ^{44}Ti

for a series of target nuclei between ^{12}C and ^{48}Ca [102]. These
back-angle oscillations are very large and not accounted for by a
standard optical-modes. The location and size of the peaks and minima
in the angular distribution agree remarkably well with those of the
square of one Legendre polynomial, P_{l_r} (cos θ) , for a given excitation energy [107]. The relevant angular momentum l_r(E) changes smoothly with energy, and each l_r appears to resonate around some specific energy. When these "resonance energies" are plotted as a function of $l_r(l_r + 1)$, straight lines are obtained as shown in fig. 12.
[107]. From the slope of these lines, an effective mass equal to about
(3.2 - 5.0) nuclear masses is obtained. All these facts point to the
existence of quasi-molecular rotational states, with an α - cluster
rotating around a core. An important element in favour of this interpretation is that the same values of l_r are found for a given exci-

tation energy, for elastic and inelastic α - scattering to the 0^+
levels at 0 , 3.35 and 7.30 MeV in ^{40}Ca [110]. From these data,
one obtains the following rough estimates [107]

$$\frac{\Gamma_\alpha^\uparrow (0)}{\Gamma_{tot}} \simeq 0.07 \quad , \quad \frac{\Gamma_{\alpha'}^\uparrow (3.35)}{\Gamma_{tot}} \simeq 0.01 \quad , \quad \frac{\Gamma_{\alpha''}^\uparrow (7.30)}{\Gamma_{tot}} \simeq 0.004 \quad .$$

$$(6.4)$$

This interpretation in terms of quasi-molecular states is quite attractive and appears to meet the various criteria associated with intermediate structure. It should be substantiated, however, by more detailed calculations and by observation of the structure from other channels. We note that a similar model has been proposed in ref. [43],
where quartet states observed in medium-weight nuclei are interpreted
as α - particle resonant states.

6. h. Heavy Ion Reactions

In the reaction ^{12}C + ^{12}C , three resonance like bumps appear
[4] at the same center of mass energies ($\simeq$ 6 MeV) in several outgoing
channels (n, α, γ and p). The first two bumps are associated with
states with well-defined angular momentum and parity (2^+ and 4^+) .
Surprizingly, the total width is nearly equal to the sum of the partial
widths [4], so that the spreading width of the corresponding doorway
states would be very small. These states were interpreted [28] as
quasi-molecular states, i.e. as quasi-bound states (resonances) in a
potential well describing the relative interaction between the two
colliding ions. A repulsive potential was assumed for small separation
distances. Marked structure is also found in the ^{16}O + ^{16}O elastic
scattering [89]. They appear too strong to be due to fluctuation phenomena. The structure can be fitted with suitable optical-model potentials. Most of these show a repulsion at short distances [21], [29],
but a ℓ - dependence of the imaginary part also provides a good fit
[33].

The physical interpretation of these data is delicate. Should
they be looked at in terms of doorway states (quasi-molecular states)
or of diffractional models? The answer is difficult in view of the
absence of a reliable theoretical framework. It was recently pointed
out by Arima, Scharff-Goldhaber and McVoy [6] that the real part of
the potentials of refs. [57], [89], which were determined phenomenolo-
gically from ^{16}O + ^{16}O , produce quasi-molecular states. Their energies form a good rotational spectrum. This "rotational band" appears
to be distinct from the ground state rotational band of ^{32}S [6].

The existence of quasi-molecular states in heavy-ion scattering probably depends upon the existence of a short-range repulsion in the optical-model potential. This repulsion is based on theoretical arguments rather than experimental evidence. For instance, a repulsion is obtained by Greiner et al. [61] in an adiabatic approximation which, however, is probably valid only for much larger energies [51]. Besides the potential scattering explanation of the structure, other explanations have been proposed, based on the excitation of quasi-bound states in an inelastic channel [70], [93]. It appears difficult to distinguish experimentally between these interpretations, unless one becomes able to compute the widths of the levels. In conclusion, intermediate structure is probably present in some heavy-ion collisions, but its proper interpretation is still somewhat uncertain.

6.i. Intermediate Structure in the Residual Nucleus

We have seen that intermediate structure is interpreted in terms of the configuration mixing between a simple state and many complicated states. This type of phenomenom of course occurs below particle threshold [23], [77], and can be observed in the spectrum of emitted particles for a fixed incident energy. It is in this way that, for instance, quartet states were seen in medium-weight nuclei. In these cases, isolated simple states are seen as intermediate structure in the residual nucleus. In many cases, however, the spectrum of emitted particles displays no striking feature. It looks like the evaporation spectrum predicted by the statistical model with, however, often an excess of high-energy particles. This excess is interpreted in terms of the pre-compound emission model [62], which is discussed in another session of the present Conference.

7. Partial Widths Correlation due to Direct Reactions

We have seen in section 5 that an isolated doorway state common to two channels gives rise to a localized correlation between the partial widths of these channels. Correlations between partial widths of particle channels have up to now been explicitly observed only in very few cases [97], and have been interpreted in terms of common doorways. If the "giant" doorway state ϕ_d^c in eq. (3.7), with the sum over j now extending to all bound states, could be identical to the giant doorway state $\phi_d^{c'}$ in another channel, a correlation would even exist for all energies [86]. It has been suggested [113] that this may occur for the channels corresponding to the levels of the same rotational band in strongly deformed residual nucleus. There exists indirect evidence for this [86].

Direct reaction processes also lead to channel-channel correlations, as shown for instance in refs. [68], [86]. Direct reactions are due
to the continuum-continuum coupling (2.9). If this coupling is treated
in first order perturbation theory, the following expression for the
partial width can be obtained from eq. (2.15) [86] :

$$
\gamma_{\lambda c} = (2\pi)^{\frac{1}{2}} \{ <\chi_E^c \mid V \mid \Omega_\lambda > + \sum_{c'} \int_{\varepsilon c'}^{\infty} dE' \ (E^+ - E')^{-1} \ V_{EE'}^{cc'}
$$

$$
< \chi_{E'}^{c'} \mid V \mid \Omega_\lambda > \} \qquad\qquad\qquad\qquad (7.1)
$$

where Ω_λ is the quasi-bound state responsible for the resonance.
Clearly, the second term in the curly brackets gives rise to correla-
tions between $\gamma_{\lambda c}$ and $\gamma_{\lambda c'}$. The existence of this correlation is
indirectly demonstrated by the success of the distorted-wave Born approximation [86] and must be taken into account as a correction to the
Hauser-Feshbach formula in the calculation of the fluctuating part of
a cross section [74].

Correlations between neutron and radiative widths have been observed in many cases, for mass numbers 90 < A < 112 (mainly), and
35 < A < 65 , 136 < A < 207 . Moreover, correlations are often seen
between the radiative widths and the spectroscopic factors of the final state. A related phenomenon is the existence of anomalous capture,
i.e. of a bump (or at least an anomalous shape, "pygmy resonance" in
the high energy part of the photon spectrum, mainly for 40 < A < 65
[11]. Lane [80] has proposed a theoretical interpretation of the correlations between neutron and photon widths in the frame of R - matrix
theory; alternative treatments were proposed by Beer [12] (using
Feshbach's projection operator formalism), and by Mahaux [86]. The radiative width is given by an expression similar to eq. (7.1) [86]

$$
\gamma_{\lambda\gamma} = (2\pi)^{\frac{1}{2}} \{ <\Psi_f \mid EM \mid \Omega_\lambda > + \sum_{c'} \int dE' \ (E^+ - E')^{-1}
$$

$$
< \Psi_f \mid EM \mid \chi_{E'}^{c'} > < \chi_{E'}^{c'} \mid V \mid \Omega_\lambda > \} \qquad\qquad . \qquad\qquad (7.2)
$$

Clearly, the second term in the curly brackets implies a correlation
between $\gamma_{\lambda\gamma}$ and $\gamma_{\lambda c}$ mainly in the vicinity of the 3 s , 3 p and
4 s giant resonances (because there χ_E^c is large inside the nucleus)
and for those final states which have a large spectroscopic factor

(because $< \Psi_f \mid EM \mid \chi_{E'}^{c'} > $ is then large).

We note that the privileged configuration which is responsible for the correlation is the channel wave function χ_E^c itself. Since the phenomenon is most pronounced for values of A associated with single-particle resonances, one can extract this single-particle resonance from χ_E^c [130] and treat it like a bound state, in a separate D - space. This is essentially what is made in R - matrix theory. The correlation can then be interpreted in terms of a common doorway state. The proper interpretation of the pygmy resonance is not yet well understood. Experimental data are even sometimes in apparent contradiction with each other [25].

8. Conclusions

We have seen that intermediate processes are those whose interpretation imply that some configurations are treated on a separate footing. As examples, we discussed the neutron strength function, the intermediate structure, the identification of the doorway states and the correlations between partial widths of different channels. This list is by no means exhaustive. Indeed, any calculation in nuclear physics consists in treating some set of configurations explicitly, while including the effect of the remaining configurations in some "effective" residual interaction. Hence, the projection operator formalism (section 2) is useful in many branches of nuclear physics. Here, we discussed only a few applications which appear to fit the topic of this Conference.

References

1. AGASSI, D., Ann. Phys. (N.Y.) $\underline{65}$ 212 (1971)

2. AFNAN, I.R., Phys. Rev. $\underline{163}$ 1016 (1967)

3. ALLEN, B.J., MACKLIN, R.L., FU,C.Y. and WINTERS,R.R., to be published.

4. ALMQVIST, E., BROMLEY, D.A. and KUEHNER, J.A., Phys. Rev.Letters $\underline{4}$ 515 (1960)

5. ARIMA, A., GILLET, V. and GINOCCHIO, J., Phys. Rev. Letters $\underline{25}$ 1043 (1970)
   ARIMA, A. and GILLET, V., Ann. Phys. (N.Y) $\underline{66}$ 117 (1971)

6. ARIMA, A., SCHARFF-GOLDHABER, G. and McVOY, K.W., Phys. Letters $\underline{40B}$ 7 (1972)

7. AUERBACH, N., HÜFNER , J., KERMAN, A.K. and SHAKIN, C.M.,Revs. Mod. Phys. $\underline{14}$ 48 (1972)

8. AXEL, P., MIN, K.K. and SUTTON, D.C., Phys. Rev. $\underline{2C}$ 689 (1970)

9. BAGLAN, R.J., BOWMAN, C.D. and BERMAN, B.L., Phys. Rev. $\underline{3}$ C 2475 (1971)

10. BALDWIN, G.C. and KLAIBER, G.S., Phys. Rev. $\underline{73}$ 1156 (1948)

11. BARTHOLOMEW, G.A., International Symposium on Neutron-Capture Gamma Ray Spectroscopy (IAEA, Vienna) 1969, p. 553

12. BEER,M., Ann. Phys. (N.Y.) $\underline{65}$ 181 (1971)

13. BENETSKIJ, B.A., NEFEDOV, V.V., FRANK, I.M. and STRANICH, I.V., Nuclear Structure Study with Neutrons (Budapest 1972), contr. C 8, to be published

14. BENÖHR, H.C. and WILDERMUTH, K., Nucl. Phys. A$\underline{128}$ 1 (1969)

15. BERES, WP. and DIVADEENAM, M., Bull. Am. Phys. Soc., II, $\underline{17}$ 579 (1972)

16. BERES, W.P. and DIVADEENAM, M., Phys. Rev. 7 C 862 (1973)

17. BERES, W.P. and DIVADEENAM, M., Phys. Rev. Letters $\underline{25}$ 596 (1950)

18. BERGERE, R., and CARLOS, J.P., private communication (Saclay, 1972)

19. BETHE, H.A. and PLACZEK, G., Phys. Rev. $\underline{51}$ 450 (1937)

20. BJØRNHOLM, S., Proceedings of the EPS-Comference on Nuclear Physics, Aix-en-Provence, 1972, vol.1

21. BLAIR, J.A., Nuclear Reactions Induced by Heavy Ions (North-Holland Publ. Comp., Amsterdam) 1970, editors R. Bock and W.R. Hering, p.1

22. BLOCH, C., Many-Body Description of Nuclear Structure and Reactions (Academic Press, New York), 1966, edited by C. Block, p. 394

23. BLOCH, C., CINDRO, N., and HARAR, S., Progress in Nuclear Physics $\underline{10}$ (1969), editors D.M. Brink and J.H. Mulley

24. BLOCK, B. and FESHBACH, H., Ann. Phys. (N.Y.) $\underline{23}$ 47 (1963)

25. BOLLINGER, L.M., LOPER, G.D. and THOMAS, G.E., Phys. Rev. Letters (in press)

26. BRACK, M. and LEDERGERBER, to be published

27. BRINK, D.M., Nucl. Phys. 4 215 (1957)

28. BROMLEY, D.A., KUEHNER, J.A. and ALMQVIST, E., Phys. Rev. 123 878 (1961)

29. BROMLEY, D.A., Nuclear Reactions Induced by Heavy Ions (North-Holland Publ. Comp., Amsterdam) 1970, editors R. Bock and W.R. Hering, p. 27

30. BROWN, G.E. and GREEN, A.M., Nucl. Phys. 75 401 (1966)

31. BROWN, R.E. and TANG, Y.C., Phys. Rev. 176 1235 (1968)
    THOMPSON, D.R. and TANG, Y.C., Nucl. Phys. A106 591 (1968)
    TANG, Y.C. and BROWN, R.E., Phys. Rev. 4C 1979 (1971)
    BROWN, R.E. and TANG, Y.C., Nucl. Phys. A170 225 (1971)
    BROWN, R.E., REICHSTEIN, I. and TANG, Y.C., Nucl Phys. A178 145 (1971)
    THOMPSON, D.R. and TANG, Y.C., Phys. Rev. 4C 306 (1971)
    THOMPSON, D.R., TANG, Y.C. and BROWN, R.E., Phys. Rev. 5C 1939 (1972)

32. BUCK,B. and PEREY, F.G., Phys. Rev. Letters 8 444 (1962)

33. CHATWIN, R.A., ECK, J.S., ROBSON, D., and RICHTER, A., Phys. Rev. C1 795 (1970)

34. COCEVA, C., CORVI, F., GIACOBBE, P. and STEFANON, M., Phys. Rev. Letters 25 1047 (1970)

35. CUGNON, J., Nucl. Phys. A175 113 (1971)

36. DANOS, M. and GREINER, W., Phys. Rev. 138 B876 (1965)

37. DAVIDSON, A.M., Nucl. Phys. A180 208 (1972)

38. DIVADEENAM, M. and BERES, W.P., private communication to H.W.Newson (ref. 99.)

39. DIVADEENAM, M., BERES, W.P. and NEWSON, H.Q., Ann. Phys. (N.Y.) 69 428 (1972)

40. DIVADEENAM, M., private communication to H.W. Newson,(ref.99.)

41. DIVADEENAM, M. and BERES, W.P., Statistical Properties of Nuclei (Plenum Press, New York), 1972, p. 579

42. DOVER,C.B., LEMMER, R.H. and HAHNE, F.J.W., Ann. Phys. (N.Y.) 70 458 (1972)

43. DUDEK, A. and HODGSON, P.E., Journal de Physique 32 C 6 - 185 (1971)

44. DUMITRESCU, O. and KÜMMEL, H., Ann. Phys. (N.Y.) 71 556 (1972)

45. ELWYN, A.J., MONAHAN, J.Q., LANE, R.O. and LANGSDORF, A., Nucl. Phys. 59 113 (1964)

46. FARRELL, J.A., KYKER Jr., G.C., BILPUCH, E.G. and NEWSON, H.W., Phys. Letters 17 286 (1965)

47. FESHBACH, H., Ann. Phys. 19 287 (1962)

48. FESHBACH, H., Ann. Phys. 5 357 (1958)

49. FESHBACH, H., KERMAN, A.K. and LEMMER, R.H., Ann. Phys. (N.Y.) 41 230 (1967)

50. FESHBACH, H., PORTER, C.E. and WEISSKOPF, V.F., Phys Rev. 96 448 (1954)

51. FLIESSBACH, T., Z. Physik 242 287 (1971); Z. Physik 247 117 (1971); Nucl. Phys. (in press)

52. FUBINI, A., BLONS, J., MICHAUDON, A. and PAYA, D., Phys. Rev. Letters 20

53. GILLET, V., MELKANOFF, M. A. and RAYNAL, J., Nucl. Phys.A97 631 (1967)

54. GIRAUD, B., LE TOURNEUX, J. and WONG, C.W., Phys. Letters 32B 23 (1970)

55. GIRAUD, B. and ZAIKINE, D., Phys. Letters 37B 25 (1971)

56. GLÖCKLE, W., Nucl. Phys. A175 337 (1971) and refs. contained therein

57. GOBBI, A., MAURENZIG, P. R., CHUA, L., HADSELL, R., PARKER, P.D., SACHS, M.W., SHAPIRA, D., STOKSTAD, R., WIELAND, R. and BROMLEY, D.A., Phys Rev. Letters 26 396 (1971)

58. GOLDHABER, M. and TELLER, E., Phys. Rev. 74 1046 (1948)

59. GOOD, W.M. and KIM, H.M., Phys. Rev. 165 1329 (1968)

60. GOSWAMI, A. and GRAVES, R. D., Phys. Letters 39B 499 (1972)

61. GREINER, W. and SCHEID, W., Journal de Physique 32 C 6 - 91 (1971)

62. GRIFFIN, J.J., Phys. Rev. Letters 17 478 (1966)

63. HAMAMOTO, J., Nucl. Phys. A126 545 (1969)
    HAMAMOTO, J., Nucl. Phys. A135 576 (1969)
    HAMAMOTO, J., Nucl. Phys. A141 1 (1970)

64. HAY, W.D., Thesis, Oxford 1969

65. HODGSON, P.E., Nuclear Reactions and Nuclear Structure (Clarendon Press, Oxford), 2971. p.558

66. HORIUCHI, H., Prog. Theor. Phys. 41 705 (1969)

67. HUBER, M.G., DANOS, M., WEBER, J.J. and GREINER, W., Phys. Rev. 155 1073 (1967)

68. HÜFNER, J., MAHAUX, C, and WEIDENMÜLLER, J.A., Nucl. Phys. A105 489 (1967)

69. IACHELLO, F., Ann. Phys. (N.Y.) 52 16 (1969)

70. IMANISHI, B., Phys. Letters 27B 267 (1968)

71. JACKSON, H.E. and TOOHEY, R.E., to be published

72. JACKSON, J.E. and STRAIT, E.N., Phys. Rev. $\underline{4}$C 1314 (1971)

73. JEUKENNE, J.P. and MAHAUX, C., Nucl. Phys. A$\underline{136}$ 49 (1969)

74. KAWAI, M., KERMAN, A.K. and McVOY, K.W., Ann. Phys. (N.Y.) $\underline{75}$ 156

75. KERMAN, A.K., RODBERG, L. and YOUNG, J.E., Phys. Rev. Letters $\underline{11}$ 422 (1963)

76. KUO, T.T.S., BLOMQVIST, J. and BROWN, G.E., Phys Letters $\underline{31}$B 93 (1970)

77. LANDE, A., and BROWN, G.E., Nucl. Phys. $\underline{75}$ 344 (1966)

78. LANE, A.M., Isospin in nuclear Physics (North-Holland Publ. Comp., Amsterdam), 1969, edited by D.H. Wilkinson, p.509

79. LANE, A.M. and THOMAS, R.G., Revs. Mod. Phys. $\underline{30}$ 257 (1958)

80. LANE, A.M., Phys. Letters 31B 344 (1970); Ann. Phys. (N.Y.) $\underline{63}$ 171 (1971)

81. LEJEUNE, A. and MAHAUX, C., Z. Physik $\underline{207}$ 35 (1967)

82. LYNN, J.E., Nuclear Structure (IAEA, Vienna) 1968, p. 463

83. MACDONALD, W.M., Nucl. Phys. $\underline{54}$ 393 (1964); Nucl. Phys.$\underline{56}$ 647 (1964)

84. MAHAUX, C. and WEIDENMÜLLER, H.A., Nucl. Phys. A$\underline{94}$ 1 (1967)

85. MAHAUX, C., Statistical Properties of Nuclei (Plenum Press, New York), 1972, edited by J.B. Garg, p. 545

86. MAHAUX, C., Nuclear Structure with Neutrons, Budapest 1972, (in press)

87. MAHAUX, C. and WEIDENMÜLLER, H.A., Shell-Model Approach to Nuclear Reactions (North-Holland Publ. Comp., Amsterdam), 1969

88. MAHAUX, C., Nuclear Isospin (Academic Press, New York), 1969, edited by J.D. Anderson, S.D. Bloom, J. Cerny and W.W. True, p.337

89. MAHER, J.V., SACHS, M.W., WEIDINGER, A. and BROMLEY, D.A., Phys. Rev. $\underline{188}$ 1665 (1969)
    SINGH, P.P., SINK, D.A., SCHWANDT, P., MALMIN, R.E. and SIEMSSEN, R.H., Phys. Rev. Letters $\underline{28}$ 1714 (1972)

90. MEKJIAN, A. and MACDONALD, W.M., Nucl. Phys. A$\underline{121}$ 385 (1968)

91. MEYER, V., private communication (1972)

92. MICHAUDON, A., Statistical Properties of Nuclei (Plenum Press, New York), 1972, edited by J.B. Carg, p. 149

93. MICHAUD, G.J. and VOGT, E. W., Phys. Rev. C$\underline{5}$ 350 (1972)

94. MIDDLETON, R., GARRETT, J.D. and FORTUNE, H.T., Phys. Rev. Letters $\underline{24}$ 1436 (1970)

95. MIDDLETON, R., GARRETT, J.D. and FORTUNE, H.T., Phys. Rev. Letters $\underline{27}$ 950 (1971)

96. MIGNECO, E. and THEOBALD, J.P., Nucl.Phys. A$\underline{112}$ 603 (1968)

97. MITCHELL, G.E., BILPUCH, E.G., MOSES, J.D., PETERS, W.C. and PROCHNON, N.H., Statistical Properties of Nuclei (Plenum Press, New York), 1972, edited by J.B. Garg, p. 299

98. MONAHAN, J.E. and ELWYN, A.J., Phys. Rev. Letters $\underline{20}$ 1119 (1968) ELWYN, A.J. and MONAHAN, J. E., Nucl. Phys. A$\underline{123}$ 33 (1969)

99. NEWSON, H.W., Statistical Properties of Nuclei (Plenum Press, New York), 1972, edited by J.B. Garg, p. 309

100. NEWSON, H.W.,BLOCK, R.C., NICHOLS, P.F., TAYLOR, A. and FURR, Ann. Phys. (N.Y.) $\underline{8}$ 211 (1959)

101. NOBLE, J.V., Phys. Rev. Letters $\underline{28}$ 111 (1972)

102. OESCHLE, H., SCHRÖTER, H., FUCHS, H., BAUM, L., GAUL, G., LÜDECKE, H., SANTO, R. and STOCK, R., Phys. Rev. Letters $\underline{28}$ 694 (1972)

103. PAYNE, G.L., Phys. Rev. $\underline{174}$ 1227 (1968)

104. RAJEWSKI, J. and KIRSON, M.W., Phys Letters $\underline{38}$B 162 (1972)

105. RICHARD, P., MOORE, C.F., ROBSON, D. and FOX, J.D., Phys. Rev. Letters $\underline{13}$ 343 a (1964)

106. RICHTER, A., GROSSE, E., HÜFNER, J., WEIDENMÜLLER, H.A. and TEPEL, J.W., Phys. Letters $\underline{38}$B 349 (1972)

107. RINAT, A.S., Phys. Letters $\underline{38}$B 281 (1972)

108. ROBSON, D., Phys. Rev. $\underline{137}$ B 505 (1965)

109. SAITO, S., Prog. Theor. Phys. $\underline{41}$ 705 (1969)

110. SCHMEING, H. and SANTO, R., Phys. Letters $\underline{33}$B 219 (1970)

111. SCHWARTZ, R.B., SCHRACK, R.A. and HEATON, H.T., Bull. Am. Phys. Soc. $\underline{16}$ 495 (1971)

112. SEIBEL, F.T., BILPUCH, E.G. and NEWSON, H.W., Ann. Phys. (N.Y.) $\underline{69}$ 451 (1972)

113. SEVGEN, A., Ph. D Thesis, Yale University, 1971, and to be published

114. SHAKIN, C.M., Ann. Phys. (N.Y.) $\underline{22}$ 373 (1963)

115. SHAKIN, C.M. and WANG, W.L., Phys. Rev. Letters $\underline{26}$ 902 (1971)

116. SOBICZEWSKI, A. and BJØRNHOLM, S., to be published

117. SPECHT, H.J., KONECNY, E., HEUNEMAN, D. and WEBER, J., Prodeedings of the EPS-Conference on Nuclear Physics, Aix-en-Provence, 1972, vol. 2, communication I. 6

118. SPECHT, H.J., FRAZER, J.S., MILTON, J.C.D. and DAVIES, W.G., Physics and Chemistry of Fission (IAEA, Vienna) 1969, p. 363

119. SPICER, B.M., Advances in Nuclear Physics, vol. $\underline{2}$ (Plenum Press), 1969, edited by M. Baranger and E. Vogt, p. 1

120. STEINWEDEL, H., JENSEN, J.H. and JENSEN, P., Phys. Rev. $\underline{79}$ 1019 (1950)

121. STOLOVY, A., NAMENSON, A.I. and GODLOVE, T.F., Phys. Rev. $\underline{4}$C 1466 (1971)

122. STRUTINSKY, V.M., Nucl Phys. A$\underline{95}$ 420 (1967); Nucl. Phys. A$\underline{122}$ 1 (1968)

123. TABAKIN, F., Nucl. Phys. A$\underline{182}$ 497 (1972)

124. DE TAKACSY, N.B., Phys. Rev. $\underline{5}$C 1883 (1972)

125. TEMMER, G.M., MARUYAMA, M., MINGAY, D.W., PETRASCU, M. and VAN BREE, R., Phys. Rev. Letters $\underline{26}$ 1341 (1971)

126. TOBOCMAN, W., Phys. Rev. $\underline{182}$ 989 (1969)
SCHMITTROTH, F. and TOBOCMAN, W., Phys. Rev. $\underline{187}$ 1735 (1969)
RODJAK, D., TOBOCMAN, W. and TANDON, G.K., Phys. Rev. (in press)

127. DE TOLEDO PIZA, A.F.R. and KERMAN, A. K., Ann. Phys. (N.Y.) $\underline{43}$ 363 (1967); Ann. Phys. (N.Y.) $\underline{48}$ 173 (1968)

128. DE TOLEDO PIZA, A.F. R., Nucl. Phys. A$\underline{184}$ 303 (1972)

129. LETOURNEUX, J., Kgl. Danske Videnskab. Selskab., Mat.-Fys. Medd. $\underline{34}$ n. 11 (1965)

130. WANG, W.L. and SHAKIN, C.M., Phys. Letters $\underline{32}$B 421 (1970)

131. WANG, W.L. and SHAKIN, C.M., Phys. Rev. $\underline{5}$C 1898 (1972)

132. WEIGMANN, H., Zeits für Physik $\underline{214}$ 7 (1968)

133. WEIGMANN, H. and THEOBALD, J.P., Nucl. Phys. A$\underline{187}$ 305 (1972)

134. WEISSKOPF, V.F., Physics Today $\underline{14}$, n. 7 18 (1961)

135. WELLER, H.R., Phys. Rev. Letters $\underline{28}$ 147 (1972)

136. WILDERMUTH, K. and McCLURE, W., Cluster Representation of Nuclei (Springer, Berlin), 1966

137. WONG, C.W., Nucl. Phys. A$\underline{147}$ 545 and 563 (1970)

138. YUKAWA, T., Phys. Letters $\underline{38}$B 1 (1972)
YUKAWA, T., Nucl Phys. A$\underline{186}$ 127 (1972)

139. ZAIKINE, D., Nucl. Phys. A$\underline{170}$ 584 (1971)

NUCLEAR EQUILIBRATION PROCESSES AT MODERATE EXCITATIONS

M. BLANN[‡]

Department of Chemistry and Nuclear Structure Laboratory

University of Rochester

Rochester, New York USA 14627

1. Introduction

The equilibrium statistical model has been with us for over 35
years [52]; direct reaction models have similarly been around for many
years. Our intuition and experimental results both tell us that the
equilibrium model must become progressively poorer as excitation energy increases, and lifetimes decrease for excited nuclear states. The
question becomes one of "what happens on the way to equilibrium?" Does
the nucleus decay after achieving an equilibrium state, or before, or
a bit of both? What is the time depedence of this process, how does
this vary with mass and energy? There have been some models proposed
for answering this question in the past, but for the most part, this
question involves work dating since 1966. It is therefore a relatively
new area in nuclear physics, and one which is still developing. The
goal of this presentation is to look at several related models which
have been applied to the equilibration question, trying to give a simple qualitative description of the underlying physical assumptions of
each model, discussing contrapunctally their similarities and their
differences. Comparisons of predictions of the several models with one
another will be presented. One must return from the world of fantasy
on occasion, and so results of these models will also be compared with
experimental results.

The main point of similarity between the models to be described
is the use of free nucleon-nucleon scattering cross sections in evaluating the rate of transitions within the nucleus, corrected of course
for the Pauli exclusion principle. This implies the assumption that
all, or at least a predominant number of the interactions are binary
in nature. This assumption is made in some ignorance at the lower energies, perhaps because we do not know how to proceed to treat multipar-
ticle or collective excitations. It becomes an assumption to test by
agreement between model and experimental results. It means,for example,
that we immediately concede that the model results should not be expected to reproduce experimental results which are due to collective be-

---

‡ Supported in part by the U.S. Atomic Energy Commission

havior.

2. The Intranuclear Cascade Model

The intranuclear cascade model is the earliest model of those to
be discussed which might have been applied to the question of equilibration in nuclear reactions (cf [49], [27],[2], [46], [43], [41], [22],
[48], [36], [42],[3], [18], [19], [20]). In point of fact, this application
has not been made, with the exception of some work within the past two
years [33], [34]. The cascade model assumes a succession of two-body
interactions which is followed in three dimensional geometry. The method of calculation which has been employed to date has been one in
which the trajectories of the nucleons are followed one at a time during the cascade, until some arbitrary energy, generally considerably
above the average equilibrium value, has been attained by the nucleon.
Then the trajectories of other struck nucleons are followed one at a
time, etc. While the time evolution of the reaction could be extracted
from such an approach, this has not been done.

As time and computers have evolved, the cascade programs have become more complex with respect to the physics going into the calculations. One aspect of this is the treatment of the nuclear potential
well and the nucleon density distributions. Figure 1, from the work of
Chen et al. [18] , illustrates some of these distributions. The early
square well (uniform) distribution is shown, compared with the Fermi
distribution from electron scattering data [35] and with a step distribution which was used to approximate the Fermi distribution in the program due to Chen et al . Bertini has approximated the Fermi distribution with a three step potential [3]. I shall return to a discussion of
the nucleon density distribution later on with respect to other models,
and then present results calculated with the cascade model.

The cascade model is the only one of those to be discussed which
predicts angular distributions. In the medium energy range, it does
not predict them very well [34]. A great simplification results if one
is willing to abandon the geometric information, substitute angle averaged distributions for nucleon-nucleon scattering and attempt an application of phase space arguments to the equilibration process. This
is basically the idea behind the other models to be discussed.

3. The Harp-Miller-Berne (HMB) Model
3.1. Physical Description and Formulation

The physical ideas of the HBM model [32] are illustrated in fig.2.
Consider the initiation of the reaction at τ_0, as shown on the left

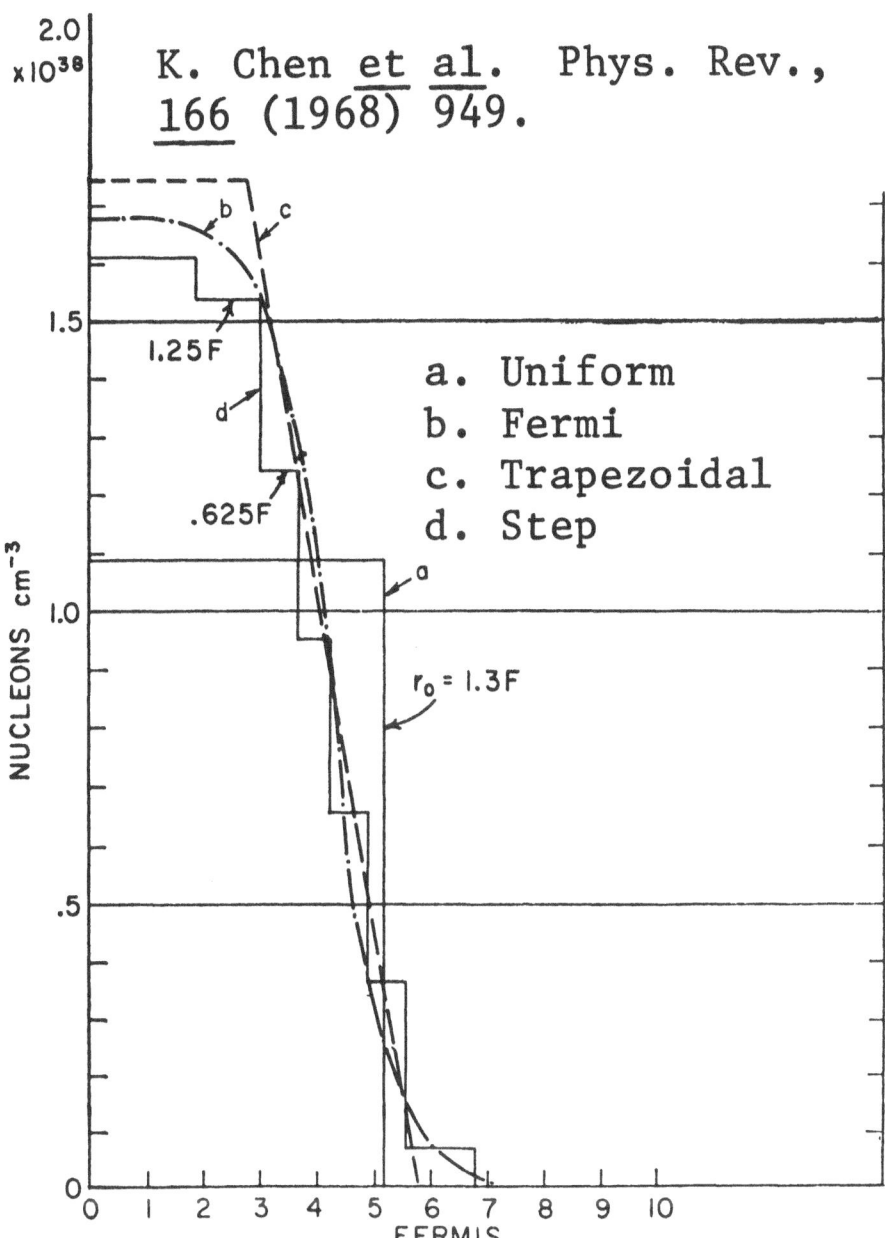

Fig. 1. Nuclear density distributions. The familiar square
well, or constant density distribution is shown compared with
the Fermi and trapezoidal distributions consistent with
electron scattering data. The step distribution used in the
cascade calculation of Chen et al is also indicated. (From
ref. [18])

46

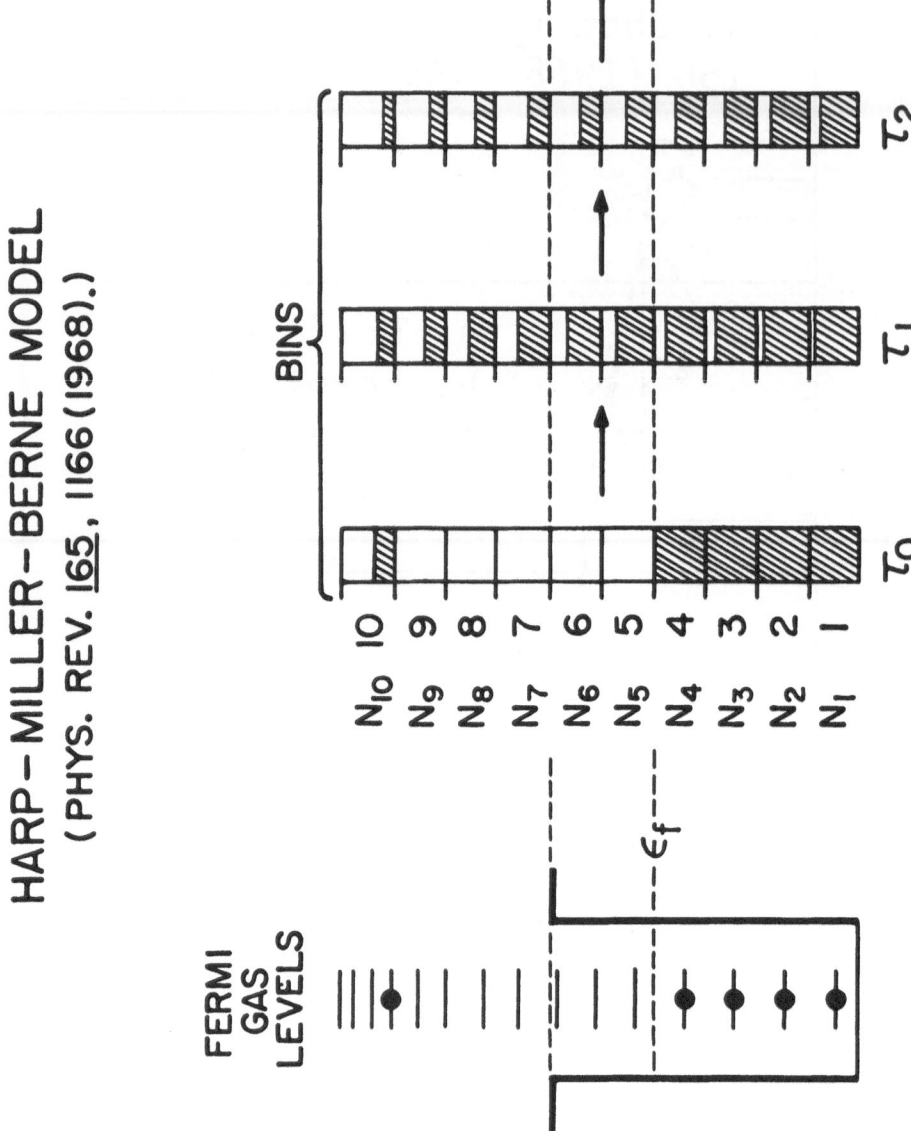

Fig. 2. Pictorial representation of the equilibration model
of refs. [32] and [33]

of the figure. Energy bins of some width, e.g. 1 MeV, are defined, and
the number of available single particle levels in each bin are computed
and stored; a Fermi gas distribution has been used in calculations co-
vered in publications to date, however, Nilsson levels could presumably
be substituted as well for calculations at low excitations. The calculation can, and has been applied either beginning with some initial
arbitrary population of excited particles and holes, or with a nucleus
in the ground state as shown here. The fractional occupation of each
bin is followed in the calculation as a function of time. For a given
incident nucleon the rate of allowed transitions with all nucleons in
the nucleus is computed, as is the rate of emission of the excited particles into the continuum. Again, n-n scattering cross sections (for
velocity vectors at $90°$) are used for calculating the two body transition rates with each energy partition being assigned equal a-priori
probability, and inverse cross sections and the free particle phase
space factors are used to compute transmission rates into the continuum. Therefore a statistical argument is applied as to the number of
ways the particle may be emitted vs. the number of ways it may make an
internal transition, and the cross section is divided proportionately.

After computing the relative probabilities of scattering into and
out of each bin and of emission from bins above the particle binding
energies, populations of all bins are changed accordingly, as shown in
the center of this figure. Now for the particles occupying each bin,
the earlier calculation must be repeated, so that all possible ways of
scattering into and out of each bin must be computed and, once more,
the populations must be changed accordingly. The solution of the equilibration problem in this model rests in computer solution of a set of
coupled differential equations, accomplishing the operations described.
Harp and Miller have recently extended this treatment to a two Fermion
gas [32]. Some details of the calculation are presented below.

(1) It is assumed that interactions within the nucleus arise from
nucleon-nucleon scattering processes; thus two nucleons are always involved, going from two initial states to two final states.

(2) The transition probabilities are all dependent only on energies of the particle or particles involved.

(3) The transition probabilities vary slowly with energy over some
interval $\Delta\varepsilon$ so that a constant value of the transition probabilities
may be used for all levels within a given "bin".

A grouping of levels within the nucleus is then defined with a total number of states within the i-th group given (for protons) by

$$g_i^P = \int_{\epsilon_i^P - 1/2\Delta\epsilon}^{\epsilon_i^P + 1/2\Delta\epsilon} \rho_p(\epsilon)\,d\epsilon \tag{1}$$

where $\rho(\epsilon) = 4\pi V(2M)^{3/2}\epsilon^{1/2}/h^3$ is the density of nuclear translational states, with V the nuclear volume and M the nucleon mass.
This definition applies to all "bins", whether the nucleon energy is
less than or greater than the Fermi plus binding energies.

A grouping of number of states for a nucleon outside the nucleus
is defined by

$$g_{i'}^P = \int_{\epsilon_{i'}^P - 1/2\Delta\epsilon}^{\epsilon_{i'}^P + 1/2\Delta\epsilon} \rho'_p(\epsilon)\,d\epsilon \tag{2}$$

where $\rho'(\epsilon') = \left[4\pi\Omega(2M)^{3/2}/h^3\right]\epsilon'^{1/2}$

Occupation numbers for the "bins" or energy subgroups are defined
as

$n_i g_i = N_i =$ the total number of occupied states within the i-th
group.

The master equations which describe the relaxation process of the
proton Fermi gas in the two-gas model are given by:

$$\frac{dN_i^P}{dt} = \sum_{jkl} \omega_{kl\to ij}^{PP} g_k^P g_l^P g_j^P n_k^P n_l^P (1-n_i^P)(1-n_j^P) - \omega_{ij\to kl}^{PP} g_j^P g_k^P g_l^P n_i^P n_j^P (1-n_k^P)(1-n_l^P)$$

$$\times \delta(\epsilon_i^P + \epsilon_j^P - \epsilon_l^P - \epsilon_k^P)$$

$$+ \sum_{jkl} \omega_{kl\to ij}^{PN} g_k^P g_l^N g_j^N n_i^N n_k^P (1-n_i^P)(1-n_j^N) - \omega_{ij\to kl}^{PN} g_j^N g_l^N g_k^P n_i^P n_j^N (1-n_k^P)(1-n_l^N)$$

$$\times \delta(\epsilon_i^P + \epsilon_j^N - \epsilon_k^P - \epsilon_l^N)$$

$$- n_i^P \omega_{i\to i'}^P g_{i'}^P \delta(\epsilon_{i'}^P - \epsilon_i^P + \epsilon_i^P + \epsilon_f^P + BE_p) \tag{3}$$

$$\frac{dN_{i'}^P}{dt} = n_i^P g_i^P \omega_{i \rightarrow i'}^P \cdot g_{i'}^P \cdot \delta(\varepsilon_{i'}^P, -\varepsilon_i^P + \varepsilon_f^P + BE_P) \qquad (4)$$

$$(i = 1, \ldots, \varepsilon_f^P + E^* \qquad i' = 1, \ldots, E^* - BE_P)$$

Symbols in Eqs. (3) and (4) are defined as follows:

$\dfrac{dN_i^P}{dt}$ — time rate of change of number of protons in the i-th subgroup

g_i^P — number of levels in the i-th subgroup

$\omega_{ij \rightarrow kl}^{XY}$ — probability per unit time that a nucleon of type x in a particular state of the i-th group scatters with a nucleon of type y in a particular state of the j-th group with particle y going to the l-th group and particle x to the k-th group.

$\omega_{i \rightarrow i'}^P$ — probability per unit time that a proton in a particular state of the i-th group escapes to the continuum.

$N_{i'}^P$ — number of escaped protons with laboratory energy ε_i^P. The delta functions are present for energy conservation.

The transition probabilities were defined as follows:

$$\omega_{ij \rightarrow kl}^{PP} = \frac{\sigma_{PP}(\varepsilon_i^P + \varepsilon_j^P) \left[2(\varepsilon_i^P + \varepsilon_j^P)/M\right]^{1/2}}{V \Sigma g_m^P g_n^P \delta(\varepsilon_i^P + \varepsilon_j^P - \varepsilon_m^P - \varepsilon_n^P)} \qquad (5)$$
$$\phantom{\omega_{ij \rightarrow kl}^{PP} = }mn$$

where $\sigma_{PP}(\varepsilon)$ is the elementary proton-proton elastic scattering cross section after removal of Coulomb effects; appropriate changes are made for PN transitions, and $\sigma_{PP} = \sigma_{NN}$. The summation in the denominator of (5) is taken only over those states which are allowed in the P-P scattering process within the nucleus.

$$\omega_{i \rightarrow i'}^P = \sigma_{inv}^P(\varepsilon_{i'}^P) \left[2\varepsilon_i^P/M\right]^{1/2}/g_i^P \Omega \qquad (6)$$

where $\sigma_{inv}^P(\varepsilon_{i'}^P)$ is the inverse cross section for the absorption by a nucleus of a proton of energy $\varepsilon_{i''}^P$, and Ω is the laboratory volume.

3.2. Results of Master Equation Approach

The calculation can be started with the system in any arbitrary initial configuration. Two classes of calculations have been investigated so far. One is a consideration of the relaxation of a gas at extremely high excitations by raising some number of particles to ε_f+B before beginning the calculation. Enough particles are "excited" in a sample calculation so as to give an excitation which exceeds the total nuclear binding energy. The relaxation process is then followed until equilibrium takes place and the number and spectrum of emitted nucleons is followed. The second type of problem considered is that of a nuclear reaction at medium excitations. In considering the case of relaxation of a one-component 100 Fermion system excited to 1054 MeV, an excitation loss of 10% and particle loss of 5% was found prior to equilibration [32]. In considering a two-component system (^{182}W) excited to1724 MeV a consistent result was found; 14% of the system excitation and 5% of the particles were emitted prior to equilibrium. Discussion of results of calculations for systems at low excitations with the HMB model will be presented in section 7.

Let me summarize several very appealing aspects of this model. For one, there are fewer assumptions than in some of the models yet to be discussed. Secondly, the model gives many important features immediately. One of these is a limit to the depth of hole excitations, which can be an important consideration. Another possibility is the use of Nilsson orbitals to determine bin occupancy, which will give some estimate of the influence of nuclear structure characteristics on particle spectra. It has been shown that this should be an important consideration in some situations [32,44]. Let us now consider other approaches to the equilibration problem, in which additional assumptions are made in order to gain in simplicity of calculation and physical clarity of the model.

4. The Exciton Model

The model of Harp, Miller and Berne is complicated by the problem of following the bin populations in time. The exciton model due to Griffin [28] avoids this by substituting densities of states characterized by particle-hole number with a statictical assumption for the population of each of these intermediate states in the equilibration sequence. A formulation for the particle spectrum which is quite simple in form may then be written. The models which are to be discussed in later sections will also use intermediate state densities to gain this simplification, and in this sense are relatives of the exciton model [28].

4.1. Physical Description and Formulation

The physical concept of the exciton model is illustrated in fig. 3. A nucleon is shown entering the nuclear potential on the left. All

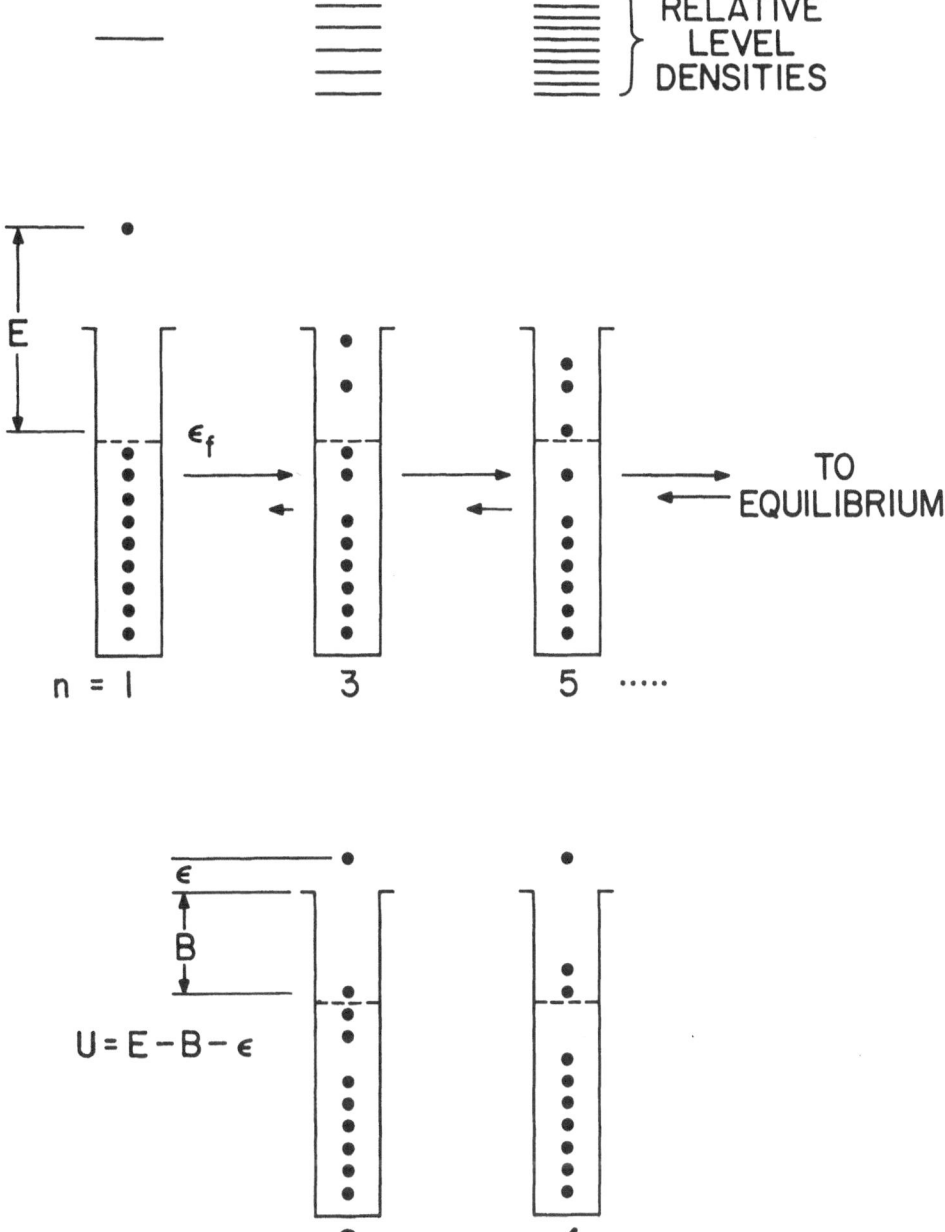

Fig. 3. Pictorial representation of the equilibration process in the exciton model

the fermions are in their ground states. It is then assumed that a series of two-body interactions occur, as is illustrated in fig. 3. The
first interaction would therefore lead to a 2p-1h (two pariicle, one
hole) state, and it is assumed that any configuration is equally likely. This, in turn, could be followed by a transition between one of
the two excited particles and a particle in the ground state, or with
the other excited particle. This could lead either to a 3p2h state,
back to the original configuration, or to a different 2p1h state. The
likehood of each occurrence will, in accord with the Golden Rule, be
assumed to be proportional to the density of the accessible final states. Since the initial simple configurations have far fewer than the
equilibrium particle-hole numbers, the level densities are rapidly increasing functions of increasing p-h number, and the system then goes
preponderately in the direction of equilibrium as is indicated in this
figure by the larger arrows in that direction. I'll return to this
point more quantitatively later.

For a given configuration, specified by the particle-hole number,
some fraction will have at least one particle with energy in excess of
its binding energy. If one can compute the fraction of particles in a
given exciton number state which are at a given energy $\varepsilon + B$ above
the Fermi energy (where B is the particle binding energy and ε is the
channel of the particle), then one can compute the relative probability
energy
of emission of a particle having such a kinetic energy from a state of
complexity as given by the exciton number. By summing over the contribution from each state, the total spectrum emitted prior to equilibration may be calculated, on a relative basis.

Note that for the simplest states there is highest probability of
having high kinetic energy particle emission as the average excitation
per particle must be highest. As the p-h number increases toward the
equilibrium value and the total number of ways the energy may be parti-
tioned between particles and holes increases exponentially, the probability that any one particle has some high energy exponentially decreases as, therefore, does the emission rate for such processes. Perhaps
in heavy ion reactions one gets initially a few lighter clusters, e.g.
alpha particles. These might then have high average excitation due to
limited numbers, and might therefore show a preequilibrium like spectrum. Telescope studies looking for this would be of interest. If the
phenomenon were observed, it might give information on the microscopic
evolution of a heavy ion reaction.

The main assumption in the Exciton model is that every configuration of each intermediate state will occur with equal apriori probabi-

53

lity during the equilibration process.

Let me present a closed form derivation of the Exciton model, followed by a discussion of the simplest predictions of the model, and a
comparison of experimental results with these predictions. This will
be followed by a pedagogical illustration of the time evolution of a
nuclear reaction according to this model. Here it will be shown that
the spectrum of emitted particles may be represented by a long and a
short component with respect to the time of emission while this is a
good approximation at lower excitations, it becomes progressively poorer at higher excitations. It will also be indicated that the Weiskopf
model is just the infinite time limit of the time dependent calculation.

It will be assumed that the fraction of n-exciton states in which
one paritcle is at an energy $\varepsilon + B$ above the Fermi energy is given
by the ratio $\rho_n(U,\varepsilon)/\rho_n(E) = \rho_{p,g}(U,\varepsilon)/\rho_{p,h}(E)$ where U is the residual
nucleus excitation if there is particle emission with channel energy ε.

Then the probability of decay from an n exciton state is given
by [9], [13]

$$P_n(\varepsilon)d\varepsilon = (2s+1) \left[\rho_n(U,\varepsilon)/\rho_n(E)\right] \frac{\Omega 4\pi p^2 dp}{h^3} \cdot \frac{\sigma v}{\Omega} \tau_n \qquad (7)$$

where τ_n is the mean lifetime of the n exciton state, the other factors being the particle spin degeneracy, the phase space and penetrability factors. The total decay expression results from summing over
the terms in eq. (7)

$$P(\varepsilon)d\varepsilon = \sum_{\substack{n=n_o \\ \Delta n=+2}}^{\bar{n}} P_n(\varepsilon)d\varepsilon \propto \frac{(2s+1)}{(gE)} \sigma\varepsilon m \sum_{\substack{n=n_o \\ \Delta n=+2}}^{\bar{n}} (\frac{U}{E})^{n-2} p(n-1)\tau_n d\varepsilon \qquad (8)$$

In writing eq. (8) down, an intermediate state density expression due
to Ericson [25] has been substituted for the particle-hole state density. We'll show later that the power series in (8) generally converges
for $n << \bar{n}$. It is implicitly assumed in (8) that the fraction of particle emission is negligible.

It was suggested [28] that τ_n be evaluated, on a relative basis,
by the Golden Rule of Fermi,

$$\lambda_{n,n'} = \frac{2\pi}{\hbar} |M|^2 \rho_n'(E) \qquad (9)$$

where $\rho_n{'}(E)$ is the density of accessible final states and $\lambda_{n,n'}$ is the transition rate from a given initial n exciton state to any of the accessible n' exciton final states. Ignorance as to $|M|^2$ makes it impossible at present to evaluate τ_n on an absolute basis.

Formulas giving the average number of accessible states for each type of transition were derived by Williams [54]:

$$\lambda_+ = \frac{2\pi}{\hbar} |\bar{M}|^2 \frac{g^3 U^2}{(p+h+1)} \tag{10a}$$

$$\lambda_- = \frac{2\pi}{\hbar} |\bar{M}|^2 g [p \cdot h \cdot (p+h-2)] \tag{10b}$$

$$\lambda_0 = \frac{2\pi}{\hbar} |\bar{M}|^2 g^2 U [\frac{3(p+h)-2}{4}] \tag{10c}$$

It may be seen from these results that $\lambda_{n+2} >> \lambda_{n-2}$ if $n << \bar{n}$, and that $\bar{n} = \sqrt{2gE}$. The latter result follows from realizing that $\lambda_{n+2} = \lambda_{n-2}$ when $n = \bar{n}$. It may also be seen that τ_n is proportional to n . This dependence of τ_n would not alter (8) very much from the form it has when τ_n is assumed to be a constant. Analyses have been made comparing these assumptions; the results are not, in my opinion, significantly different [21].

4.2. Comparisons with Experimental Results

Reference to Eq. (8) permits one to see the qualitative predictions of the exciton model. The following points might be noted:

(1) The shapes of the predicted particle spectra are independent of the single particle level density parameter g. Thus no variation of this parameter will affect the spectral distribution predicted within the approximations of level densities used. While targets of different masses are predicted to give equilibrium spectral distributions varying in shape with mass number, no such variation is predicted in the pre-equilibrium model.

(2) The spectral distribution is very sensitive to n_o, the initial exciton number of the compound state as it enters as an exponent in the power series in $(\frac{U}{E})$. If a given projectile always forms the compound state with the same n_o in any target, then pre-equilibrium spectra from all targets at a given bombarding energy should be nearly identical. This would explain the observation of Swenson et al. [50], [51] that all of their (α,p) spectral analyses gave the same level spacing parameter, independent of target mass and in contradiction to the predictions of the equilibrium statistical model.

This point is further illustrated in fig. 4. which contains data

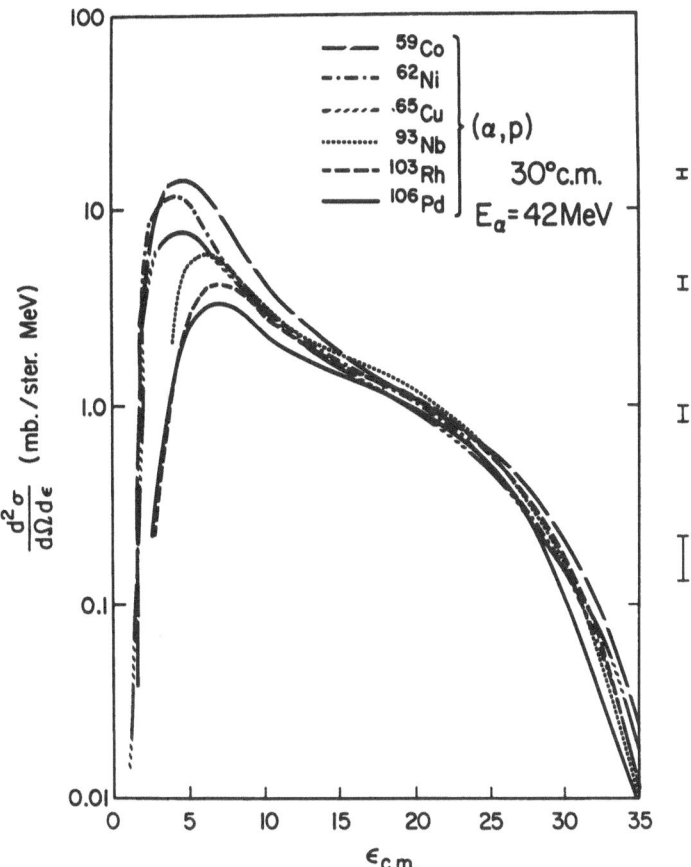

Fig. 4. Proton spectra following 42 MeV ^{4}He ion bombardment
of several targets (R.W. West, ref. [53])

due to West [53]. Note that while the equilibrium portions of the
spectra vary with mass number, the high energy regions are nearly identical. These data have not been normalized.

(3) For low (U/E), the leading term of the series dominates and
the value of n_o may be determined from the slope of log $P(\varepsilon)/\varepsilon\sigma$ versus log U. At higher values of U/E, higher order terms become more im-
portant, and such an analysis may no longer be valid. What about other
bombarding energies and projectiles? Analyses of a large body of data
for nucleon and α induced reactions are summarized in fig. 5. The
sources of the experimental data are indicated on the figure, followed
by the name of the person doing the analysis. The data shown for incident
α particles were analyzed from the slopes of the spectra, suitably
plotted; the data for proton induced reactions were analyzed both by
the slope method and by finding the integer value for n_o which gave
calculated spectra best reproducing the data. The indication is that
n_o is fixed, at least to within a 1 unit range, for a given projectile
type, independent of the bombarding energy and mass number of the
projectile. It would be very desirable to have data over a wider range
of bombarding energies for α and nucleon induced reactions for several
reasons. First, the independence of n_o on energy is supported by these data over only a narrow range for a given projectile type and, second, because uncertainties decrease at higher bombarding energies. It
should be stated that analyses of (α,n) spectra by Magda et al. [39],
Magda et al. [40], Grimes et al. [29] have given n_o values of 3 to 5.
The different n_o values for α particle induced reactions may result
from analyzing different regions of the spectra. Note the discrepancies
on fig. 5. for reactions on targets near shell closures: suggesting the
inadequacy of the use of an equidistant spacing model for level densities near shell closures.

4.3. Master equation approach to exciton model

In Subsection 4.1. we listed Williams' expressions for the average
density of accessible states for transitions for which $\Delta n = +2,-2$ and
0. With Eq. (9) rates are therefore specified for each transition type.
It seemed that it might be interesting from a pedagogical viewpoint to
follow the time evolution of a reaction considering both ±2 exciton
change transitions, in order to investigate several questions. One of
these is whether the closed form expression is a good approximation to
results of more detailed calculations. The other interesting point is
to emphasize that the exciton model provides a unified description of
nuclear reactions beginning with the target projectile interaction,
proceeding through the attainment of equilibrium, and finally including

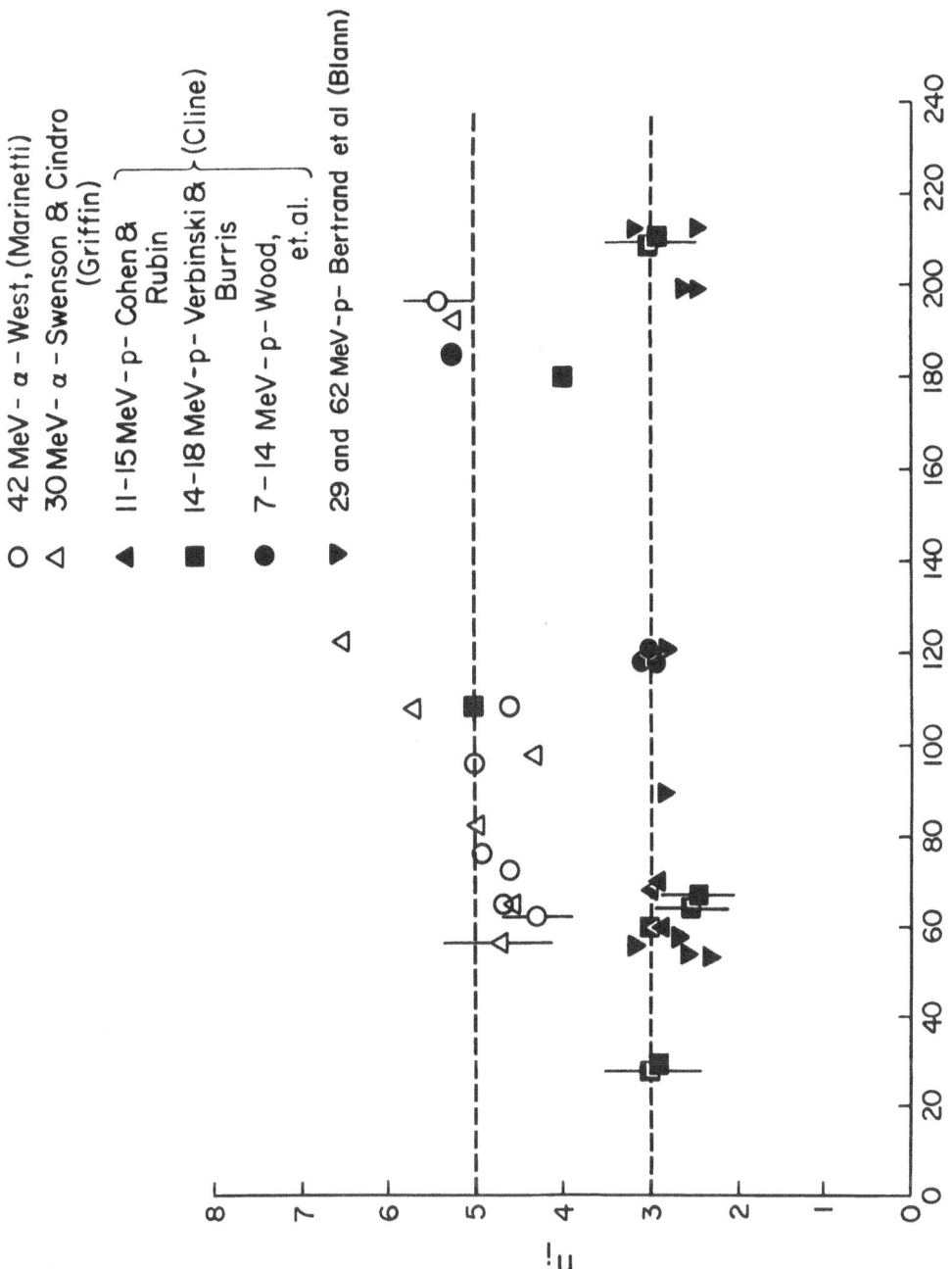

Fig. 5. Initial exciton values, n_o, deduced from particle spectral measurements for nucleon and α induced raactions on a variety of targets. Source of data and name of person doing the analysis are indicated on the figure

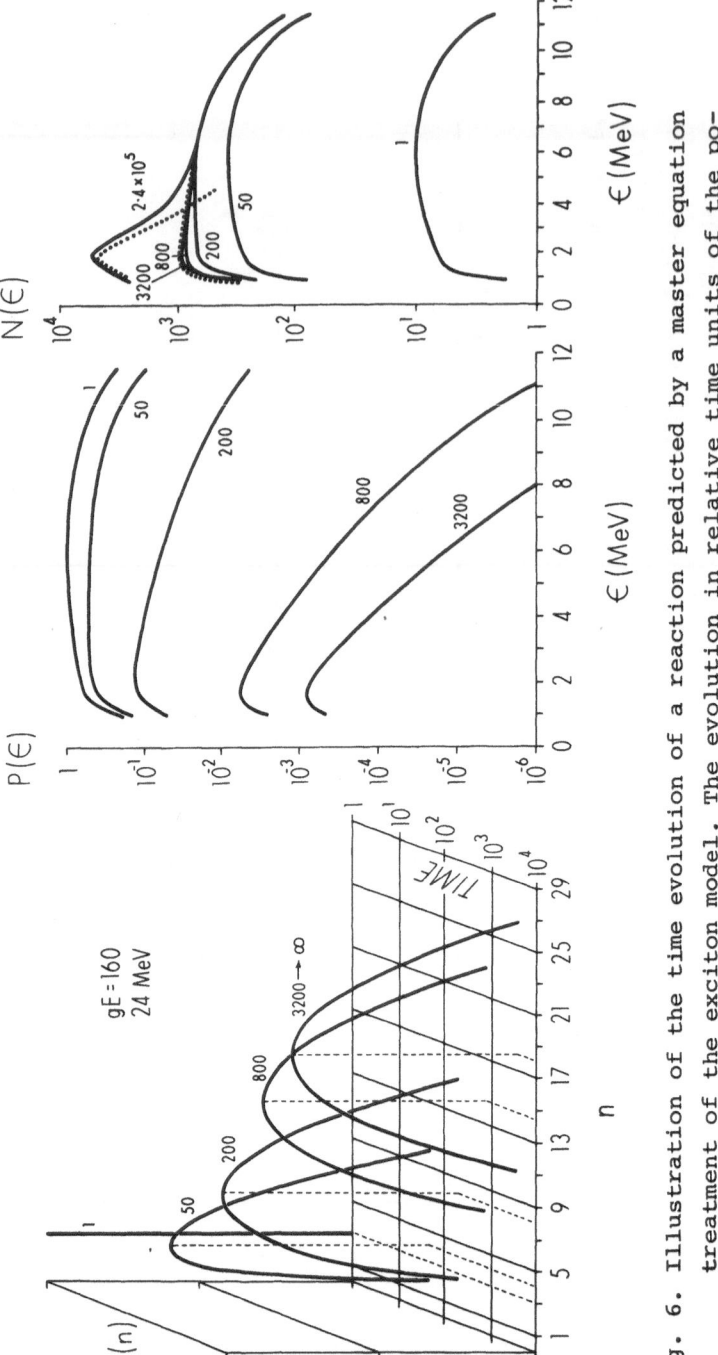

Fig. 6. Illustration of the time evolution of a reaction predicted by a master equation treatment of the exciton model. The evolution in relative time units of the population probability of different exciton numbers is shown on the three-dimensional plot. The spectral distributions at given instants of time are shown in the center. Times shown correspond to those given for the population versus exciton distributions on the left. The spectra integrated up to the indicated times are shown on the right hand side of the figure

the decay of the equilibrium compound nucleus. The Weiskopf equilibrium expression [52] is seen to be the pre-equilibrium model
expression after the passage of a time equal to the equilibration time
[32] , [21] .

The set of coupled differential equations for computer solution
are represented by the set

$$\frac{dP(n,t)}{dt} = P(n-2,t)\lambda_+(n-2,E) + P(n+2,t)\lambda_-(n+2,E) \qquad (11)$$

$$-P(n,t)\lambda_+(n,E) - P(n,t)\lambda_-(n,E)$$

By solving these equations numerically, we can follow the time evolution of a reaction. The implicit assumption has been made in these
calculations that the percent of the population undergoing particle
emission is negligible.

Results obtained using the rate expressions (9),(10a) and (10b),
are shown in figure 6.A large population of nuclei at the time of the
initial projectile-target interaction has been assumed, characterized
by some exciton number and excitation energy. For purposes of an example, a result has been taken for which n_o, the initial exciton number,
is 3 and $gE \approx 160$, where g is the single particle level density and E is
the compound state excitation energy [21] . This corresponds to an Fe
nucleus at 24 MeV of excitation. The initial population at time 1 is
shown as a delta function at n =3, a value that might be valid for a
nucleon induced reaction. A computer was then used to follow the shift
in exciton number as a function of time, using the rate expressions described and time increments about 1% of a mean transition time. The results are shown after 50,200, 800 and 3200 units of time. The distribution at 3200 units of time is indistinguishable from the equilibrium
distribution so that that distribution will continue to move out unchanged on the time axis.

At any moment of time, there is a probability of particle emission
into the continuum from the various states in the ensemble; these are
shown as particle spectra on fig. 6. It may be observed that the high
energy or "hard" spectra come early in time, whereas very little is
contributed at the higher energies from those spectra characteristic of
the ensemble after say 800 units of time. Note that the scale on the
instantaneous spectral probabilities is not shifted but represents the
relative probabilities per unit time of emission into the continuum.
Note that the scale is logaritmic so that after 3200 time increments
the probability of emission of a 3 MeV neutron is reduced by a factor

of 10^3 from its value after only 50 time units. The reduction factor for 10 MeV neutrons is 10^7. This, of course, represents the effects of sharing the excitation between many more particles and holes at later times.

Now what would be observed of the particle spectrum was measured with continuous integration as a function of time, as with a pulse height analyzer? After fifty time units one would see roughly fifty times the instantaneous spectrum shown here for t=50; at 200 time units one would observe the spectrum at 50 units plus the spectrum of all neutrons emitted in the 150 units region between 50 and 200 . The time integrated spectra are also shown in fig. 6.

It may be seen that the "hard components" of the spectrum seem to cease increasing after 800 time units. Indeed, if a motion picture camera had been used to record the integrated spectra one frame for a specified number of time units, it would appear that the spectrum stopped coming in between 800 and 3200 units of time. If one then switched to a much longer time scale per frame, the "equilibrium"component could then be seen growing with time. Fig. 7. shows the present result compared with the same reaction at four times the excitation [21]. Here the division into fast and slow components is not so obvious as at lower excitations; note that the time scales are not the same for the two sets of reactions.

It can be seen that at lower excitations the reaction seems to divide into two components, a relatively long-lived one, which we have for years been calculating as the equilibrium statistical model, and a very much shorter component which we now identify as the pre-equilibrium component. Thus, at least at lower excitations, we may analyze data using a separate prescription for each component as a mathematical convenience, but realizing that a single model serves for the entire reaction.

In this section we have emphasized only a few aspects of the exciton model. It should be noted that the model has been very successful in reproducing spectral shapes and excitation functions for a great many reactions (cf [13], [39], [40] , [29] ,[8],[7], [26] , [45] , [23] ,[30] , [11]).

We have failed to note the correspondence between the Exciton model as described in this section, and the doorway state model [31]. This correspondence has been discussed by Grimes et al. [29]. Neither have we discussed the earlier work of Izumo on pre-equilibrium decay as a treatment of the excitation of a few "valence nucleons", but refer interested parties to the original references [37].

61

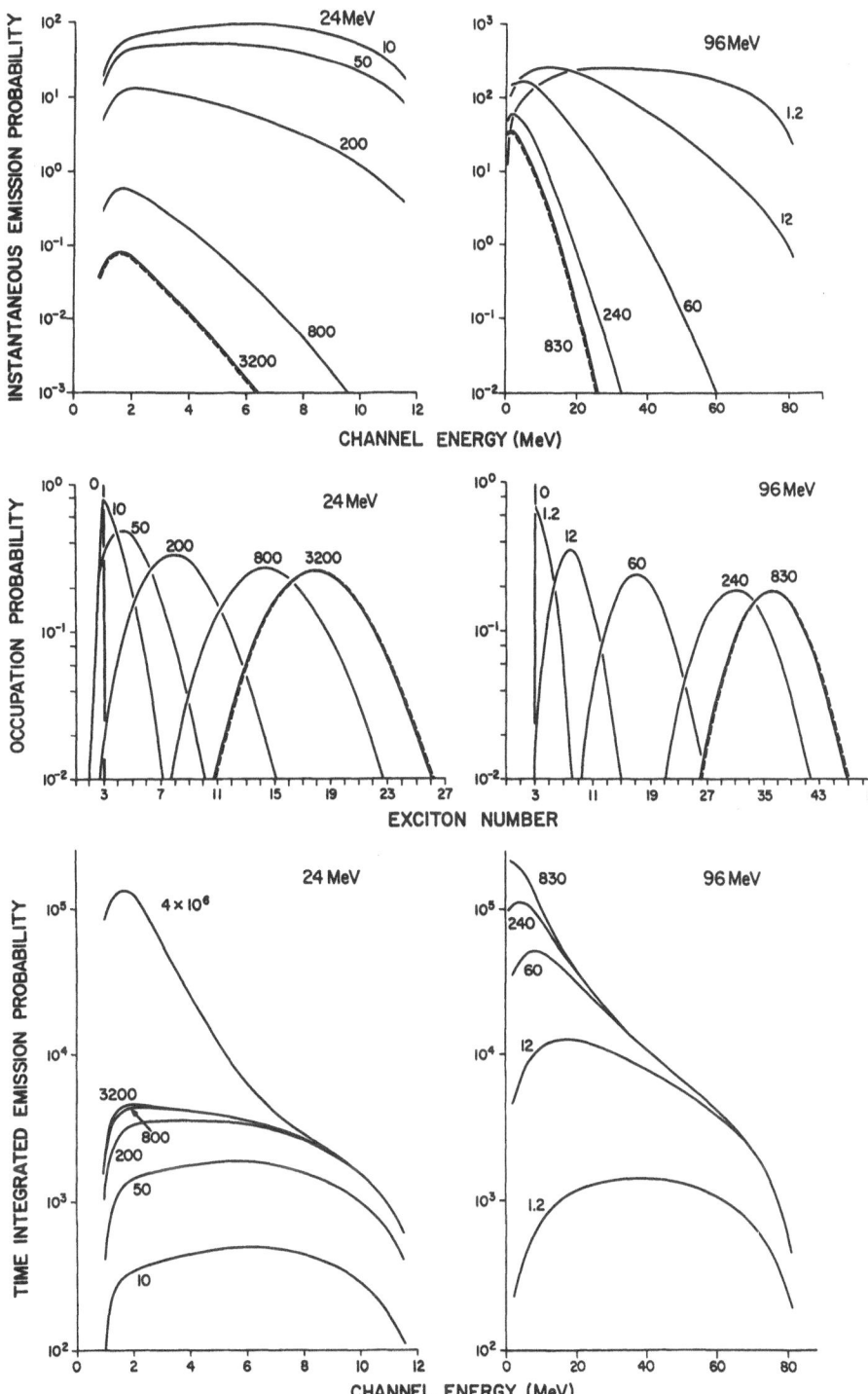

Fig. 7. Comparisons of time evolution as per exciton model
nuclei at 96 MeV and 24 MeV, as in fig. 6

5. The Hybrid Model

One shortcoming of the Exciton model, as far as it has been discussed, is that only relative spectral yields can be determined until
such time as the two body interaction matrix can be evaluated. The Milano group has been working on this aspect, and I hope that we will
hear about that work shortly [17].

At this point I would like to discuss an approach which was designed to maintain the physical transparency and simplicity of the Exciton model, while permitting the calculation of absolute spectral yields
as in the HMB model. Since ideas were borrowed from both models, and
because of the à propos translation into German and transliteration into
English this approach has been called the "Hybrid Model" [10]. This
approach permits prediction of the fraction of pre-equilibrium decay
as a function of excitation energy, target mass and charge, and projectile type. It of course remains to be seen if it permits a correct
prediction of any of these phenomena. Let me first present a derivation
of the Hybrid model, and then address myself to the latter more trou-
blesome question.

5.1. Physical Description and Formulation

The problem in calculating absolute spectral yields from the Exciton model is one of calculating the intranuclear transition rates
for the intermediate states. The simplest approach to this is as in
the Cascade and HMB models, through use of the nucleon-nucleon scattering cross sections. The mean free path of a nucleon in nuclear matter
has been calculated by Kikuchi and Kawai from free scattering cross
sections, with exclusion of those events which would leave either particle with less than the Fermi energy [38]. These mean free path values
are converted to transition rates by dividing the particle velocity by
the mean free path. The result of this exercise may be represented by
the simple binomial expression [10]:

$$\lambda_{n+2}(\varepsilon) = \left[1.4 \times 10^{21}(\varepsilon + B_x) - 6 \times 10^{18}(\varepsilon + B_x)^2\right]\sec^{-1} \qquad (12)$$

where $\lambda_{n+2}(\varepsilon)$ represents the rate at which a nucleon at energy $\varepsilon + B_x$
above the Fermi energy undergoes allowed two body collisions. The ave-
rage rate of transitions for the n exciton excited state may be represented fairly well by Blann et al. [14]

$$\lambda_n(E) = \left[1.4 \times 10^{21}E - 6 \times 10^{18}E^2\right]\sec^{-1} \qquad (13)$$

where E is the complex state excitation energy. Reference will be made
later on to both of these expressions.

As in high energy cascade calculations, it will be assumed that a
reaction proceeds through a series of particle-particle or particle-
hole interactions, in which the total particle-hole number characterizing the nuclear state may either increase by two, decrease by two, or
remain unchanged as a result of each interaction. As in earlier work,
we assume that the transitions in which the particle-hole (or exciton)
number increases by two dominate in the early stages of the equilibration process. As in Griffin's model, we assume that the intermediate
states are characterized by appropriate state density formulas and
that all states of a given exciton number may be populated with equal
a-priori probability (within limitations of energy conservation and the
Pauli principle) during the equilibration process. As in earlier treatments of the exciton modes the total particle emission probability in
a given channel energy range $P_x(\varepsilon)d\varepsilon$ is given as a sum over the contributions of the intermediate states. The sum is taken from some initial number of excitons n_o to the equilibrium number $\bar{n}$. We write the
decay probability as

$$P_x(\varepsilon)d\varepsilon = \sum_{n=n_o}^{\bar{n}} {}_nP_x \left[\rho_n(U, \ )/\rho_n(E)\right] \cdot \left[\lambda_c(\varepsilon)/(\lambda_c(\varepsilon)+\lambda(\varepsilon))\right]_{n+2} \cdot$$

$$\prod_{n'=n_o}^{\bar{n}} (1-P_{n'-2}) = \sum_{n=n_o}^{\bar{n}} {}_nP_x(\varepsilon)d\varepsilon \qquad (14)$$

where ${}_nP_x$ is the number of particles in an n-particle-hole state which
are of the type x, and $\lambda_c(\varepsilon)$ is the density of translational states
for a particle in the continuum. All other symbols are as previously
defined.

The expression in the first set of brackets is the fraction of the
n exciton state population which has one particle in a virtual level
which corresponds to a continuum energy between ε and $\varepsilon+d\varepsilon$. The expression in the second set of brackets is the ratio of the decay rate into
the continuum to the total decay rate for the unbound particle. It is
similar in form to that given in Harp et al. [33] except that the internal transition rate is taken here to depend on a particular particle. The transition rate will be simplified to the rate for interaction
with a nucleon below the Fermi energy to give a state with an additional excited particle and hole. A result using eq. (13) will also be
presented later on. The rates $\lambda_c(\varepsilon)$ of (14) are calculated as

$$\lambda_c(\varepsilon) = \sigma(\varepsilon)(2\varepsilon/M)^{1/2}\rho_c(\varepsilon)/g\Omega \qquad (15)$$

The term in the third set of brackets of (14) is the depletion factor, which reduces the population of each state according to the amount of particle emission from simpler states. With this definition $P_{n'}$ is given by

$$P_{n'} = \sum_{x=n,p} \int_{\varepsilon=0}^{\varepsilon_{max}} {}_{n'}P_x(\varepsilon)d\varepsilon \qquad (16)$$

where $P_{n'}$ is zero for the first term in the summation of (14).

The state densities used in (14) were those given by Williams [55]

$$\rho_{p,h}(E) = \frac{g(gE-\theta)^{n-1}}{p!h!(n-1)!} \qquad (17)$$

where $\theta = f(p,h)$ is a correction term for the Pauli principle. However for all systems we have investigated, we find that the Ericson expression gives essentially identical results to those resulting from use of (17).

5.2. Theoretical and Experimental Results

Comparisons of calculated and experimental (α,p) particle spectra are shown in fig. 8. The data are due to West for 42 MeV He^4 particles. The calculated results shown include the equilibrium component of the cross section(i.e. the part not decaying prior to equilibrium). Generally the preequilibrium components are reproduced as well or better than the equilibrium components.

A method of testing the predicted energy variation of fraction of preequilibrium emission (fpe) is through excitation functions. Those for ^{197}Au(α,xn) reactions are shown in figure 9; results are seen to be very good. Results for ^{51}V$(\alpha,3n)$ and $(\alpha,p3n)$ are shown in figure 10. (Bowman et al. [16], Blann et al. [15]). Here the results seem somewhat less promising. Still, it should be noted that agreement between pre-equilibrium yields and experimental results is as good as for the equilibrium yields over a very wide energy range.

Another particle spectrum is shown in figure 11. Here the ^{197}Au(p, p') spectrum for 62 MeV incident protons is shown [4]. Calculated results for a 2p1h, and for 2p initial states are shown. The 2p1h state gives a result inconsistent with experiment; the assumption of a 2p

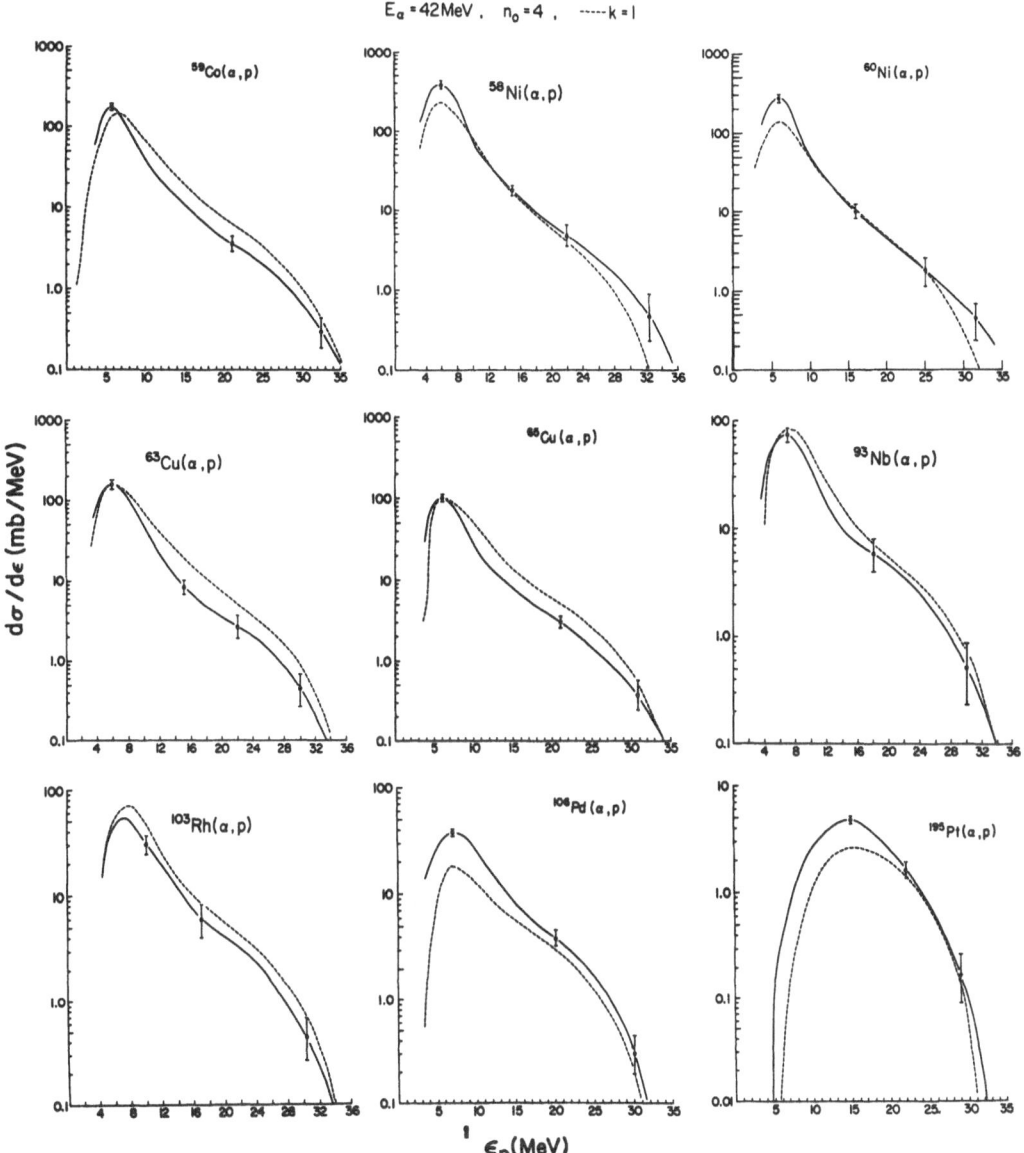

Fig. 8. Experimental and calculated (α,p) spectra on several
targets. The experimental results (ref. 53) are shown as solid curves with several error bars, and are for incident ⁴He
energies of 42 MeV. Calculated results are shown as dashed curves

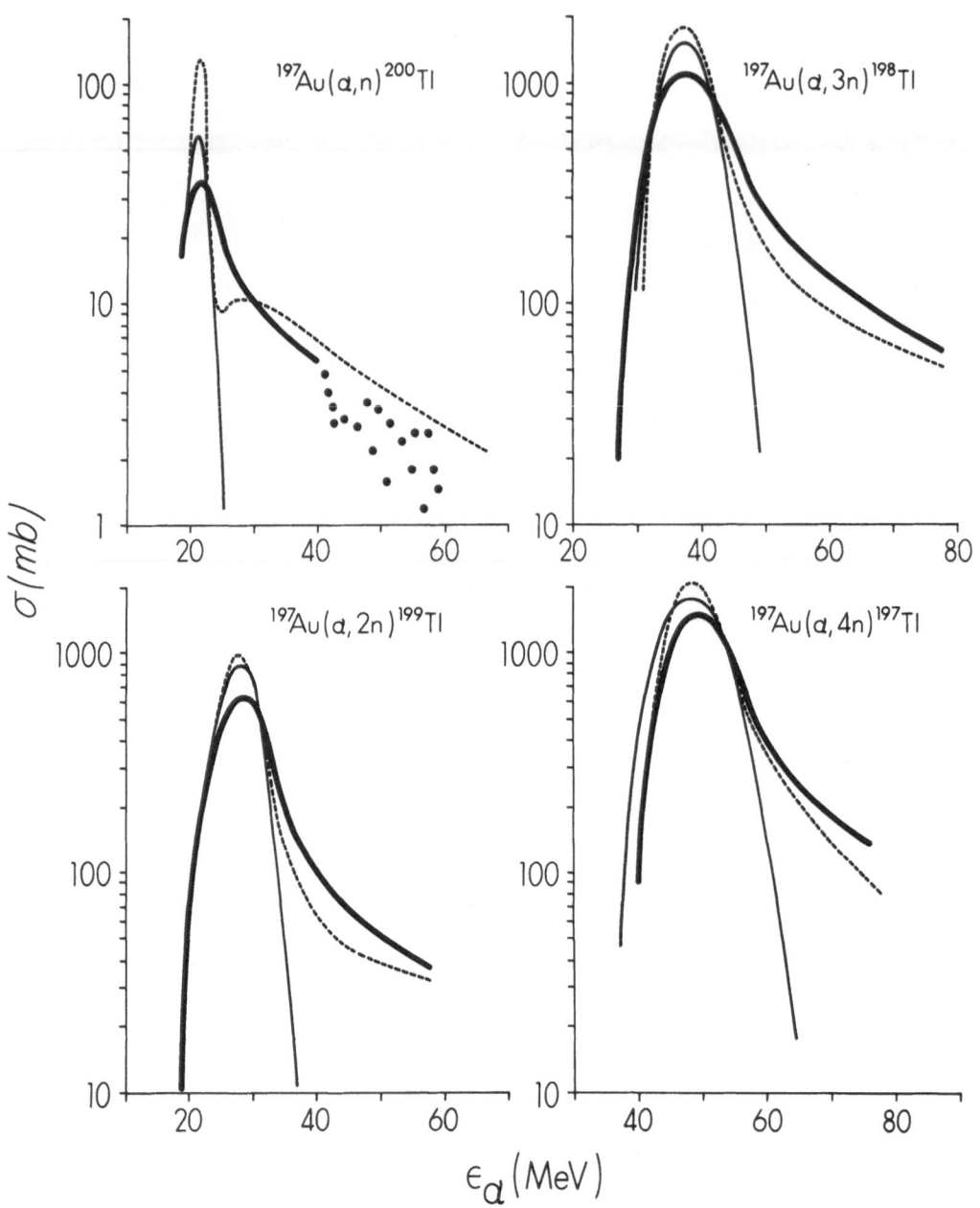

Fig. 9. Calculated and experimental excitation functions for the reactions ^{197}Au(α,xn). The heavy solid curves represent experimental yields from ref.[13]. The thin solid curves represent equilibrum statistical model calculations, and the dashed curves represent results of a combined Hybrid model - Equilibrium model calculation

67

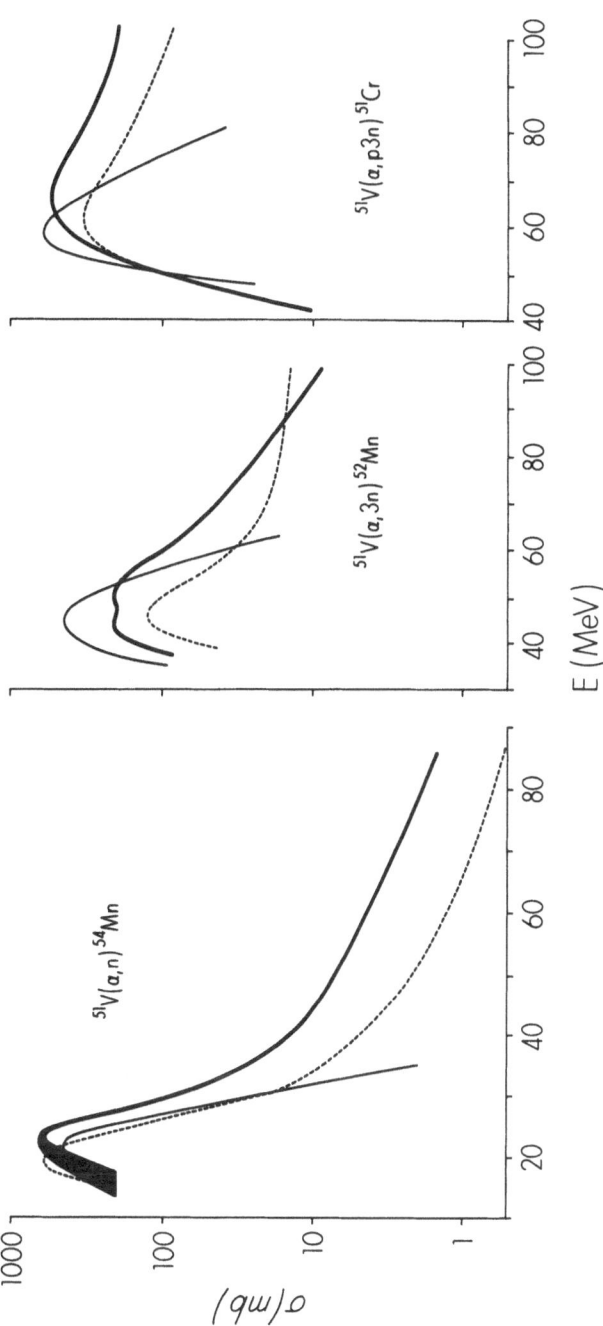

Fig. 10. Curves are as defined for fig. 9. for reactions ^{51}V
(α,xpyn). Experimental results are from ref.[16]

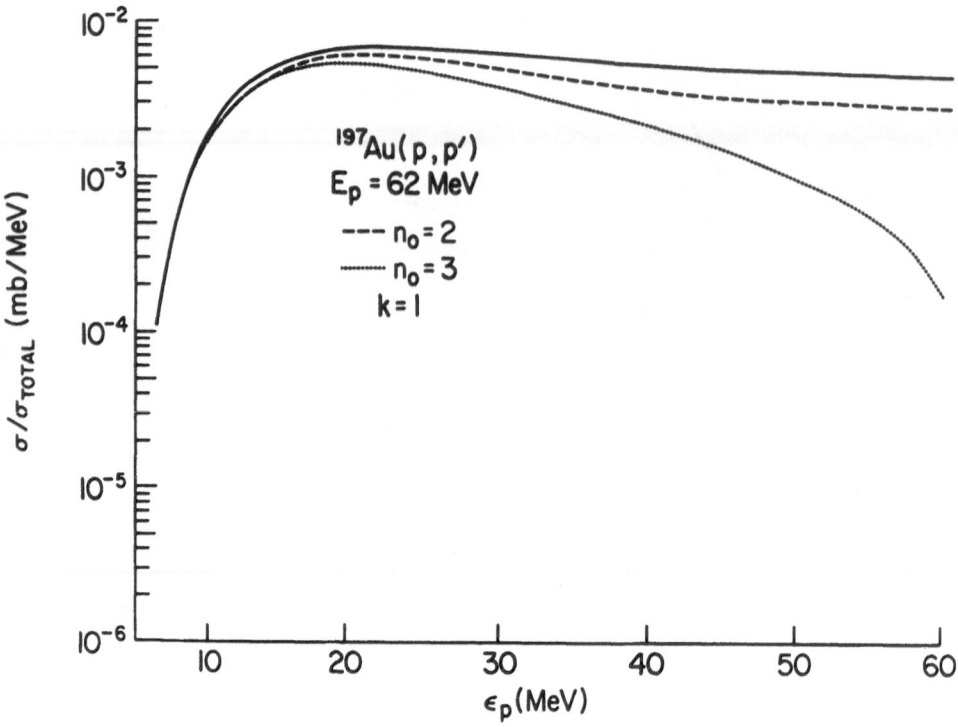

Fig. 11. Experimental and calculated ^{197}Au(p,p'7)spectrum
for 62 MeV incident protons. Experimental results
(ref. [4]) are given by the solid curve, calculated results by the dashed line for n_o=2, and by
the dotted line n_o=3

state is hard to justify. Discrepancies of this type resulted in a consideration of possible effects of the nuclear density distribution on
the preequilibrium decay process.

6. The Geometry Dependent Hybrid Model
6.1. Formulation
The nucleon density distribution in the nuclear skin can affect
the preequilibrium decay in two ways: first, the mean free path for
intranuclear transitions should be greater in the diffuse edge; second, the Fermi energy will be lower in that region, so that the hole
depth is limited. The latter effect is to limit a degree of freedom
for the energy partition. A 2p1h excited state would then give a spe-
ctras distribution more characteristic of a two exciton (2p) state.

To consider how these two effects might influence particle spectra, the Hybrid model has been reformulated as a sum of contributions
over impact parameter [13],

$$\sigma_x(\varepsilon)\,d\varepsilon \;=\; \pi \lambdabar^2 \sum_{\ell=0}^{\infty} (2\ell+1)\,T_\ell P_x(\varepsilon)\,d\varepsilon \qquad (18)$$

where P_x is the preepuilibrium probability from an analogue of eq.
(14), but calculated as a function of nuclear density. A Fermi density
distribution was assumed [35], with

$$d(R) \;=\; \bar{d}\left[\exp(R-C)/Z+1\right]^{-1} \qquad (19)$$

where $d(R)$ is the density at radius R, $\bar{d}$ is the central nuclear density, $C = 1.07\ A^{1/3}$ fm, and $Z = 0.55$ fm. The transition rates given by
eqs. (12) and (13) are modified in each region to be $\bar{d}/<d(R)>$ times
greater, where $<d(R)>$ represents the average density for the impact parameter in question.

Similarly the Fermi energy in each zone will be taken as

$$E_f(R) \;=\; E_f\left[<d(R)>/\bar{d}\right]^{2/3}$$

where E_f is the Fermi energy at central density. The single particle
level density is also taken to have a density dependence,

$$g_\nu(R) \;=\; \left[E_f/E_f(R)\right] \cdot (A/28). \qquad (20)$$

A calculation may now be performed as in eq. (14), but as a sum over
contributions from each impact parameter.

Unfortunately intermediate state densities with limits on the hole depth have not been published. We have derived expressions for the first term in the preequilibrium decay sequence [12],

$$\rho_{2p1h}(\epsilon,U) = \frac{1}{2}g^2 (E_f(R)); \quad U>E_f(R) \qquad (21a)$$

and

$$\rho_{2p1h}(E) = \frac{1}{4}g^3 E_f(R) [2E-E_f(R)]; \quad E>E_f(R) \quad (21b)$$

where this should be the most important contribution. Eq. (17) was used for all higher order states.

6.2. Applications to Preequilibrium Decay

Figure 12 shows the density distribution for ^{54}Fe, and the histogram used to approximate it for the (p,p') reaction at 62 MeV. It looks somewhat like the density distribution for one of the cascade models, except that here there will be one step for each impact parameter. The upper part of this figure shows the partial reaction cross sections in each zone, and the cross section predicted to go into preequilibrium decay.

Figure 13 shows some predicted spectra for the ^{54}Fe(p,p') reaction at three bombarding energies, compared with the experimantal results of Bertrand and Peelle [5]. The dotted curve represents the Hybrid model calculation previously described, with the assumption of a 2p1h initial state. The dot-dash curve results from increasing the mean free path in the skin region according to the reduced nucleon density, and the dashed curve results from including also a limit on hole depth as given by eqs. (21a) and (21b) for the first term in the decay sequence. Apparently both effects - mean free path and limited hole depth - are necessary to reproduce the experimental spectrum.

The density effect alone could, for this nucleus, have been reproduced within the framework of the Hybrid model by the assumption that the mean free path for a nucleon in nuclear matter is twice the value given by the calculation of Kikuchi [38].

7. Comparisons of Results Calculated with Several Models

Several models or permutations of models which could be used to calculate aspects of the nuclear equilibration process have been described. It is interesting to see how well predictions of the various models agree with one another and, to add a note of reality, with experimental results.

Figure 14 shows the ^{54}Fe(p,p') spectra calculated with the HMB mo-

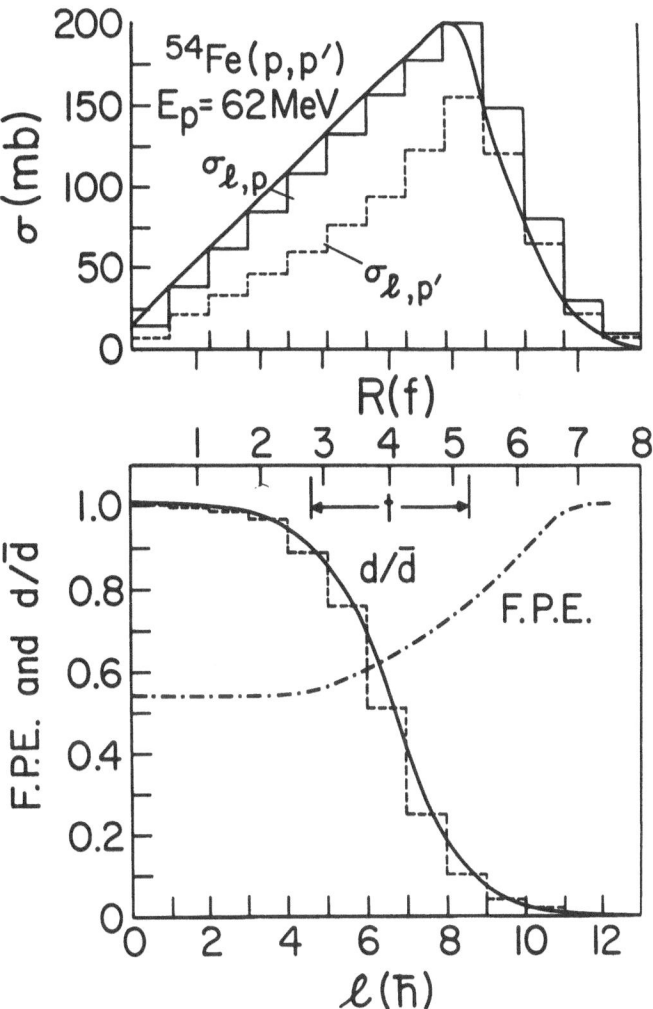

Fig. 12. The lower half of the figure gives the ratio of nuclear density at radius R (upper abscissa) to the interior density $\bar{d}$, shown as a continuous solid curve, for a mass-55 nucleus. The dashed histogram shows the average density ratios used in each reaction zone defined by the partial wave ℓ (lower abscissa). The 90-10% skin thickness t is also indicated. The fraction pre-equilibrium emission predicted for each zone is shown as a smooth dot-dashed curve. The upper curves show histograms for the total reaction cross section of each zone, and the corresponding integrated pre-equilibrium proton emission cross sections

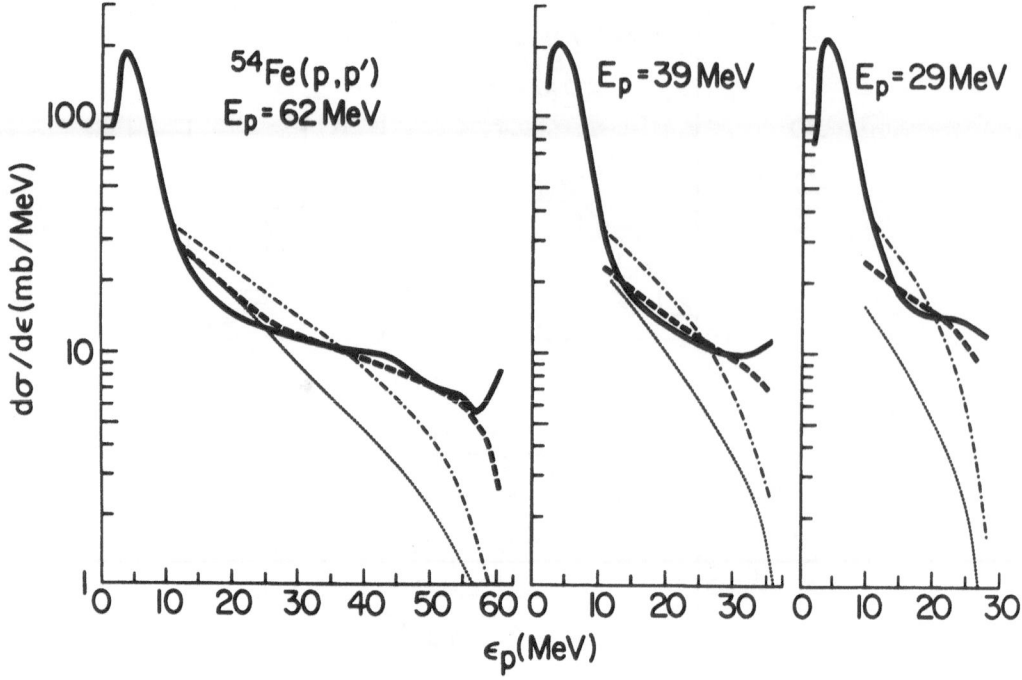

Fig. 13. Calculated and experimental (p,p') spectra on an
^{54}Fe target for incident proton energies of 62, 39
and 29 MeV. Experimental results (ref. [4]) are
represented by the heavy solid curves. Calculated
pre-equilibrium components are based on the hybrid
model (dotted curve), and on the density-dependent
models under discussion in this work. The dot-dashed
curve shows the effect of including a density-dependent mean free path. The dashed curve represents the
effect of also including the density-dependent potential

del (cf [32], [33]), and with the Hybrid [10] and geometry dependent
Hybrid (GDH) models [12]. The Hybrid model calculation was performed
with a 20 MeV Fermi energy, and a limit to hole depth for the first
term in the decay sequence. This approximately matches the conditions
which exist in the HMB model. Without doing this, the two models would
not give the same degree of agreement (e.g. see the Hybrid result of
fig. 13). Also shown on fig. 14 are results of the GDH model, and at
39 MeV bombarding energy a result with the GDH model with a Fermi energy of 40 rather than 20 MeV. In all these calculations continuum theory inverse cross sections [24] were used as in the HMB model calculations, and all results were normalized to a reaction cross section of
approximately 1050 mb, as given by optical model calculations [1].
This normalization was also performed for the HMB calculations, so that
the comparisons shown use parameters which are comparable.

The experimental cross section for 62 MeV protons on ^{54}Fe is near-
er 850 mb than 1050 mb [44], so the calculated results should perhaps,
in this case, all be decreased by 20%. However, the ^{56}Fe (p,p') spectrum is reported to be the same as that for ^{54}Fe at 62 MeV [6], and
for that nucleus the optical model reaction cross section is in better
agreement with the experimental result [44]. Since these calculations
would give the same results for ^{56}Fe as for ^{54}Fe, the downward normalization may be questionable, or in any case less than 20%. We feel that
agreement to better than 20% for these calculations is at best fortu-
ituous.

In fig. 15 similar comparisons are shown for ^{209}Bi(p,p') at ε_p =
62 and 39 MeV [5]. In this case nearly all of the (p,p') cross section
should result from other than equilibrium emission, whereas for ^{54}Fe
this is not the case. Some results not in figure 14 are shown in fig.
15. Here we have shown the Hybrid calculation in which an average state lifetime, using eq. (13) rather than a single particle lifetime as
in eq. (12) has been used. The shape and cross sections are in better
agreement with those of the HMB model at the lower particle energies.
Both calculations agree well at the higher particle energies. At 62
MeV the GDH calculations of fig. 15 were performed using optical model
inverse cross sections, with Fermi energies of 20 and 40 MeV. Results
were all normalized to an optical model reaction cross section of 2300
mb at ε_p = 62 MeV, and 2200 mb at ε_p=39 MeV; the results are within
approximately 10% of the experimental values [44]. At ε_p = 39 MeV results are shown for the GDH model only for a Fermi energy of 20 MeV,
but with both continuum theory and optical model inverse reaction cross
sections having been used.

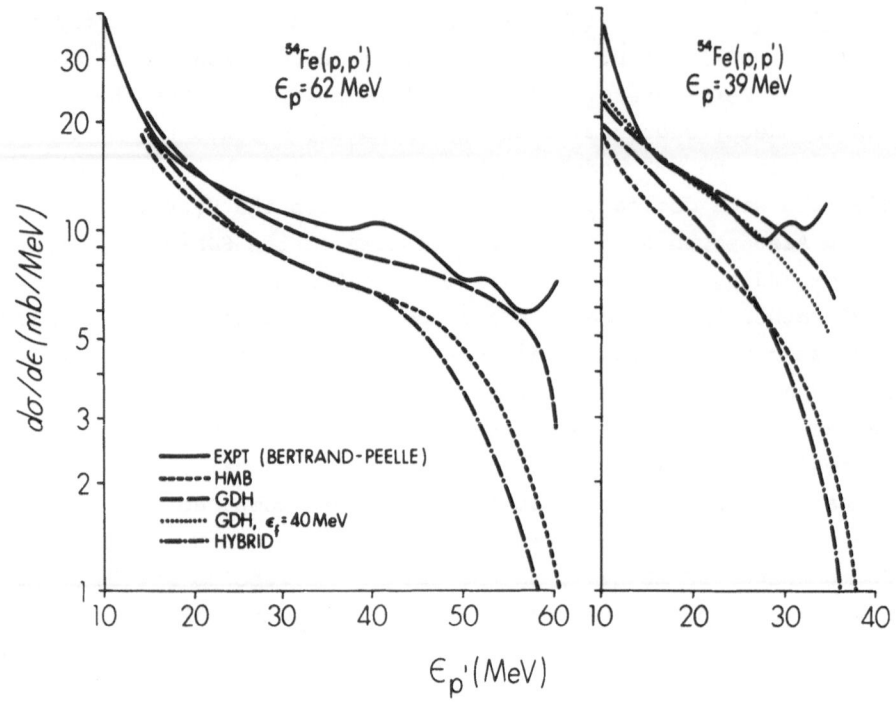

Fig. 14.   Comparison of experimental and calculated ^{54}Fe
(p,p') spectra for incident proton energies of 39
and 62 MeV. Calculations are by the model of Harp-
Miller and Berne (HMB), the Hybrid model with limit to hole depth of 20 MeV and the Geometry Dependent Hybrid model (GDH) with Fermi energies of 20
and 40 MeV

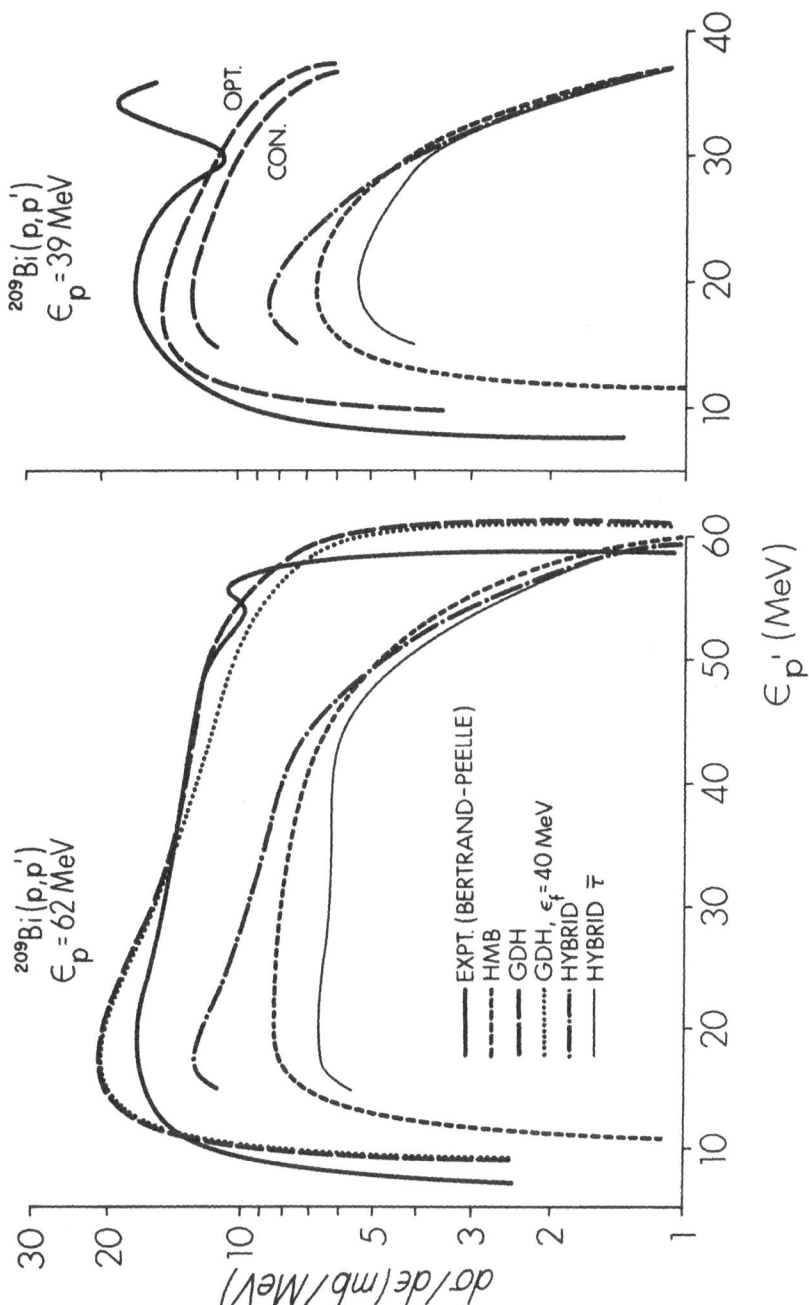

Fig. 15. Comparisons of experimental and calculated ^{209}Bi(p,p')
spectra. Incident protons were 39 and 62 MeV. Calculations are
as in fig. 14, but a Hybrid calculation using an average state
lifetime is also shown. For the calculations with GDH at 39 MeV
results using both continuum and optical model inverse cross
sections are shown

In figs 16 and 17 comparisons are shown between some of the earlier experimental (p,p') spectra and the HMB, GDH, and intranuclear cascade models (cf [3], [18-19], [20], [33], [34]). For ^{209}Bi, the GDH results are presented for both optical model and continuum theory cross sections. The cascade calculations were performed with the Brookhaven "VPOT" code, and the ORNL code due to Bertini [3]. In the Fe region the agreement between all calculations and experimantal results seems reasonable. In the ^{209}Bi region the absolute cross sections of some calculations are not in such good agreement with the experimental results, although the shapes of the HMB and cascade calculations are in excellent agreement with experiment. The conclusion as to how well results of the models agree with experimental results is somewhat subjective. The author of this work feels that the results are far better than one might have expected from an application of nucleon-nucleon scattering cross sections in nuclear matter at such low energies. The shapes of the calculated results are in quite good agreement with experimental results. There is some evidence as to the importance of considering both the nuclear density distribution and limited hole depth in calculations of this type. The GDH model suggests that the major fraction of pre-equilibrium particle emission comes from the skin region; it would be of interest to compare this result with results of an intranuclear cascade calculation. It has been suggested that the VPOT results are low in the Bi region due to excessive internal reflection in the multi-step potential [6].

8. Conclusions

Equilibration models have survived their birth and have even begun to achieve a status of use as tools in interpretation of other data. Examples of this to date have included interpretation of neutron/fission width data, fission isomer yields, and as an aid in interpreting equilibrium reaction data.

Perhaps of greater importance thus far, the ideas of pre-equilibrium models have started us to thinking differently about nuclear reactions. We are no longer resolutely glued to ideas of direct or equilibrium decay, but rather to the realization that we may have to follow the evolution of the complex state in time and that contributions to experimental particle yields may come throughout the entire lifetime of the state. Such new outlooks may lead us to fresh approaches in many areas, and I'll try to mention several shortly. As an immediate benefit of these models, we can presently explain in a semi-quantitative to quantitative fashion many phenomena which could not be explained in

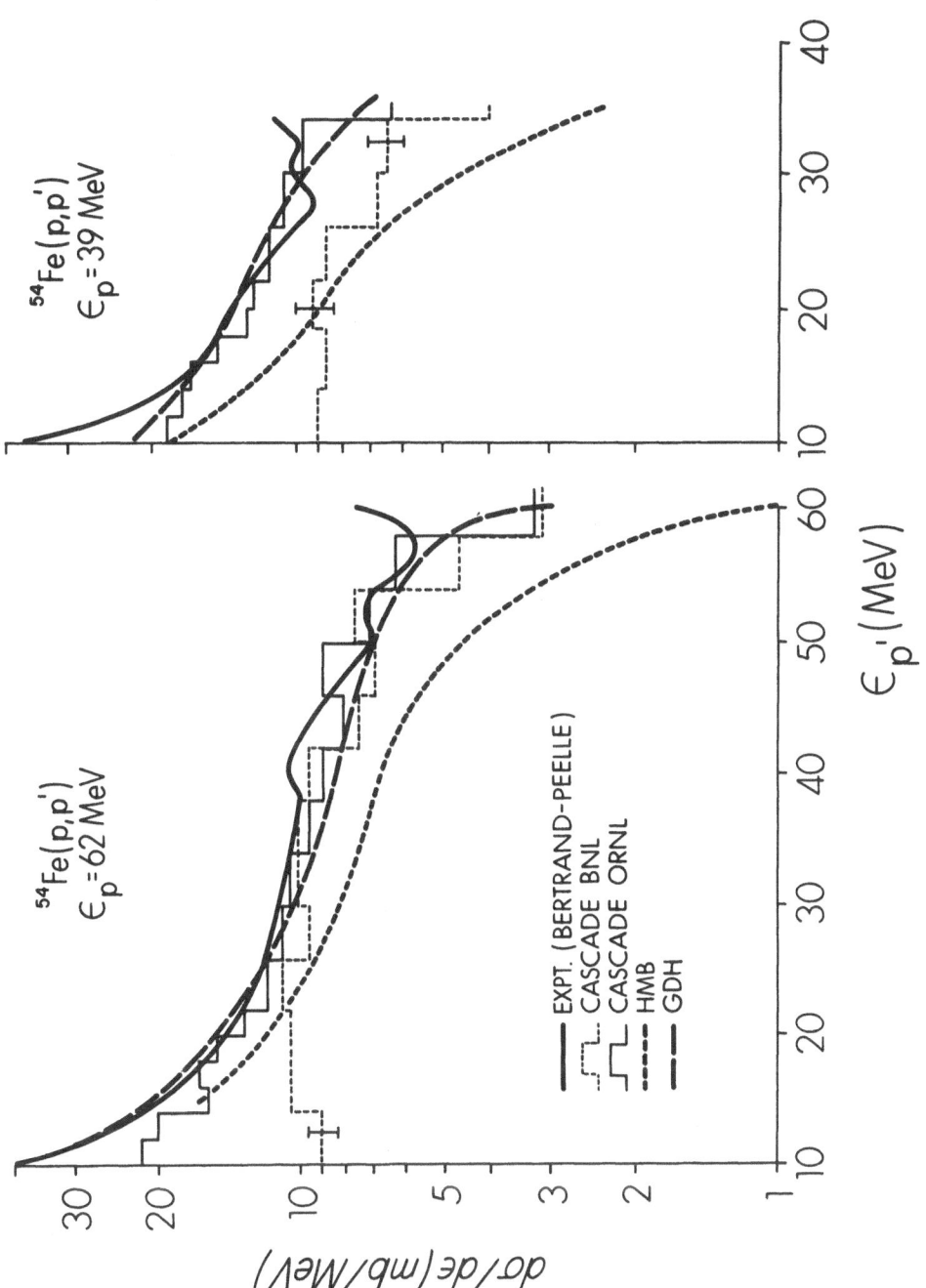

Fig. 16. Results as in fig. 14, but intranuclear cascade calculations are also presented as histograms

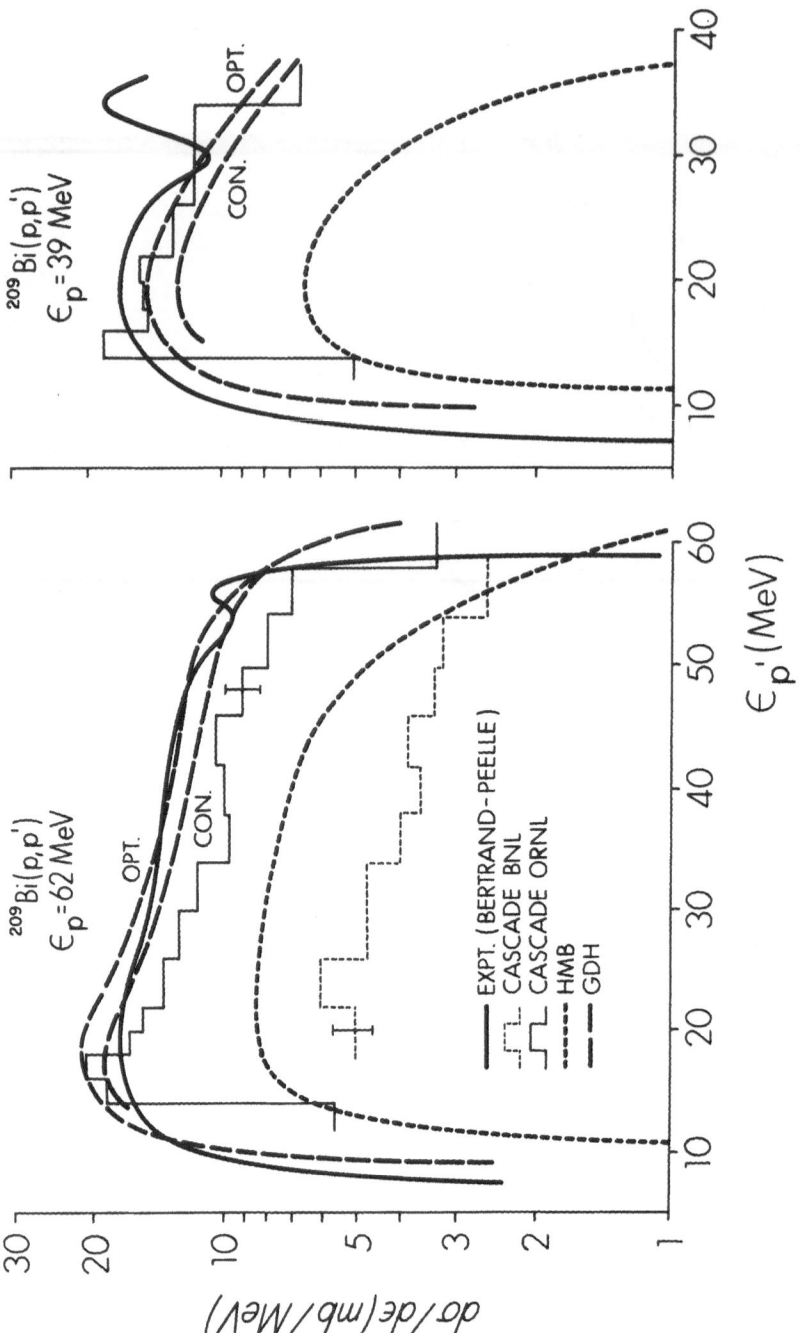

Fig. 17. Results as in figure 15, but intranuclear cascade
calculations are also presented as histograms. Calculated
results of figs 14-17, excluding GDH and Hybrid are from
unpublished results of ref. [34]

the past.

Applications of the concepts of these models may possibly be made toward an understanding of the microscopic description of the heavy ion fusion reaction. If the projectile initially divides into several aggregates, e.g. α particles, then one has a system of few "excitons" with high average energy. At sufficiently high bombarding energies (I would estimate 10 MeV per nucleon) these aggregates should then undergo preequilibrium emission. If this happens, then counter telescope studies of the types of clusters emitted and their spectra may yield information on the mode of breakup in the fusion reaction.

Work remains in understanding and being able to calculate the intermediate state dinsities with inclusion of structure and pairing effects, as demanded by the Sewer Pipe principle. Along those lines, there is room for improvement in the evaluation of intranuclear transition rates, or if you prefer spreading widths; several different approaches might be taken to this problem - and should be.

There is some evidence that the preequilibrium particles are emitted largely from the region of the nuclear "skin", and this may require further scrutiny. Related to this is the very interesting question of complex particle emission, which (in the preequilibrium case) cannot be separated from the preformation question and this again involves the "skin". A great deal remains to be done before this aspect of nuclear reactions can be said to be understood. I should mention that the Milano group has made an interesting start in this direction [45].

Each year I see more areas where the ideas of these models might be applied, and more areas where they are applied. I conclude that these models have earned their place in nuclear physics. They no doubt have a long evolutionary road ahead, and I expect that they will yield a good deal of new insight in the future.

9. Acknowledgements

The Author wishes to acknowledge the generosity of Drs. Harp, Bertrand, and Bertini in supplying, and allowing reproduction of their unpublished results on cascade calculations quoted herein, as well as the HMB model calculations. Helpful conversations in this general area with the same group is also acknowledged with great pleasure. The Author has also enjoyed instructive discussions with Profs. J.M. Miller, J.J. Griffin, N. Cindro, L. Milazzo-Colli, and E. Gadioli.

Definition of Symbols Used

Symbol	Definition
ε	particle channel energy
E	compound state excitation
x	particle type, neutron or proton
B_x	binding energy of particle x
U	residual nucleus excitation $= E-B_x-\varepsilon$
h	number of holes in excited state
p	number of excited particles
n	sum of number of excited particles plus holes, $n=p+h$
g	density of single particle levels in the excited nucleus, assumed for simplicity to be a constant at all energies
$\rho_{p+h}(E)$	density of states having p+h excited particles plus holes at excitation E, also abbreviated $\rho_n(E)$
$\rho_{p+h}(U,\varepsilon)$	dinsity of states with p+h excitons, one of which has an energy such that if emitted, the residual nucleus would have excitation U and the particle would have channel energy ε
$\sigma_x(\varepsilon)$	inverse cross section as used in statistical model
$\lambda_c(\varepsilon)$	decay constant for transitions into the continuum for a particle at excitaiton $\varepsilon+B_x$ above the Fermi energy
$\lambda_{n+2}(\varepsilon)$	decay constant for transitions of a particle at excitation $\varepsilon+B$ above the Fermi energy to create another particle-hole pair in a final state of n+2 excitons
$\lambda_n(E)$	decay constant for transitions from an n exciton state at excitation E to an n+2 exciton state
$P_x(\varepsilon)d\varepsilon$	probability of particle decay in the interval ε to $\varepsilon+d$
$_nP_x$	number of all particles in an n-particle-hole state which are of the type x
$\rho_c(\varepsilon)$	density of translational states of a particle in the continuum
Ω	volume in which free particle phase space states are normalized

References

1. AUERBACH, E.H.S., Calculations kindly supplied by Auerbach, E.H. using ABACUS code (1972).

2. BERNARDINI, G., BOOTH, E.T. and LINDENBAUM, S.J., Phys. Rev. $\underline{88}$ 1017 (1952).

3. BERTINI, H.W., Phys. Rev. $\underline{131}$ 1801 (1963).

4. BERTRAND, F.E. and PEELLE, R.W., ORNL $\underline{4469}$ and $\underline{4460}$ (1970), unpublished.

5. BERTRAND, F.E. and PEELLE, R.W., ORNL $\underline{4638}$ (1971), unpublished.

6. BERTRAND, private communication (1972).

7. BIMBOT, R. and LEBEYEC, Y., Journal Phys. Rad. $\underline{32}$ 243 (1971).

8. BIRATTARI, C., GADIOLI, E., GRASSI, A.M.-STRINI, TAGLIAFERRI, G. and ZETTA, L., Nucl. Phys. $\underline{A166}$ 605 (1971).

9. BLANN, M., Phys. Rev. Letters $\underline{21}$ 1357 (1968).

10. BLANN, M., Phys. Rev. Lett. $\underline{27}$ 337 (1971).

11. BLANN, M., Proceedings of the IV International School in Nuclear Physics (Rudziska, Poland, 1971) Warsaw University Press (Ed. W. Zych) 1971.

12. BLANN, M.,Phys. Rev. Lett. $\underline{28}$ 757 (1972).

13. BLANN, M. and LANZAFAME, F.M., Nucl. Phys. $\underline{A142}$ 559 (1970).

14. BLANN, M. and MIGNEREY, A., Nucl. Phys. $\underline{A186}$ 245 (1972).

15. BLANN, M. and MIGNEREY, A., unpublished (1972).

16. BOWMAN, W.W. and BLANN, M., Nucl. Phys. $\underline{A131}$ 513 (1969).

17. BRAGA MARCAZZAN, M.G., GADIOLI, E.-ERBA, MILAZZO COLLI, L. and SONA, P.G., to be published (1972).

18. CHEN, K., FRAENKEL, Z., FRIEDLANDER, G., GROVER, J.R., MILLER, J. M. and SHIMAMOTO, Y., Phys. Rev. $\underline{166}$ 949 (1968).

19. CHEN, K., FRIEDLANDER, G. and MILLER, J.M., Phys. Rev. $\underline{176}$ 1208 (1968).

20. CHEN, K., FRIEDLANDER, G., HARP, G.D. and MILLER, J.M., Phys. Rev. $\underline{C4}$ 2234 (1971).

21. CLINE, C.K. and BLANN, M., Nucl. Phys. $\underline{A172}$ 225 (1971).

22. COMBE, J., Nuovo Cimento $\underline{3}$ 5182 (1956).

23. DEMEYER, A., CHERY, R., CHEVARIER, A., TOUSSET, J. and TRAN MINH-DUC, J. de Phys. $\underline{31}$ 225 (1970); DEMEYER, A., CHERY, R., CHEVARIER, N., CHEVARIER, A.,TOUSSET, J. and TRAN MINH-DUC, J. de Phys. $\underline{31}$ (1970) 847; CHEVARIER, N., CHEVARIER, A., DEMEYER, A. and TRAN MINH-DUC, Report LYCEN/7092, unpublished.

24. DOSTROVSKY, I., FRAENKEL, Z. and FRIEDLANDER, G., Phys. Rev. $\underline{116}$ 683 (1959).

25. ERICSON, T., Adv. Physics $\underline{9}$ 423 (1960).

26. GADIOLI, E., Nuovo Cim. Lett. $\underline{3}$ 515 (1972).

27. GOLDBERGER, M.L., Phys. Rev. $\underline{74}$ 1268 (1948).

28. GRIFFIN, J.J., Phys. Rev. Letters $\underline{17}$ 478 (1966).

29. GRIMES, S.M., ANDERSON, J.D., POHL, B.A., McCLURE, J.W. and WONG, C., Phys. Rev. $\underline{C4}$ 607 (1971); $\underline{C3}$ 645 (1971).

30. GUAZZONI, P., IORI, I., MICHELETTI, S., MOLHO, N., PIGNANELLI, M. and SEMENESCAU, G., Phys. Rev. $\underline{C4}$ 1092 (1971); GADIOLI, E., IORI, I., MOLHO, N. and ZETTA, L., Phys. Rev. $\underline{C4}$ 1412 (1971).

31. FESCHBACH, H., KERMAN, A.K. and LEMMER, R.H., Ann. Phys. (N.Y.) $\underline{41}$ 230 (1967).

32. HARP, G.D., MILLER, J.M. and BERNE, B.J., Phys. Rev. $\underline{165}$ 1166 (1968).

33. HARP, G.D. and MILLER, J.M., Phys. Rev. $\underline{C3}$ 1847 (1971).

34. HARP, G.D., BERTRAND, F.E. and BERTINI, H.W., unpublished, private communication (1972).

35. HOFSTADTER, R., Ann. Rev. Nucl. Sci. 295 (1957).

36. IVANOVA, N.S. and PIANOV, I.I., J. Exptl. Theoret. Phys. USSSR $\underline{31}$ 416 (1956) (Soviet Phys. JETP $\underline{4}$ (1957)) 367.

37. IZUMO, K., Progr. Theoret. Phys. (Kyoto) $\underline{26}$ 807 (1961); IZUMO, K., "Direct Interactions and Nuclear Reaction Mechanisms", New York, Gordon and Breach 312 (1963), IZUMO, K., Nucl. Phys. $\underline{62}$ 673 (1962).

38. KIKUCHI , K. and KAWAI, M., "Nuclear Matter and Nuclear Reactions", North Holland Publishing Co. p. 40 (1968).

39. MAGDA, M.T., ALEVRA, A., INGRID R. LUKAS, PLOSTINARU, D., ELENA TRUTIA and MOLEA, M., Nucl. Phys. $\underline{A140}$ 23 (1970).

40. MAGDA, M.T., ALEVRA, A., DUMITRESCU, R., LUKAS, I.R., PLOSTINARU, D., TRUTIA, E., CHEVARIER, N., DEMEYER, A. and TRAN MINH-DUC, "Analysis of Precompound processes in (α,n) Reactions" presented at International Conference on Statistical Phenomena, Albany, New York, August 32-27, 1971 (Proceedings to be published).

41. MEADOWS, J.W., Phys. Rev. $\underline{98}$ 744 (1955).

42. METROPOLIS, N., BIVINS, R., STORM, M., TURKEVICH, A., MILLER, J.M. and FRIEDLANDER, G., Phys. Rev. $\underline{110}$ 185 (1958); $\underline{110}$ 204 (1958).

43. McMANUS, H., SHARP, W.T. and GELLMAN, H., Phys. Rev. $\underline{93}$ 924A (1954).

44. MENET, J.J., GROSS, E.E., MALANIFY, J.J., and ZUCKER, A., Phys. Rev. $\underline{C4}$ 1114 (1971).

45. MILAZZO-COLLI, L. and BRAGA-MARCAZZAN, M.G., Phys. Lett. $\underline{36B}$ 447 (1971), $\underline{38B}$ 155 (1972).

46. MORRISON, G.C., MUIRHEAD, G. and ROSSER, W.G.V., Phil. Mag. $\underline{44}$ 1326 (1953).

47. ROSENZWEIG, N., Phys. Rev. <u>105</u> 950 (1957), <u>108</u> 817 (1957).

48. RUDSTAM, G., Ph. D. thesis, University of Uppsala, Uppsala, Sweden, (1956), unpublished.

49. SERBER, R., Phys. Rev. <u>72</u> 114 (1947).

50. SWENSON, W. and CINDRO, N., Phys. Rev. <u>123</u> 910 (1961).

51. SWENSON, L.W. and GRUHN, C.R., Phys. Rev. <u>146</u> 886 (1966); see also

52. WEISKOPF, V.F., Phys. Rev. <u>52</u> 295 (1937).

53. WEST, R.W., Phys. Rev. <u>141</u> 1033 (1966).

54. WILLIAMS, F.C., Jr., Physics Letters <u>31B</u> 184 (1970).

55. WILLIAMS, F.C., Jr., Nucl. Phys. <u>A166</u> 231 (1971).

56. WILLIAMS, F.C., Jr., A. MIGNEREY and BLANN, M. (unpublished).

PRE-EQUILIBRIUM EMISSION IN NEUTRON AND PROTON
INDUCED REACTIONS

E. GADIOLI

Istituto di Scienze Fisiche dell'Università, Milano

and

I.N.F.N., Sezione di Milano

L. MILAZZO-COLLI

C.I.S.E., Segrate (Milano)

and

Istituto di Scienze Fisiche dell'Università, Milano

1. Introduction

The mechanism for nuclear reactions called pre-equilibrium (PE)
emission, proposed a few years ago by Griffin [13], can be well classified as an intermediate process for particle emission. This mechanism takes into consideration the emission of particles during the nuclear cascade initiated in a nucleus by a projectile. The salient feature of this process is the statistic hypothesis done while calculating both the cascade development and particle emission. This makes
the whole description of the mechanism very simple and easy to compare
with experiments.

Evidence that this model reproduces very satisfactorily the
spectra of nucleons and alpha particles coming from reactions induced
by ∿5 to ∿65 MeV nuclear projectiles in a variety of targets as well
as the slopes, beyond the respective maxima, of the excitation functions of (α,×n), (α,p×n), (d,×n), (p,×n), (p,p×n) reactions, has been
given by several authors (cf [5],[15],[2]).

In the following report it will be shown that the PE model can also
satisfactorily. predict the absolute cross sections of many reactions
induced by neutrons and protons in target nuclei spanning the entire
periodic table.

This report is the result of a joint research of the authors with
several colleagues in Milan: C. Birattari, M.G. Braga -Marcazzan, E.
Gadioli-Erba, A.M. Grassi-Strini, P.G. Sona, G. Strini and G. Tagliaferri and with D. Seeliger of Dresden University. Many topics discussed
here were published or are being published in expanded versions which
contain a much more detailed reference list to earlier work in the
field.

2. Neutron Induced Reactions

Several reactions induced by neutrons in the energy range 10-20 MeV cannot be explained by means of an evaporative calculation. These are, in particular, (n,p) reactions on nuclei heavier than A = 100, a part of the emitted spectrum in (n,n') reactions, (n,α) reactions on nuclei heavier than 140 , and also a part of the emitted spectrum in the latter for masses between 100-130.

In order to apply the PE emission model to the analysis of these reactions we have mainly followed the formulation given by Williams [19].

This process is described as particle emission from the different stages of a cascade, each stage being characterized by the number $\underline{n}$ of excitons, i.e. excited particles and holes. Therefore the final cross-section will be represented as a sum of many terms:

$$\frac{d\sigma}{d\varepsilon} = \sigma_R \sum_{n_o}^{\bar{n}} n \frac{dW_c^n}{d\varepsilon} \tau_n \qquad (1)$$

$$(\Delta n = +2)$$

where ε is the kinetic energy of the outgoing particle, σ_R is the re-action cross-section, $\frac{dW_c^n}{d\varepsilon}$ the decay rate for emission of the conside-red particle from the states with $\underline{n}$ excitons and τ_n the lifetime of these states. n_o is the number of excitons which characterizes the ini-tial configuration.

The calculation of $\frac{dW_c^n}{d\varepsilon}$ is performed using the statistical hypo-theses. In the case of a decay of a state named $\underline{k}$ to a given final sta-te we have

$$\frac{\Gamma_k}{\hbar} = \frac{\langle D_k \rangle}{2\pi\hbar} \sum_l (2l + 1) T_l^k \qquad (2)$$

where $\langle D_k \rangle$ is the average level spacing of the considered levels $\langle D^k \rangle = \frac{1}{\rho_k}$; ρ_k is the level density of states having $\underline{n}$ excitons.

This function has been calculated by Ericson [9]

$$\rho_{p,h} = \frac{g(gE)^{p+h-1}}{p!h!(p+h-1)!} \qquad (3)$$

where $\underline{p}$ and $\underline{h}$ are, respectively, particle and hole numbers, $\underline{n} = \underline{p} + \underline{h}$, and $\underline{g}$ is the density of single nucleon states at the Fermi energy.

Considering the transition to all final states in the interval $d\varepsilon$ we obtain the expression

$$\frac{dw_c^n}{d\varepsilon} = \frac{\rho_{p-1,h}(U)}{2\pi\hbar\rho_{p,h}(E)} \sum_1 (2l + 1) \; T_l^{p,h} \qquad (4)$$

where U and E are, respectively, the effective excitation energies of residual and compound nucleus (the pairing energy Δ is taken into account). Here we can, as usual, substitute $\sum_1 (2l + 1) \; T_l^{p,h}$ by the inverse cross-section.

The final step is to calculate τ_n. To do that, we assume that the probability per unit time for particle emission is small compared with the probability per unit time for an exciton-exciton interaction inside the nucleus which increases the exciton number from $\underline{n}$ to $\underline{n+2}$ (this can be shown to be true around 14 MeV incident particle energy). The value of τ_n has been calculated by Williams to be:

$$\tau_n = \frac{1}{W_{eq}^n} = \frac{\hbar(n+1)}{2\pi |M|^2 g^3 E^2} = \frac{n+1}{2W_{eq}^1} \qquad (5)$$

where $|M|^2$ is the matrix element for an exciton-exciton interaction and $\frac{1}{W_{eq}^1}$ is the life-time of an one-exciton state.

Substituting (4) and (5) in formula (1) we finally obtain for the emission of a nucleon ($n_o = 3$):

$$\frac{d\sigma}{d\varepsilon} = (2S+1)\sigma_R \frac{m\varepsilon\sigma_{inv}(\varepsilon)}{4\pi^3\hbar^2 |M|^2 g^4 E^3} \sum_{\frac{1}{3}n}^{\bar{n}} (\frac{v}{E})^{n-2} (n+1)^2 (n-1) (\frac{n\pm 1}{n}) \qquad (6)$$

where the sign in the last factor is "+" in the (n,n′) case and "−" in the (n,p) case. This factor takes into account charge conservation.

All the terms of this formula are known within reasonable limits, except the value of $|M|^2$ which can be calculated following the method of Goldberger and of Kikuchi and Kawai [12] (taking the free-particle value for the nucleon-nucleon scattering cross-section inside the nucleus) assuming a dominant two-body interaction and including the Pa-

uli principle and momentum conservation. In this way the mean free path
of a fast nucleon in nuclear matter can be calculated and its value is
related to $|M|^2$.

At low excitation energies, $|M|^2$ is found to be dependent on A as
A^{-3} and, in first approximation, independent of energy.

The theoretical value deduced in this way is $|M|^2 = 36 \, A^{-3} \, MeV^2$.
We think, however, that this value could be overestimated, the effective two body interaction in nuclear matter is expected to be much lower
than in free collisions.

2.1. (n,p) Reactions [6]

It is now interesting to compare theoretical predictions with experimental results.

We start with the observation that the reactions (n,p) on nuclei
with mass >100, at neutron energy of ∿14 MeV cannot be explained by
means of the evaporation mechanism.

This fact is evidenced both by spectrum shape and cross-section
values [8].

A comparison of spectrum shape and absolute cross-section values,
as predicted by the pre-equilibrium emission process with experimental
data can be easily performed using the closed form expression [6].

The few existing spectra from (n,p) reactions on target nuclei of
A∿100 are on CsI (fig.1), Rh, Ta and Au (fig.2).

The CsI measurement was done at different incident energies; the
spectra show that for the highest incident energies (18, 19.6 , 21.5
MeV) all the emission seems to be due to a pre-equilibrium process,
while at lower energies the same process accounts for at least 50%.
In the CsI case we also have a way of testing the energy dependence of
the cross-section shown in fig.3 on a relative scale. The agreement is
quite satisfactory. In the other three cases (Rh, Ta and Au) the agreement with pre-equilibrium spectrum is also good. Some deviation at lower energy is probably due to deuteron emission.

To compare the absolute value of (n,p) cross-sections with the
predictions of the model, we have used the measurements made at 14 MeV
incident neutron energy by means of the activation technique. The reasons were the following:

1) there exist many measuremants. About 75 nuclei have been examined;

2) the isotope is always well defined and the contribution from
   other reactions is excluded;

3) experimental value gives the angle integrated cross-section

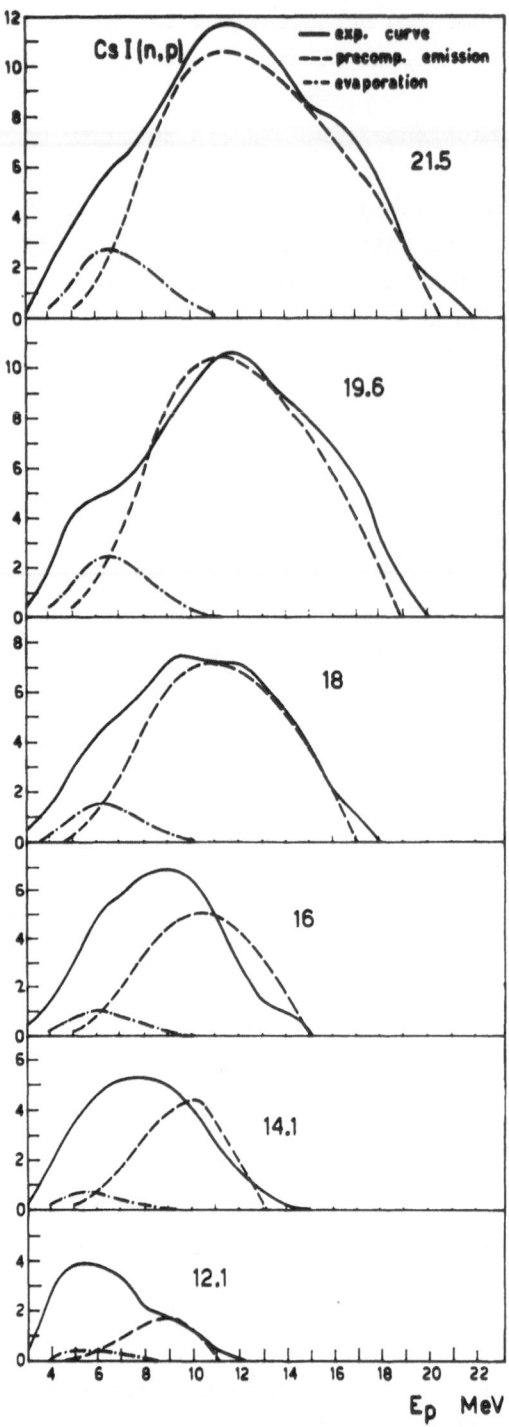

Fig. 1. Comparison between the experimental spectrum for protons emitted in CsI (n,p) reaction (full line) at various energies and the calculated evaporative and precompound contributions (broken lines)

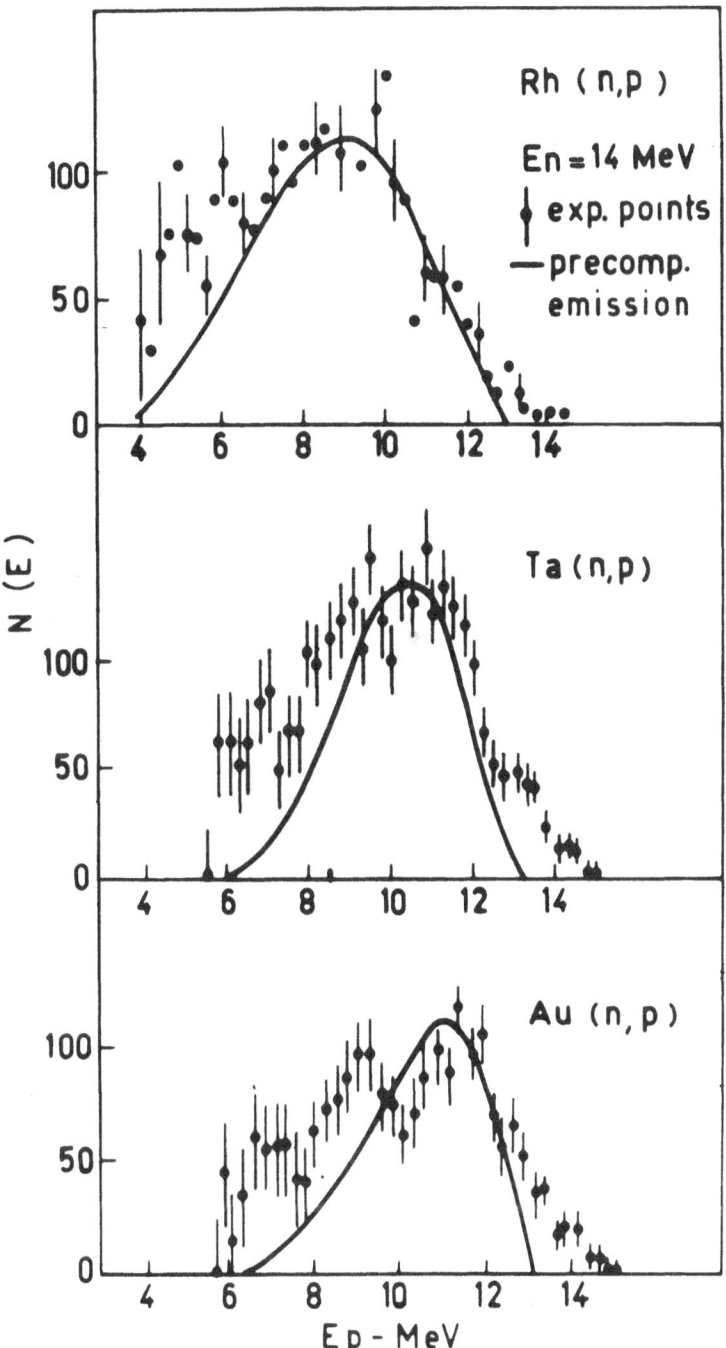

Fig. 2. Comparison between experimental proton spectra and calculated precompound emission in relative units for (n,p) reactions on Rh^{103}, Ta^{181}, Au^{197}

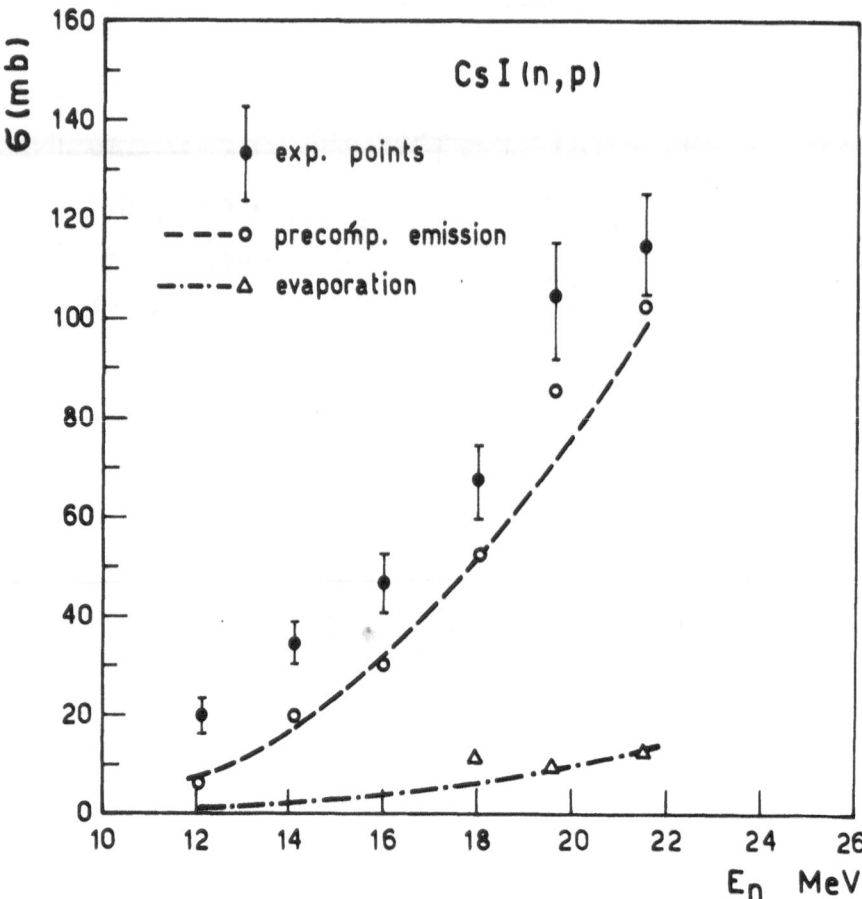

Fig. 3. Cross-section values for the different contributions
to the CsI (n,p) reaction as a function of incident
neutron energy

which is directly comparable with theory.

It is useful to observe the term $|M|^2 g^4/A = \alpha$ which appears in the denominator of the formula (6). Here α is a constant whose theoretical value is 0,0015 MeV^{-2}. Using the experimental values for $\sigma(n,p)$, we can obtain an empirical value of α for each examined nucleus and see what it comes out to be, both in absolute value and in its A-dependence (we recall that it should be constant).

The result of this comparison is shown in fig. 4. The histogram gives the distribution of the α values obtained in this way. The average α value is 3.3 10^{-4}MeV^{-2}, that is ~4 times smaller than the theoretical prediction based on free nucleon-nucleon scattering.

The rather large dispersion of the α values around the most probable one follows the experimental errors which affect the activation measurement.

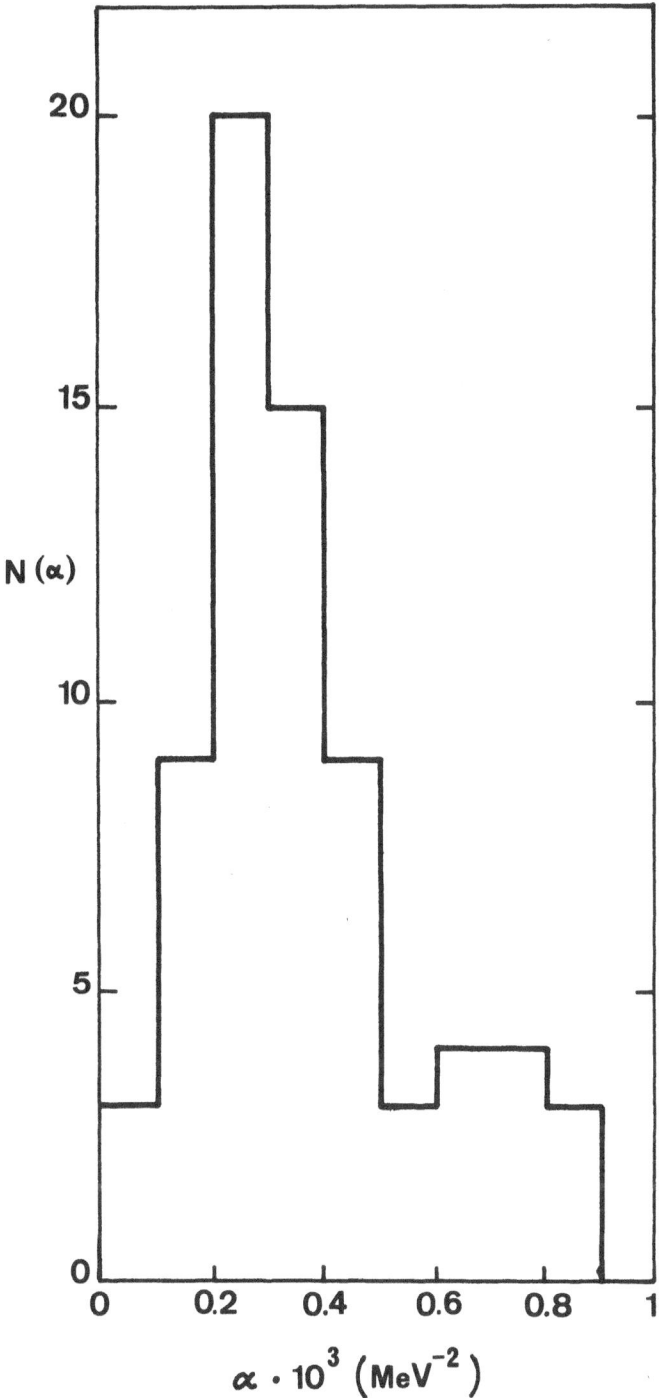

Fig. 4. Distribution of the α values extracted from (n,p) reaction cross-section values

To show the A dependence of the calculated values in a more direct way, we have calculated the ratio $R = \sigma_{exp}(n,p)/\sigma_{calc}(n,p)$ using for the α value the result of the former figure, that is 3.3 10^{-4} MeV^{-2}. This ratio as a function of A is shown in fig. 5. The distribution of points around the line R=1 is almost the same in all the mass range, showing that α has no important variation with A.

2.2. (n,n') Reactions

Since we have seen that pre-equilibrium emission is present in (n,p) reactions and we have learnt that their cross-sections can be predicted on the basis of formula (6) generally within a factor of two, we can ask whether the same model can be applied to neutron emission, that is to (n,n') reactions.

We expect here that the evaporative contributions will be always present, for all the mass values, because here we do not have the effect of the Coulomb barrier suppressing the emission of slow particles.

We expect, therefore, that pre-equilibrium emission, if it exists, should give only a contribution to a total spectrum, which should also display evaporation effects.

Very recent measurements of (n,n') spectra at 14 MeV neutron energy have been obtained by D. Seeliger et al. [18] in Dresden, with the time of flight technique. They have measured neutron spectra emitted by 20 elements, all over the periodic table.

For all the nuclei the spectrum of emitted neutrons consists of a large amount of slow neutrons and a long tail of higher energy neutrons.

The lower energy yield can be attributed to evaporation, and gives the largest contribution to the total cross-section value, while the higher energy part of the spectrum can be attributed to pre-equilibrium emission. A reasonably good fitting of the spectrum shape can be done using these two emission mechanisms.

In this way, the cross-section value of the part due to pre-equilibrium emission can be extracted from the integrated spectra, and these values compared with the prediction of formula (6). The comparison is shown in fig. 6. The agreement in this case is better than for (n,p) reactions which is very probably due to smaller errors in the experimental measurements. The mass range is larger here than in the (n,p) case, showing that α is essentially constant with A.

The remaining discrepancies, never worse than a factor of two, could be due to the use of parameters, like Δ (pairing energy) and $\underline{a}$ (level density parameter) the value of which was obtained from analyses of compound nucleus (CN) evaporation data, i.e. data taken at equilibrium, when many nucleons are excited. We cannot be sure that the same

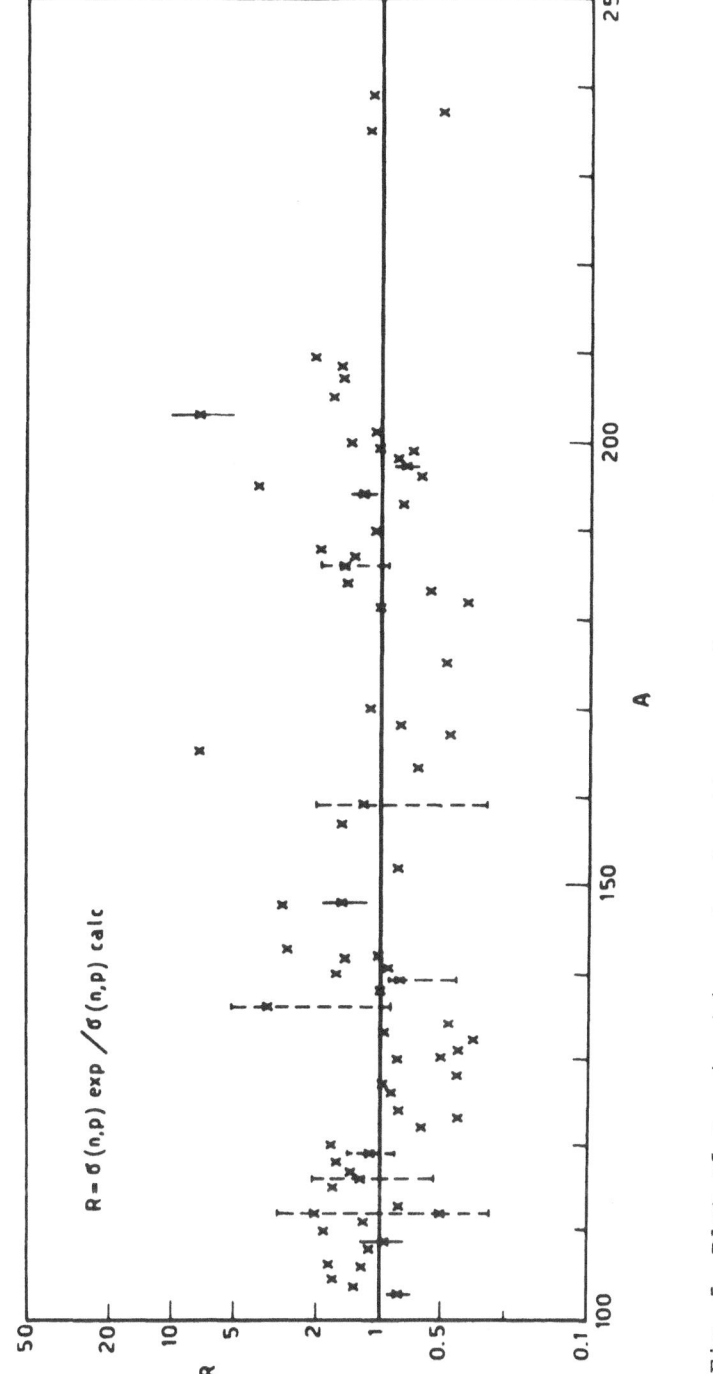

Fig. 5. Plot of $\sigma_{exp}(n,p)/\sigma_{calc}(n,p)$ ratios as a function of mass number A. When various measurements exist the average experimental cross-section value is reported. Full and dotted lines represent respectively, the experimental error when only one measurement exists and the maximum spread of experimental values when more than one measurement is known (taken as examples in some cases)

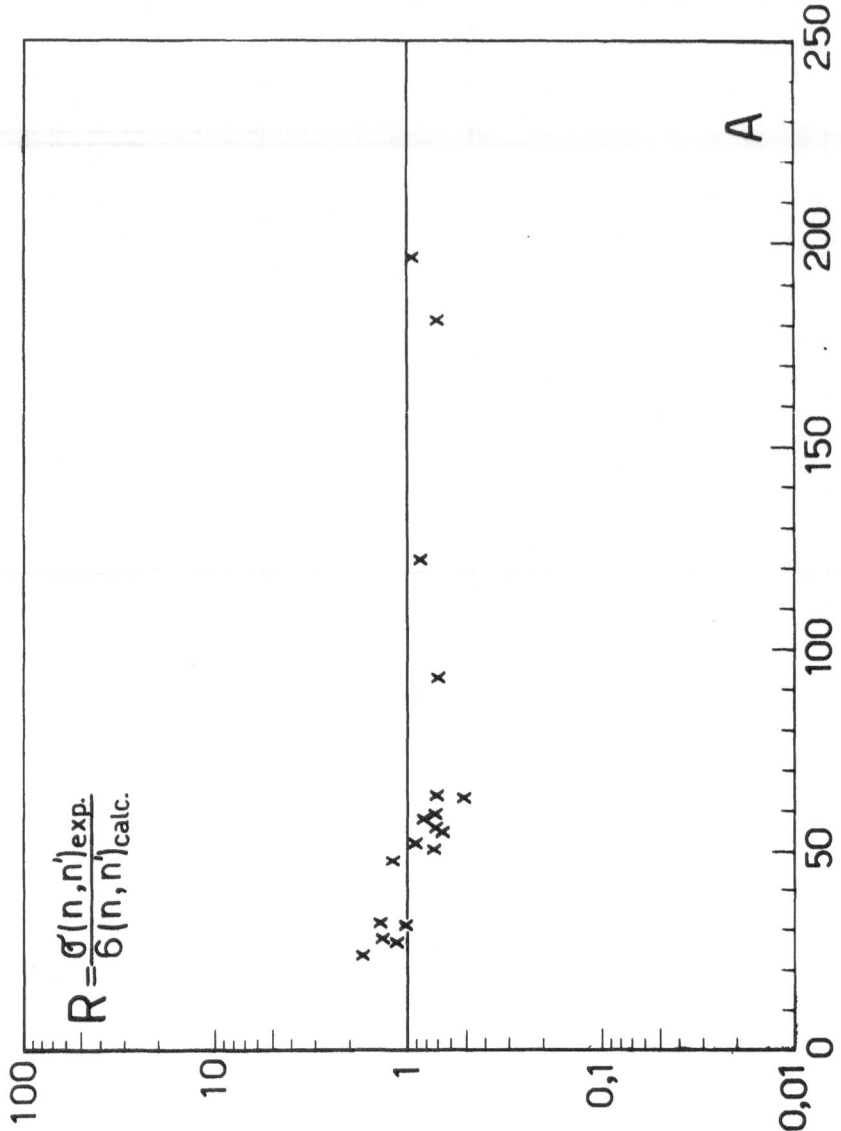

Fig. 6. Plot of the $\sigma_{exp}(n,n')/\sigma_{calc}(n,n')$ ratio as a function of mass number A

set of parameters could describe the average properties of a nucleus when only few nucleons are excited, as it is implied by the PE model.

2.3. (n,α) Reactions

It has been shown that the PE model can be applied to (n,α) reactions on heavy nuclei, emphasizing that there are two ways of emitting alpha particles by a pre-equilibrium mechanism [15].

These two ways can be described as follows:

1) We can assume that there is a certain probability that an α particle
could exist in the nucleus in its ground state. Therefore, in the
same way as we calculate cross-sections of (n,p) or (n,n') reactions we can calculate cross-sections of (n,α) reactions. The exciton
number characterizing the initial configuration is therefore $n_0=3$
also in this case.

2) The other possibility is that an alpha can be formed during the development of the cascade. In this case, the first state from which
an alpha can be emitted must contain at least four excited particles, that is 7 excitons. A formula similar to (6) is obtained:

$$\frac{d\sigma}{d\varepsilon} = \sigma_R \frac{(2S+1)m\varepsilon\sigma_{inv}(\varepsilon)g_R}{32\pi^2\hbar^2|M|^2 g_C^8 E^6} \cdot$$

$$\cdot \sum_{7n}^{\bar{n}} \frac{1}{n} (\frac{g_R U}{g_C E})^{n-5} (n+1)^2 (n-1)^3 (n-3)^2 (n-2) (n-4) (n-5), \qquad (7)$$

$$(\Delta n=+2)$$

where g_R and g_C are the densities of single nucleon states at the Fermi
energy for the residual and compound nuclei , respectively.

The two cases give markedly different spectra. A critical examina-
tion of (n,α) spectra found in the literature, shows, however, the existence of two distinct processes.

In heavy nuclei, A>140, at energies lower than 22 MeV, we have
found the emission mechanism 1).

An example is shown in fig. 7 for the reaction ^{181}Ta (n,α). In
this figure the alpha spectrum was calculated for both mechanisms and
it is clear that the experimental spectrum agrees well only with the
one calculated according to (6)(mechanism 1) and not with the other.

The same kind of agreement is found for 17 other nuclei; a few
examples are shown in fig. 8.

When we look at smaller A values, we have to consider also the second possibility, that is the process described by formula (7). This is
shown in fig. 9 where alpha spectra from CsI are shown. In this last
case the measurements have been taken at varying neutron energies, so
that we can see that process 2) increases with energy faster than process 1. These are angle integrated measurements.

The next step is to try the fitting of absolute values of cross-
sections. Of course, in the case of alpha emission, this is not simple.

First we consider the case of heavy nuclei, where the spectrum
shows that mechanism 1 is to be preferred. Even assuming that alpha

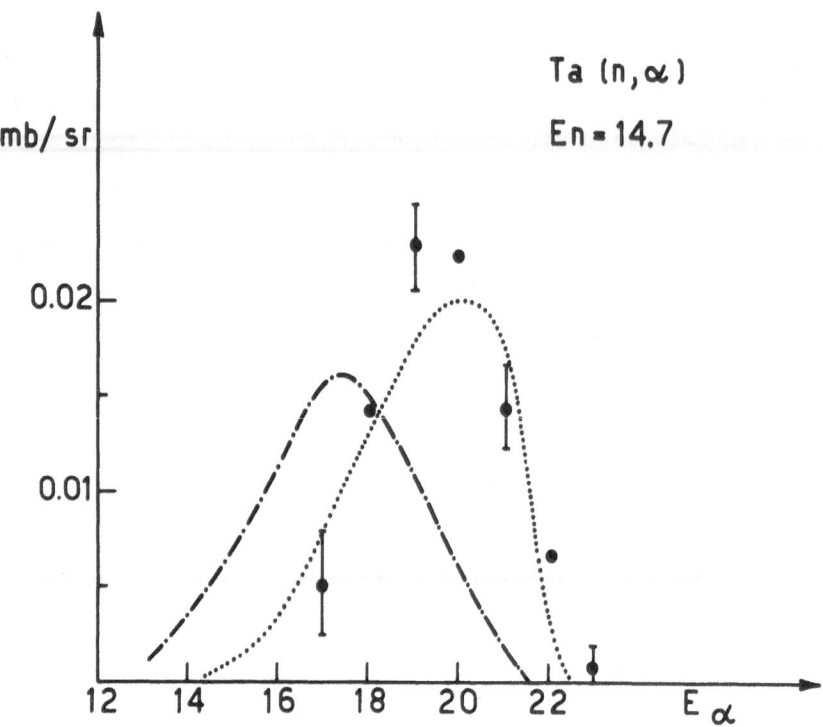

Fig. 7. Comparison between the experimental alpha spectrum
   and the calculated precompound emission with $n_o=3$
   (dots) and $n_o=7$ (dot and dash) for the reaction
   Ta (n,α)

particles are performed in the target nucleus, the probability for the
incoming neutron or proton to strike an alpha, once inside, is less
than one. Therefore, we have an unknown coefficient, smaller than one,
to be introduced in formula (6). Also the interpretation of $|M|^2$ is
less clear.

   Anyway, using the same value of $|M|^2$ as for proton emission, we
have made the same type of analysis performed for (n,p) and (n,n')
cross-sections for (n,α) cross-section measurements, made by activation
technique on heavy nuclei. Beginning with A∿130 we have ∿45 values of
(n,α) cross-sections at 14 MeV neutron energy. Though they show rather
large experimental uncertainty, the use of these data presents the
same advantages we discussed in the (n,p) cases.

   Fig. 10 shows the ratio $R=\sigma_{exp}(n,\alpha)/\sigma_{calc}(n,\alpha)$ for these nuclei.

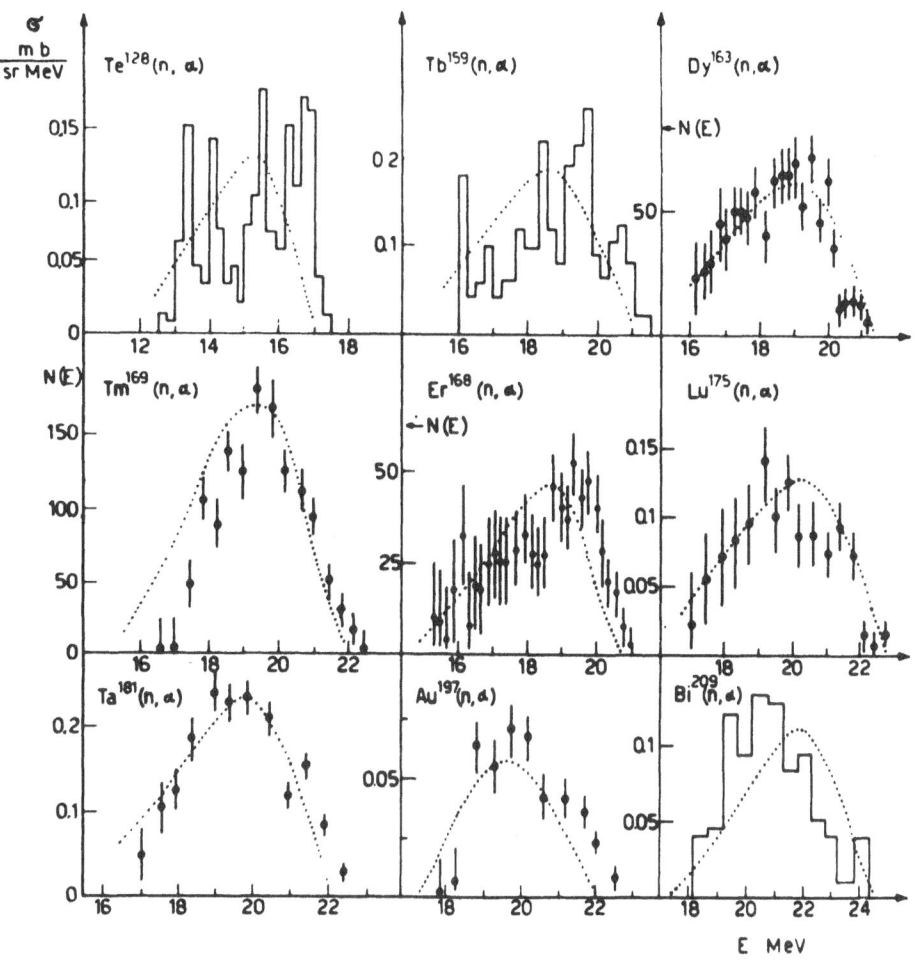

Fig. 8. Comparison of some experimental alpha spectra at
En   14-15 MeV with the calculated ones;"a pre-
formed alpha" precompound calculation ($n_o=3$)
was used

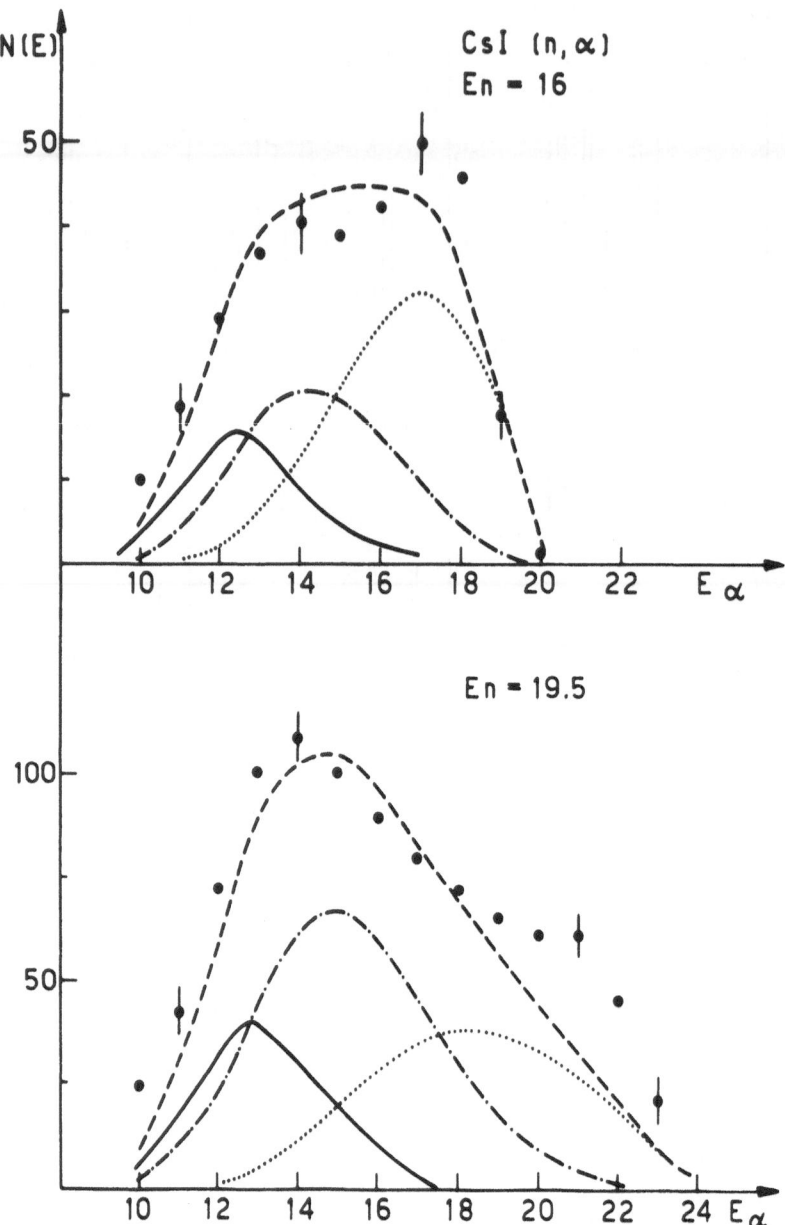

Fig. 9. Analysis of the experimental α spectra for the CsI
(n,α) reaction at 19.5 and 16 MeV incident neutron energy
respectively. The calculated spectrum is given as a sum of
evaporative and precompound emission contributions. The
precompound contributions calculated with $n_o=3$ and with
$n_o=7$ are marked with dots and dash and dots in the figure,
respectively

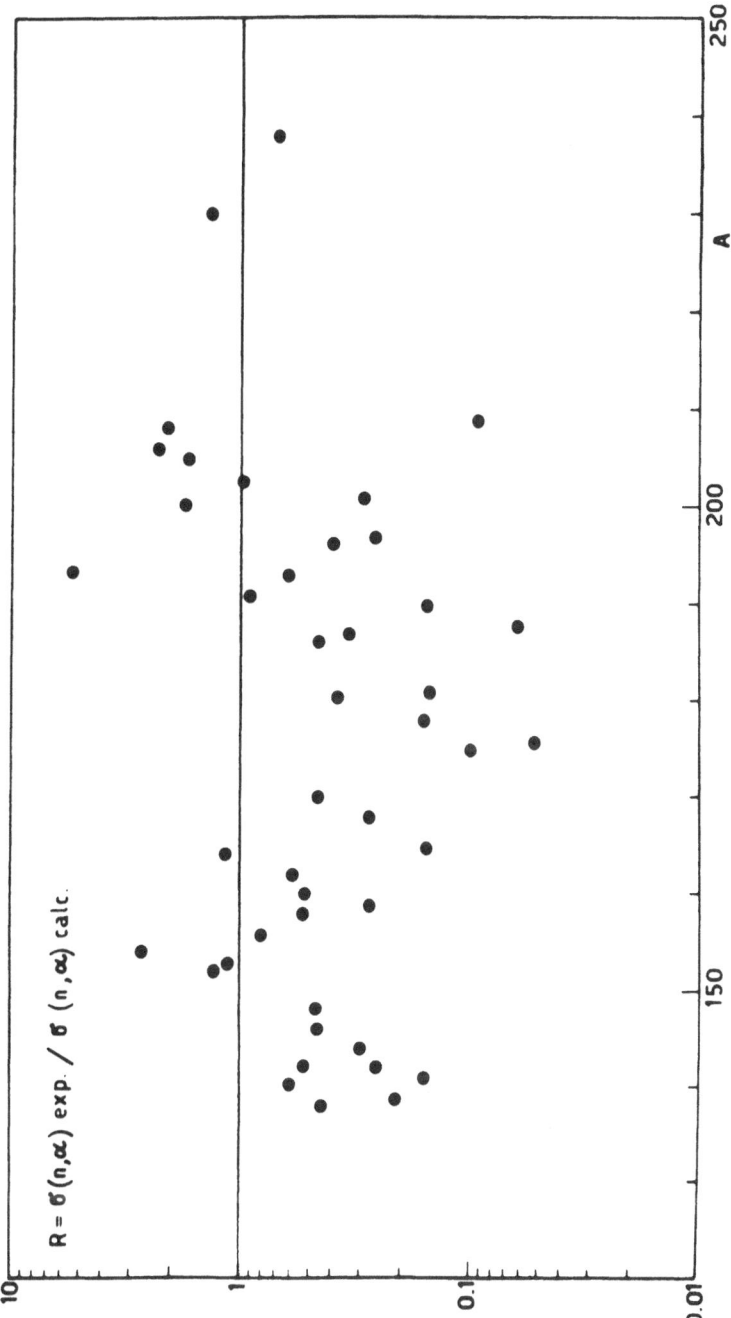

Fig. 10. Plot of the $\sigma_{exp}(n,\alpha)/\sigma_{calc}(n,\alpha)$ ratio as a function of mass number A

We see here that the points are much more scattered than in the (n,p) case and that their value is generally less than one, confirming that the probability to strike an alpha particle is less than one. This probability seems to exhibit an A dependence.

The most interesting point is that this calculation predicts at least the correct order of magnitude of the cross-sections, while other theories, for example the evaporation, in this mass range often predict cross-sections several orders of magnitude too low.

A pre-equlibrium formula applied to the case of the CsI (n,α) reaction for the part described by formula (7),(that is for process 2), allows, as we have seen, a very good description of the spectrum shape, but gives a too small value for the cross-section. In this case the ratio R (for the process 2) is $\sim$100.

The fact that the calculation gives a lower value is not extremely surprising because we calculate here only the statistical probability to have an alpha particle formed during the development of the cascade. If, due to a realistic interaction not taken into account in the model, this probability is higher, a term larger than one must appear in the formula. It is surprising that this term should be so large.

3. Proton Induced Reactions

In the analysis of proton induced reactions that we have considered so far, the excitation energy of the composite nucleus is much higher than in the neutron case. Therefore, some of the approximations that have been introduced in the preceding paragraphs are no more valid.

It is no more possible to neglect particle emission during the development of the cascade. In order to reproduce the experimental data it is, then necessary to take into account, at each stage of the equilibration process, the competition between particle emission into the continuum and intranuclear interactions that increase the exciton number.

Also, the assumption of $|M|^2$ being approximately energy independent is no more valid. The theoretical estimate of W_{eq}^1 , that will be denoted here by $\bar{W}$, at energies 20 MeV higher than the Fermi energy, has a linear dependence on the nucleon energy (see fig. 11) and this means that $|M|^2 \alpha 1/E$.

3.1. Excitation Functions of (p,n) Reactions [3]

The analysis of excitation functions of (p,n) reactions, where the emission of one neutron is not accompanied by other particles, may be taken as a comparatively simple process, which offers a mean to study the competition between particle emission and cascade development of

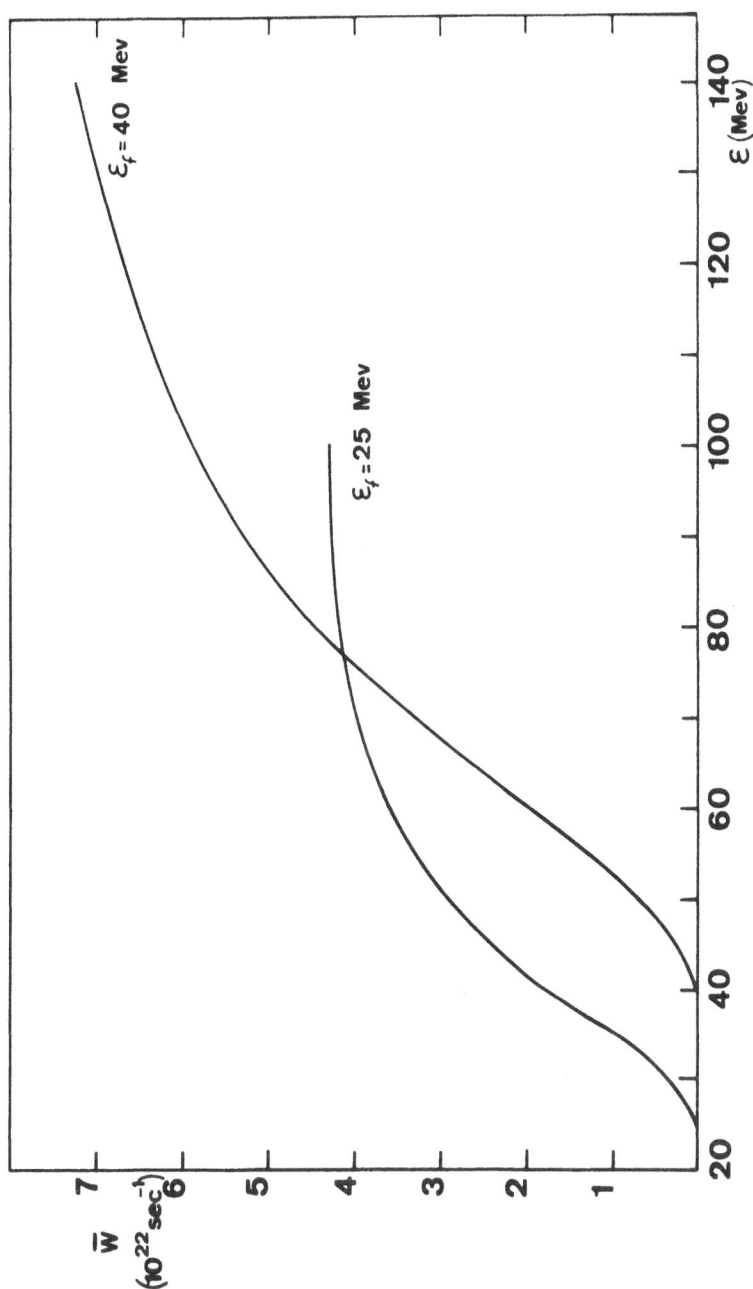

Fig. 11. Predicted values of $\bar{W}$ as a function of the kinetic
energy ε for two values of the Fermi energy

the initial exciton configuration. The reasons why such reactions appear suitable for this purpose are, briefly, the following: (i) the formalism required for the theoretical treatment of the model is simpler when nucleons, rather than more complex particles are involved, as shown in Section 2; (ii) the experimental cross-section measurements, using conveniently chosen target elements and radioactive measurement techniques, are reliable and precise; (iii) the excitation functions beyond their maxima measure directly the pre-equilibrium neutron emission yield, and since the neutrons have to carry away enough energy to leave the nucleus unable to emit further particles, their emission is expected to take place quite early in the equilibration process, when the excitation energy is shared by very few excitons; in this situation the experimental results can be related to a well defined step in the development of the nuclear interaction; (iv) finally, the fact of dealing with the first stages of the equilibration process avoids the amplification of errors introduced by approximations.

The (p,n) excitation functions grow steadily starting from threshold, reach a maximum and beyond the maximum have a long tail, slowly varying with the proton energy (approximately as E_p^{-2}).

At each proton energy the cross-section is given by

$$\sigma_{\alpha\beta} = {}_E\sigma_{\alpha\beta} + {}_P\sigma_{\alpha\beta} \tag{8}$$

where α and β denote the entrance and final channels respectively; ${}_E\sigma_{\alpha\beta}$ is the contribution to the cross-section due to neutrons evaporated by compound nucleus (CN) at equilibrium, ${}_P\sigma_{\alpha\beta}$ is the contribution due to neutrons emitted in the pre-equilibrium decay of the composite nucleus. The two contributions are assumed to sum incoherently due to the relatively low energy resolution of the incident proton beam ($\gtrsim 300$ keV in the best cases) and the fact that transitions to many final levels are measured.

Up to and including the maxima of the excitation functions the evaporation of neutrons from the CN at equilibrium is by far preponderant. On the contrary, the long tail of the excitation functions is entirely accounted for by pre-equilibrium emissions.

Under the usual assumption that at each step of the equilibration process every possible decay mode of an n-exciton state is equally probable, the two contributions are given by

$$E\sigma_{\alpha\beta} \approx \sigma_{CN}(\alpha) G_N(\beta) \approx \sigma_R(\alpha)\{ \prod_{\substack{j=n_0 \\ (\Delta n=+2)}}^{\bar{n}} \frac{W_{eq}^j}{W_{eq}^j + W_c^j} \} G_N(\beta) \tag{9}$$

and

$$P^{\sigma}_{\alpha\beta} \simeq \sigma_R(\alpha)\left\{\frac{W^{n_o}_c}{W^{n_o}_c+W^{n_o}_{eq}} \frac{W^{n_o}_{c\beta}}{W^{n_o}_c} + \sum_n^{\bar{n}} \left[\prod_{n_o}^{n-2} j \frac{W^j_{eq}}{W^j_c+W^j_{eq}}\right] \frac{W^n_c}{W^n_c+W^n_{eq}} \frac{W^n_{c\beta}}{W^n_c}\right\}$$

$$(\Delta n=+2)$$

(10)

here $\sigma_{CN}(\alpha)$ is the compound nucleus formation cross-section, $\sigma_R(\alpha)$ the reaction cross-section for the incident particle, n_o the initial exciton number, W^n_c the probability per unit time of emission of one particle into the continuum from an n-exciton state, W^n_{eq} the probability per unit time of a two-body interaction inside the nucleus whereby the exciton number is increased from n to n+2 and $\bar{n}$ is the average exciton number at equilibrium. $G_N(\beta)$ is the branching ratio for neutron evaporation from the CN into the β channel, whose expression is reported in several publications [16] and $W^n_{c\beta}/W^n_c$ is the branching ratio for the pre-equilibrium emission considered.

W^n_c and $W^n_{c\beta}$ can be calculated as in the case of neutron induced reactions (see formula (2)), using the detailed balance principle. Assuming further that pre-equilibrium emission of particles other than protons and neutrons can be neglected, one gets:

$$W^n_c = \frac{(2S+1)}{\pi^2\hbar^3} \frac{mA}{A+1} \frac{1}{\rho_n(E)} \left\{\int_0^{E_{max,N}} \sigma_{inv,N}(\varepsilon)\varepsilon\rho_{n-1,N}(U)d\varepsilon + \right.$$

$$\left. + \int_0^{E_{max,P}} \sigma_{inv,P}(\varepsilon)\varepsilon\rho_{n-1,P}(U)d\varepsilon \right.$$

(11)

and

$$W^n_{c\beta} = \frac{(2S+1)}{\pi^2\hbar^3} \frac{mA}{A+1} \frac{1}{\rho_n(E)} \int_{E_1}^{E_{max,N}} \sigma_{inv,N}(\varepsilon)\varepsilon\rho_{n-1,N}(U)d\varepsilon.$$ (12)

A is the mass number of the target nucleus, ε the energy of the emitted nucleon of spin s and mass m, $\rho_n(E)$ the density of the n-exciton states of the composite nucleus and E its effective excitation energy, $\rho_{n-1,N}(U)$ and $\rho_{n-1,P}(U)$ are the densities of the (n-1)-exciton states of the residual nuclei and $E_{max,N}$ and $E_{max,P}$ their maximum effective excitation energies, after the emission of one neutron or proton, respectively; $\sigma_{inv,N}(\varepsilon)$ and $\sigma_{inv,P}(\varepsilon)$ are the inverse neutron and proton cross-sections respectively. Finally, E_1 denotes the minimum energy that a neutron should carry away in order to make emission of further particles from the residual nucleus impossible.

For the density of n-exciton states formula (3) is employed. As in the case of neutron induced reactions $n_0=3$. In this case, since for charge conservation the initial 2p-1h configurations can only be either of the 2 proton-1 proton-hole or of the 1 proton, 1 neutron - 1 neutron -hole type, in the expression of $W_C^{(3)}$ and $W_{C\beta}^{(3)}$ the densities of exciton states become

$$\rho_3'(E) = \frac{3}{8}\,\rho_{2p-1h}(E); \quad \rho_{2,N}'(U) = \frac{1}{4}\,\rho_{1p-1h,N}(U); \quad \rho_{2,P}'(U) = \frac{1}{2}\,\rho_{1p-1h,P}$$

$$\rho_{2,P}'(U) = \frac{1}{2}\,\rho_{1p-1h,P}(U).$$

Similar corrections for $n > 3$ have been neglected on account of their smallness.

The relation which connects W_{eq}^n and W_{eq}^1 is again given by (5).

The theoretical estimate of W_{eq}^1 (to be denoted by $\bar{W}$ calculated according to Kikuchi and Kawai [12] as a function of the nucleon kinetic energy ε is shown in fig. 11 for two different values of the Fermi energy. ε is measured from the bottom of the Fermi sea, and is given by the relation $\varepsilon = E_{inc} + B_{inc} + \varepsilon_f$ where E_{inc} and B_{inc} are, respectively, the incident proton energy and its binding energy in the composite nucleus. However, as it was already assessed in the preceding section, the $\bar{W}$ values thus obtained are likely to be overestimates of the transition probability W_{eq}^1.

To obtain corrected W_{eq}^1 values, a relation of the type $W_{eq}^1 = f(\varepsilon)$ $\bar{W}$, with $0 < f(\varepsilon) < 1$, could be used. As a first approximation, the function $f(\varepsilon)$ may be replaced by a constant C, should the analysis of the experimental data suggest a single value for C over a sufficiently large energy interval. This appears to be the case for (p,n) excitation-functions with $C \sim 0.7$ and $\bar{W}$ calculated for $\varepsilon_f = 40$ MeV.

The comparison of the theory with three excitation functions measured with the AVF cyclotron of Milan University in the energy interval $5 \lesssim E_p \lesssim 45$ MeV is shown in fig. 12.

All the parameters entering eqs. (9) and (10) and the reaction cross-section have been taken from available compilations (for further details see ref. [3]). The only free parameter is the C factor entering the evaluation of the W_{eq}^n.

105

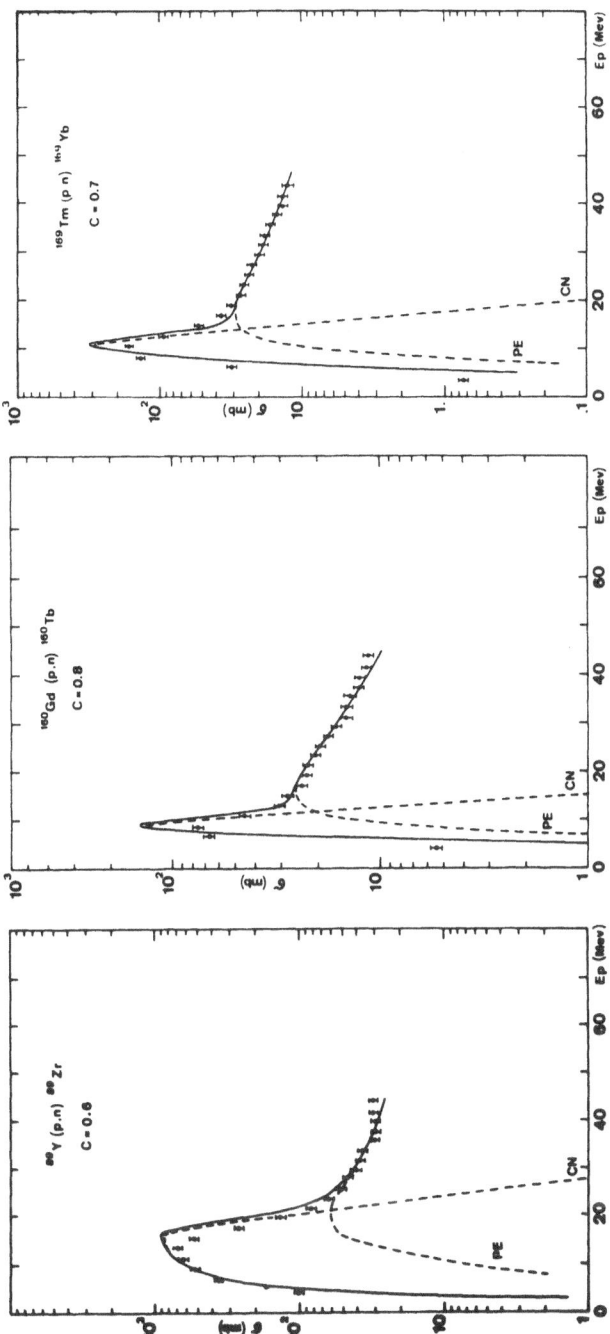

Fig. 12. Excitation functions of ^{89}Y, ^{160}Gd and ^{169}Tm(p,n)
reactions. The solid curve is the theoretical es-
timate with the C values shown; the contributions
due to neutron evaporation from CN and PE emission
are given by dashed lines

The overall agreement is very satisfactory, the C factor being about the same for all the considered reactions: C = 0.7 ± 0.1 (for a Fermi energy ε_f = 40 MeV).

The theoretical expressions above reported succeeded in reproducing satisfactoraly also the other (p,n) excitation functions reported in the literature (on ^{63}Cu, ^{68}Zn, ^{69}Ga, ^{127}I, ^{142}Ce), which have been measured over energy intervals (5≲Ep≲60-80 MeV) extending well into the pre-equilibrium emission region [3].

3.2. Excitation Functions of (p,xn) and (p,pxn) Reactions

The PE Emission Model has been further applied to the analysis of (p,xn) and (p,pxn) excitation functions in ^{181}Ta. The analysis extends up to 45 MeV incident proton energy. The extension of the analysis at higher energies likely requires substantial refinements of the model, as it will be discussed in section 3.3.

The ^{181}Ta nucleus has been chosen since for this nucleus the (p, 3n), (p,4n), (p,pn), (p,p2n), (p,p3n) excitation functions have been measured and in three cases measurements done in two different laboratories are reported [2], [17], [4].

In the case of heavy nuclei the analysis is simplified by the fact that only neutrons are evaporated by the CN at equilibrium. For this reason the pre-equilibrium emission should give a major contribution not only to the tails at high energy of (p,xn) excitation functions but also to the (p,p xn) excitation functions.

In the present analysis we assumed for the proton reaction cross-section the value $\sigma_R = \sum_{x,y} \sigma(p,xpyn)$. This value was calculated by assuming for the (p,n) and (p,2n) cross-sections values estimated on the basis of the comparison with experimental data on neighbouring nuclei.

The values so obtained in the interval 30-45 MeV are slightly smaller than the ones estimated by the optical model calculations used in the analysis of (p,n) data [14]. As a consequence, and to be consistent with previous analyses, the C value used in the present analysis had to be slightly reduced (C∿0.6).

The analytical formulae that must be used are somewhat complicated and will not be explicitly reported.

In all the calculations no free parameters have been introduced. For the parameter C the value 0.6 was assumed according to the new choice of the proton reaction cross-section. The other parameters have been taken from literature as explained in ref. [2].

3.2.1. (p,3n) and (p,4n) Excitation Functions

The two excitation functions have been measured both by Rao and

Yaffe [17] and in our laboratory [2]. The absolute value of the (p,3n) excitation function measured by Rao and Yaffe is higher than the one measured in Milan by about 50%. In the case of the (p,4n) excitation function the agreement is good.

The comparison between the measured excitation functions and the calculated ones is shown in fig. 13.

As one can see the calculations appear to reproduce satisfactorally the experimental data in both cases.

3.2.2. (p,pn) and (p,p2n) Excitation Functions

Both excitation functions have been measured by Rao and Yaffe [17]. In the case of the (p,pn) reaction, only the fraction of the total cross-section feeding the isomeric level of ^{180}Ta at 212 keV was measured. Since this level has spin 1^+ as compared with the spin 8^+ of the ground state, it can be assumed that for the considered proton energies the measured cross-section is a major fraction of the total cross-section. The comparison between the measured excitation functions and the calculated ones is much less satisfactory than in the case of (p,3n) and (p,4n) reactions as shown in fig. 14. The shape is reasonably well reproduced but the calculated absolute value is smaller by approximately a factor 2-3. No reasonable choice of the parameters could improve the agreement.

It has to be stressed that in our approach the reaction path including the emission of deuterons has not been considered. At present, it is difficult to estimate correctly the yield of deuteron emission in the frame-work of the PE model, it can be, however, safely stated, looking at experimental deuteron spectra that the above approximation should introduce an error in theoretical estimates not larger than about 10%.

3.2.3. (p,p3n) Excitation Function

The excitation function has been measured both by Rao and Yaffe [17] and in our laboratory [4] and the agreement between the two measurements is good. The comparison between the measured and the calculated excitation function is satisfactory as shown in fig. 15.

3.2.4. Complementary Data: Proton and Deuteron Spectra from Gold

The proton and deuteron spectra form proton bombardment of ^{181}Ta, at comparable energies, have not been measured; however, measurements on neighbouring nuclei have been reported by Bertrand and Peelle [1].

The proton spectra form proton bombardment of gold at 28.8 and 61.5 MeV are shown in fig. 16. The PE decay model predicts reasonably well the shape of the proton spectrum at 28.8 MeV but fails to reproduce its absolute value by a factor ∿3 (the total theoretical cross-

108

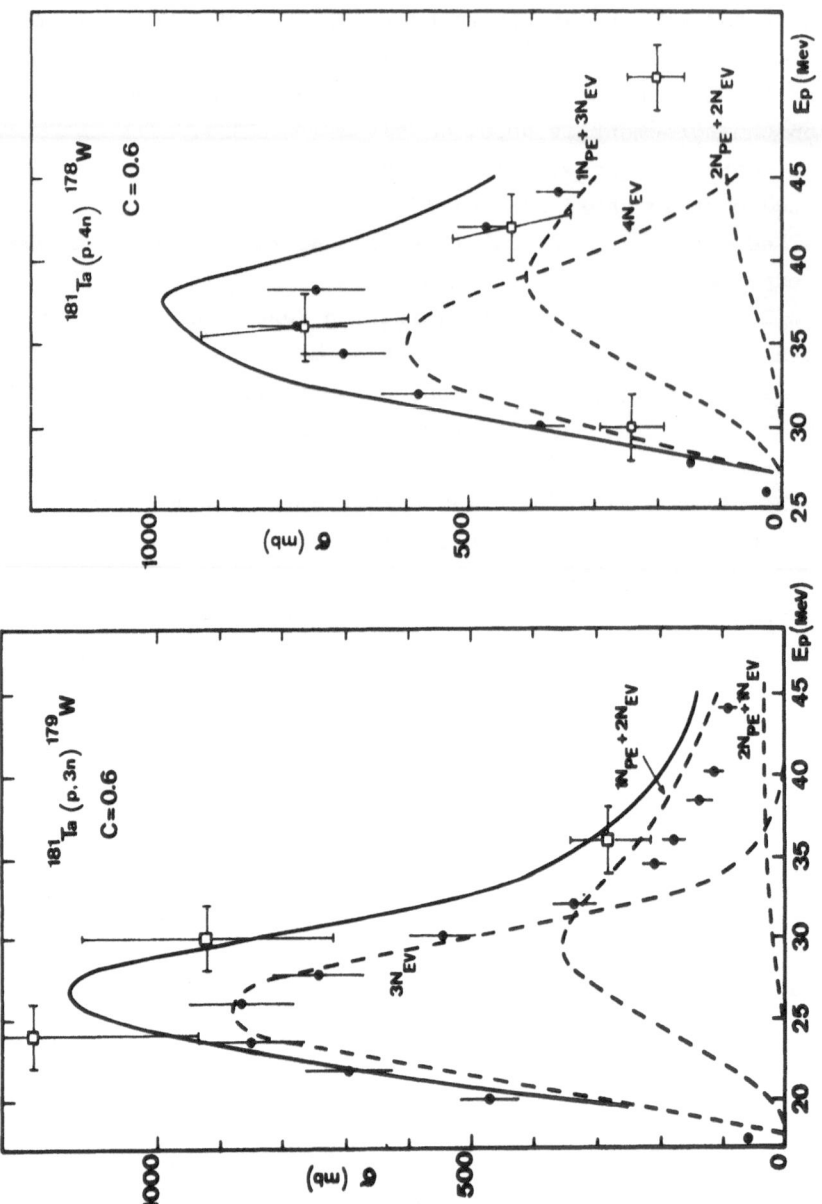

Fig. 13. Excitation functions of ^{181}Ta (p,3n) and (p,4n) reactions. The experimental cross-sections are from ref. [13],
(open squares) and ref. [2] (black circles). The solid curve
is the theoretical estimate. The dashed curves give the contributions of particular reaction paths. Here and in the
following, the index EV below N (or P) indicates that the neutron (or the proton) has been evaporated by the CN at equilibrium; the index PE indicates that it has been emitted in
a Pre-Equilibrium process

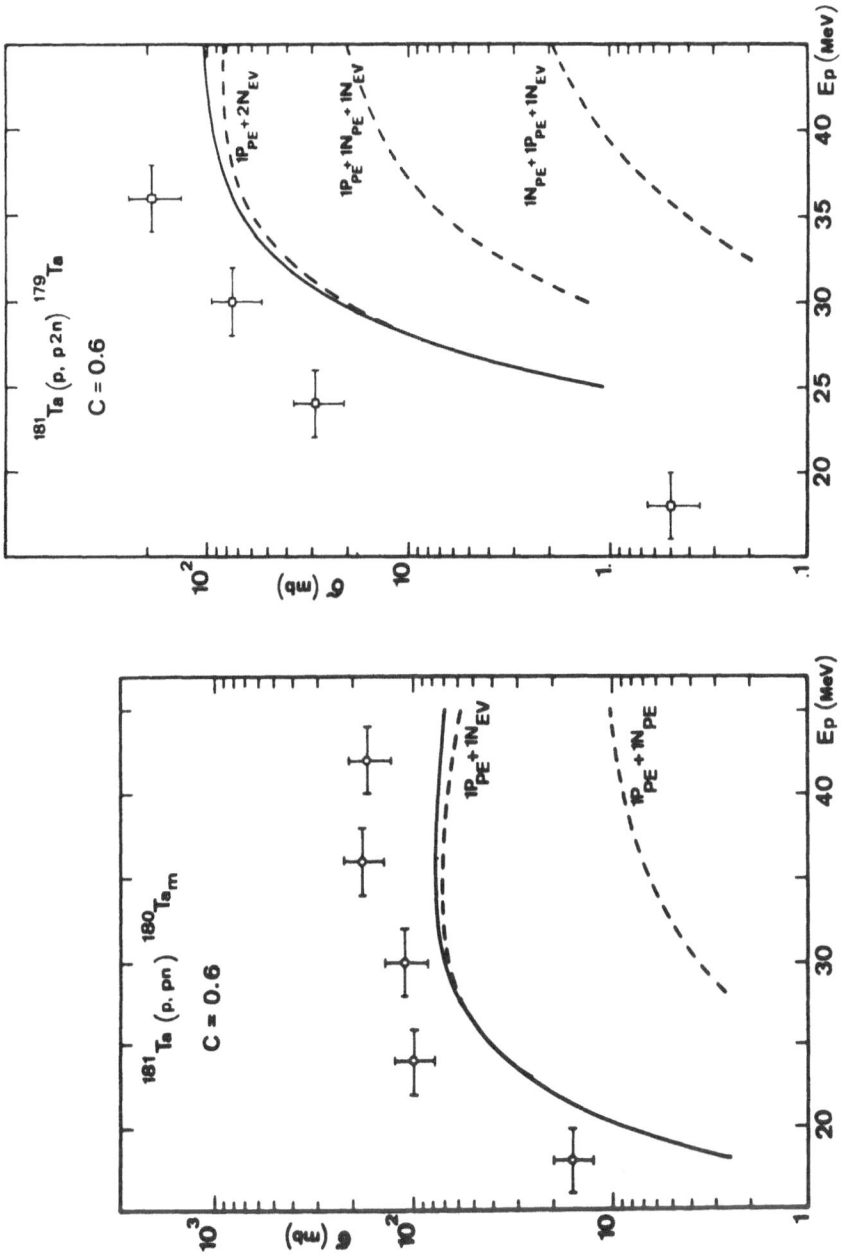

Fig. 14. Excitation functions of ^{181}Ta (p,pn) and (p,p2n).
The solid curve is the theoretical estimation; the
dashed curves give the contributions of particular
reaction paths

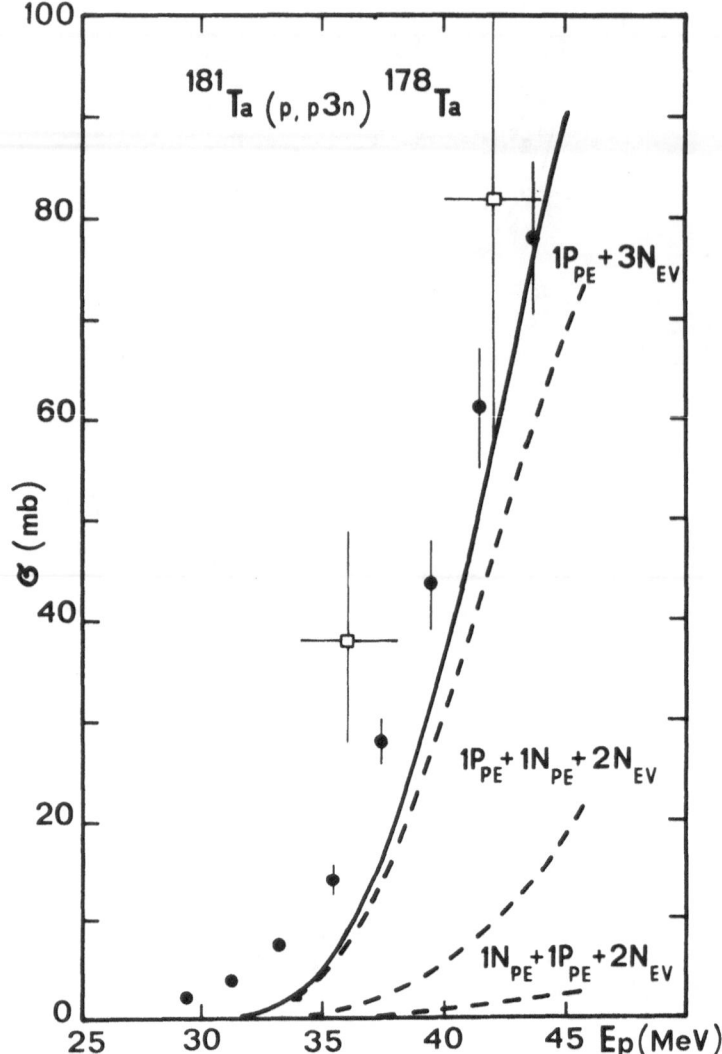

Fig.15 Excitation function of ^{181}Ta(p,p3n) reaction. The experimental cross sections are from ref (17), open squares, and ref (4), black circles.
The solid curve is the theoretical estimate.
The dashed curves give the contribution of particular reaction paths

111

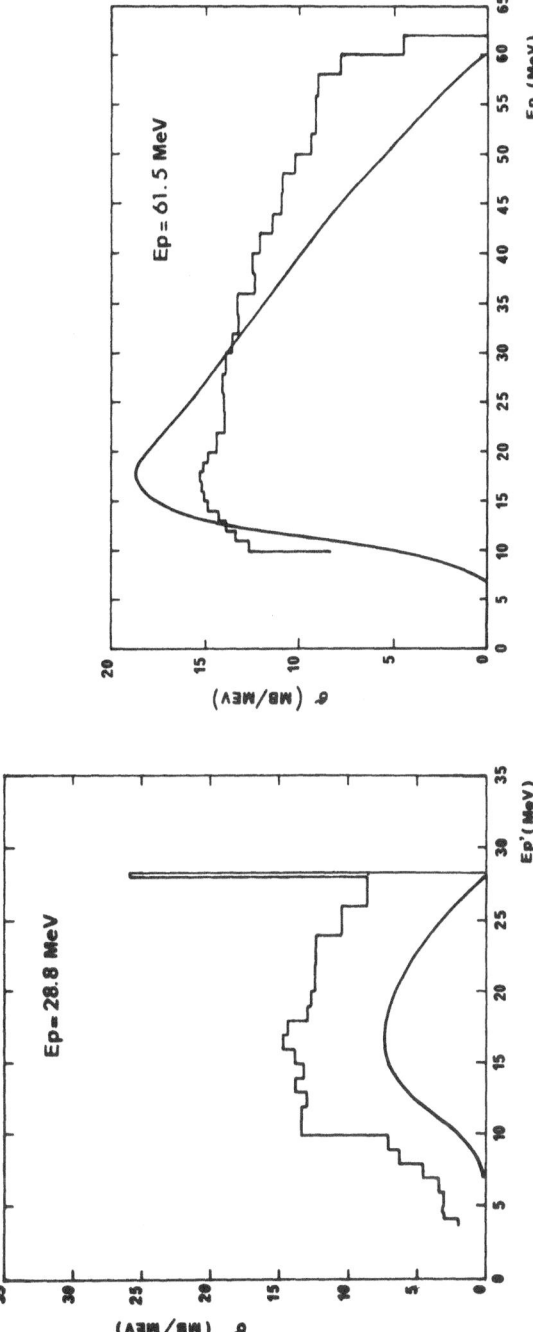

Fig. 16. Angle integrated proton spectra from proton bombardment of gold at 28.8 and 61.5 MeV incident proton energies. The histograms are experimental results from ref. [1], the solid lines the theoretical extimates. The parameters entering the theoretical formulae are from ref. 1o with C = 0.7. The experimental histograms could slightly underestimate on a relative basis the soft contribution to the proton spectrum as compared with the hard one. In fact no measurements have been done at angles greater than ≈125° and the integrals over angle were performed using a trapezoidal quadrature where the cross-section at 180° was assumed equal to the cross-section at the largest angle. On the other hand the peak of experimental spectra shifts at lower particle energies when increasing the emission angle

112

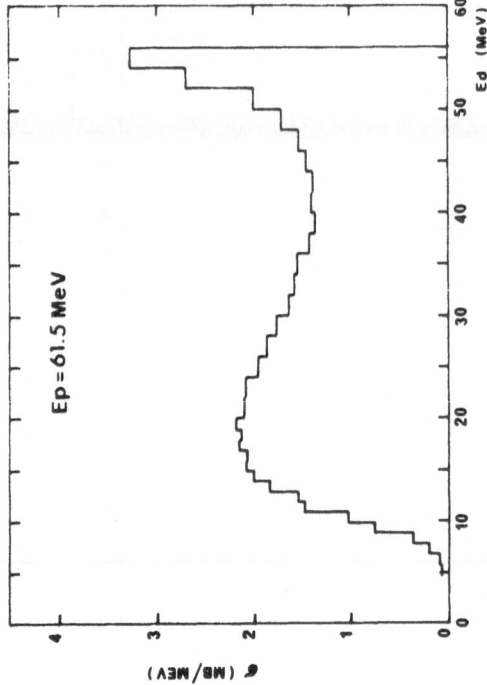

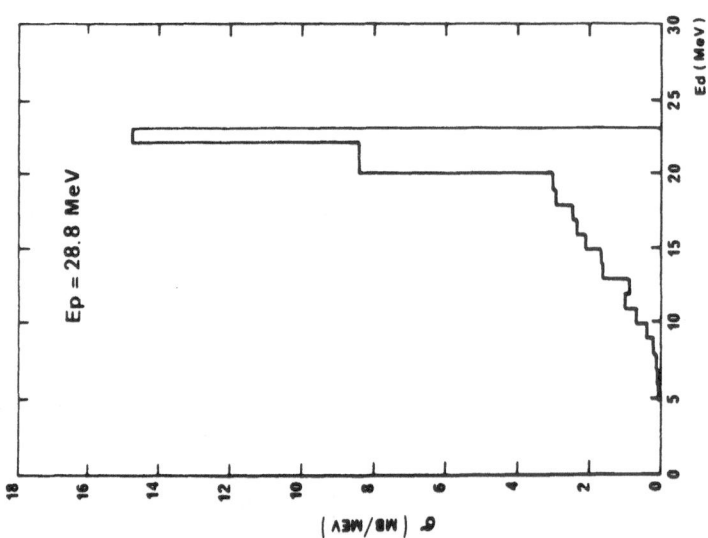

Fig. 17. Angle integrated deuteron spectra from proton bombardment of gold at 28.8 and 61.5 MeV incident proton energies. The data are from ref. [1]

section is ∿94 mb, the experimental one ∿260 mb). At the same energy
the neutron yield is correctly predicted by the model (see fig 13).
At 61.5 MeV the absolute theoretical value is ∿550 mb, the experimental
one ∿660 mb, so the disagreement found at the lower energy is strongly
reduced. At the high energy part of the spectrum, however, the PE emission model predicts cross-sections lower by a factor 3 than the experimental ones.

As an example, in the interval 50≤Ep'≤60 MeV the theoretical cross
-section is ∿27.5 mb, while the experimental value is ∿100 mb. The absolute value of the neutron production cross-section is, on the contrary, correctly predicted by the model also in the hardest part of
the spectrum of emitted neutrons as demonstrated by the success of the
model to reproduce the (p,n) cross-section data (see section 3.1. and
ref. [3]).

The deuteron spectra at 28.8 and 61.5 MeV are shown in fig. 17.
As it is difficult to calculate correctly with PE model the absolute
cross-section for the emission of complex particles (see sect. 2.3),
we shall, in this case, give only a qualitative discussion.

The 28.8 MeV deuteron spectrum shows a huge peak at the highest
available energy. This spectrum shape cannot be predicted by the PE model. If one assumes that the PE contribution to the deuteron spectrum
should have a shape similar to the calculated proton spectrum, one can
roughly estimate a ∿12 mb contribution due to PE decay, while a contribution of ∿16 mb cannot be accounted for by the model. The deuteron
spectrum at 61.5 MeV shows a clear superposition of two effects.

In this case assuming a PE contribution to the spectrum not much
different in shape from the proton one, a ∿63 mb contribution to the
total cross-section, due to PE decay and a ∿23 mb contribution, which
cannot be accounted for by the model, can be estimated.

The above discussion seems to indicate that the total yield of
proton and deuteron emissions which cannot be described as due to PE
decay amounts roughly to ∿182 mb at 28.8 MeV and ∿133 mb at 61.5 MeV,
estimates that are in reasonable agreement with the ones obtained by
the analysis of (p,p×n) excitation functions.

In the case of proton spectra the contribution that we think could
not be described by PE decay extends from 0 to ∿20-25 MeV of excitation energy of residual nucleus. This explains why (p,pn) and (p,p2n)
cross-sections could not be correctly predicted by the model, while
the (p,p3n) cross-section is correctly predicted. Another experimental
information that suggests the presence of two different reaction mechanisms comes from the angular distribution of emitted particle spectra.

Both in the case of gold and bismuth, Bertrand and Peelle data [1] show that the contribution which cannot be described by PE decay model is strongly localized at forward angles lower than 75°; the PE decay contribution, though forward peaked, is dominant at greater angles.

3.3. Discussion

The analysis of (p,n) and (p,×n) excitation functions in medium and heavy nuclei seems to demonstrate that the PE decay model can correctly predict the yield and spectral shape of emitted neutrons. On the other hand the analysis of (p,p×n) excitation functions, the spectrum shape and angular distribution of emitted protons and deuterons indicates the presence of effects in proton and deuteron final channels that cannot be described by PE decay model. In the case of deuterons the effect is likely to be the well known pick-up process, in the case of protons, as shown in the case of bismuth by Bertrand and Peelle, one could invoke an inelastic scattering which excites preferentially groups of levels of collective nature.

In any case, the PE decay model up to ∿50 and perhaps 70 MeV excitation energy accounts for about 90% of total proton reaction cross-section.

Finally, if one assumes a ∿10% contribution to the total proton reaction cross-section which cannot be described by PE decay model, the σ_R to be introduced in the calculations should be ∿90% of the values we used. The use of these new values does not modify any of our conclusions and only improves the fit to (p,×n) excitation functions.

Any refinement of the PE model introduced in order to try to reproduce the most energetic part of the proton or deuteron spectra should explain why such enhancements of high energy particle yields are not seen in the neutron decay.

It could be remarked that, at not too low excitation energies, (i) the use of the equidistant spacing model to calculate level densities is not much justified in the case of PE decay, since the excitation energy is shared among few excitons and (ii) the hole-hole and particle-particle scattering do not behave in a fully symmetric way.

These considerations throw some doubt on the validity, at high energies, of the expression (5) which gives explicitly the dependence of W_{eq}^n on n and implicitly its dependence on energy. Detailed calculations to clarify this point are in progress [11]. Preliminary results seem to assess that at excitation energies lower than ∿50 MeV the energy dependence of W_{eq}^n as implied by (5) is correct and that the ratios W_{eq}^n/W_{eq}^{n+2} are not appreciably altered; however, the relationship between

W^3_{eq} and W^1_{eq} does change. For these reasons a more appropriate approach to evaluate W^n_{eq} should not affect our general conclusions and the goodness of the fits we have shown, but the numerical value of the constant C could be modified.

4. Final Conclusions and Lifetime of an Exciton in the Nucleus

The α-value deduced in Section 2 from (n,p) cross-sections at 14 MeV, allows to estimate that at an excitation energy $E \approx 21$ MeV,

$$\frac{\pi |M|^2 g^3 E^2}{2\hbar} = \frac{\pi \alpha A E^2}{2\hbar g} = W^3_{eq} \approx 0.5 \ 10^{22} \ sec^{-1}.$$

Also the fit to (p,n) excitation functions allows essentially to estimate the value of W^3_{eq} (E) for $20 \lesssim E \lesssim 50$ MeV, since for $E \gtrsim 20$ MeV the first term in the curly bracket in formula (10) is dominant; from this analysis one obtains W^3_{eq} (E=21 MeV) $\gtrsim$ 0.6 x 10^{22} sec^{-1}.

The agreement between the two values is indeed striking on account of the differences in theoretical handling and the uncertainties affecting several experimental data and calculation parameters.

An interesting result of the above analyses is the estimation of the lifetime ($\tau_1 = 1/W^1_{eq}$), in the nucleus, of a particle (hole) at an energy E above (below) the Fermi energy. $\tau_1 \approx 10^{-22}$ sec at $E \approx 21$ MeV and its value decreases to $\tau_1 \approx 0.3 \ 10^{-22}$ sec at $E \approx 52$ MeV.

The corresponding mean free paths are equal to ~ 9 and 4 fm, respectively.

It is interesting to point out that the lifetimes and mean free paths do not depend on A but must be seen as a property of nuclear matter.

References

1. BERTRAND, F.E. and PEELLE, R.W., Reports ORNL-4460 and ORNL 4638, unpublished.
2. BIRATTARI, C., GADIOLI, E., STRINI-GRASSI, A.M., TAGLIAFERRI, G., STRINI, G. and ZETTA, L., Nucl. Phys. A116 , 605 (1971).
3. BIRATTARI, C., GADIOLI, E., GADIOLI-ERBA, STRINI-GRASSI, A.M., STRINI, G. and TAGLIAFERRI, G., Nucl. Phys., A 2o1 (1973) 579
4. BIRATTARI, C., GADIOLI, E., STRINI-GRASSI, A.M., STRINI, G. and TAGLIAFETTI, G., Nuovo Cim. Lett. 7 (1973) 1o1
5. BLANN, M. and LANZAFAME, F.M., Nucl. Phys. A142, 559 (1970);
   BLANN, M., "Equilibration Processes in Nuclear Reactions at Moderate Excitations", Univ. of Rochester Report UR-3591-20, unpublished and bibliography therein,
   CLINE, C.K. and BLANN, M., Nucl. Phys. A172, 226 (1971);
   DEMEYER, A. et al., Journal de phys. 31, 225 (1970), ibidem 31, 847 (1970);
   CLINE, C.K., Nucl. Phys. A174, 73 (1971).
6. BRAGA-MARCAZZAN, M.G., ERBA-GADIOLI, E., COLLI-MILAZZO, L. and SONA, P.G., Phys. Rev. C6 (1972) 1398
7. DOSTROVSKI, E., FRAENKEL, Z. and FRIEDLANDER, G., Phys. Rev. 116, 683 (1960).
8. ERBA, E., FACCHINI, U. and MENICHELLA-SAETTA, E., Nuovo Cimento 22, 1237 (1961).
9. ERICSON, T., Advan. Phys. 9, 425 (1960).
10. GADIOLI, E., IORI, I., MOLHO, N and ZETTA, L., Phys. Rev. C4, 1412 (1971).
11. GADIOLI -ERBA, E., private communication.
12. GOLDBERGER, M.L., Phys. Rev. 74, 1268 (1948).
    KIKUCHI, K. and KAWAI, M., Nuclear Matter and Nuclear Reactions, North Holland Publ. Comp., Amsterdam, 1968.
13. GRIFFIN, J.J., Phys. Rev. Lett. 17, 478 (1966); Phys. Lett. 24B, 5 (1967); BLANN, M., Phys. Rev. Lett. 21, 1357 (1968).
14. MANI, G.S., MELKANOFF, M.A. and IORI, I., Report CEA 2379 (1963) unpublished.
15. COLLI-MILAZZO, L. and BRAGA-MARCAZZAN, M.G., Phys. Lett. 36B, 447 (1971); ibidem 38B, 155 (1972).
16. PORILE, N.T., Low Energy Nuclear Reactions, In Nuclear Chemistry, ed. L. Yaffe, Academic Press, New York 1968.
17. RAO, C.L. and YAFFE, L., Can. J. Chem. 41, 2516 (1963).
18. SEELIGER, D., private communication.
19. WILLIAMS, F.C., jr., Phys. Lett. 31B, 184 (1970).

THE INTERMEDIATE STATE IN FISSION AND SHAPE ISOMERS*

DANIEL SPERBER

Department of Physics,Rensselaer Polytechnic Institute

Troy, Now York 12181 U. S. A.

1. Introduction

The purpose of the present paper is to survey different versions
for the calculation of 1) isomer ratios for shape (fission isomers),
2) the ratio between the cross section for isomeric (fission) and
prompt fission and the ratio between the cross section for neutron
evaporation to prompt fission. The models discussed for the evaluation
of the above observable quantities are based on the statistical compound nucleus model. Although the models have much in common they differ in detail. Certain models are time dependent but no explicitly
energy dependent. Others are energy dependent but time dependent. An
energy dependent model is discussed in detail. In this model the neutron evaporation is discussed in a rigorous way. Also in the latter
calculation there are no adjustable parameters. All parameters are
obtained from other sources. The predictions of the theory are compared with experiments. Very good agreement between theory and experiment
is obtained.

As early as 1964 Flerov [11] found a spontaneous fissioning isomeric state (such isomers are referred to as shape isomers in the present
paper) in the odd americium isotopes. Traditionally isomeric state have
been attributed to the existence of a low lying state which differs in
spin by a few units of h from the spin of the ground state. (Such isomers are reffered to as spin isomers.) The isomers discovered by Flerov
[11] are at an excitation of about 3 MeV. The existence of a spin isomer at such high excitations would imply that the spin of the isomer
is so high as to make it difficult to reconcile with present nuclear
models. Many other fissioning isomers have been discovered since Flerov's [11] work. In fig. 1 the "island of isomers" can be seen.

In section 2 the theory leading to a double humped potential and
the existence of shape isomers is reviewed. In section 3 various statistical theories of shape isomers are discussed. In section 4 the time and energy dependent model for the study of shape isomers is presented. A comparison between theory and experiment is found in section
5 and the meaning of this comparison is discussed in section 6.

---

*Work supported by the United States Atomic Energy Commission

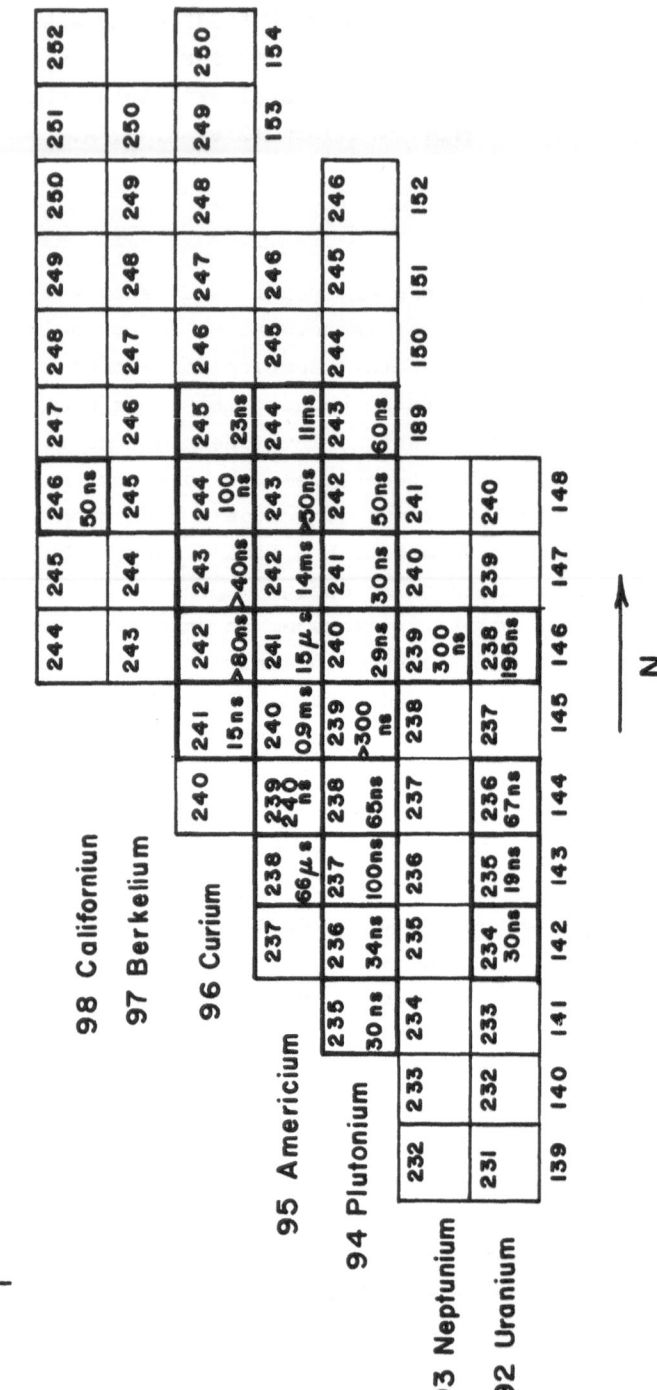

Fig. 1. Island of shape isomers

2. Theory of Potential Surfaces

The nature of the isomeric states can be explained using the Strutinsky [33] model for the calculation of the energy as a function of
deformation. According to this model the potential energy as a function of deformation has more than one minimum and one maximum. So far
attention has been focused on potentials with two minima and two maxima. In fig. 2 a schematic plot of energy versus deformation is found.
The isomeric state is interpreted as the ground state in the second
well. The transition between the isomeric and ground state is slow because of barrier penetration and not due to spin differences. For spin
isomers one has a long lived excited state and the isomer state differ
in spin. For shape isomers one has a long lived excited state since
the ground state differ in shape. It should be clear that the existence of a double humped potential does not necessarily imply the existence of a shape isomer. For example, if the second barrier is very low
the states in the second will fission , or if the barrier between the
wells is low the state in the second well decays very fast to the gro-
und state of the first well.

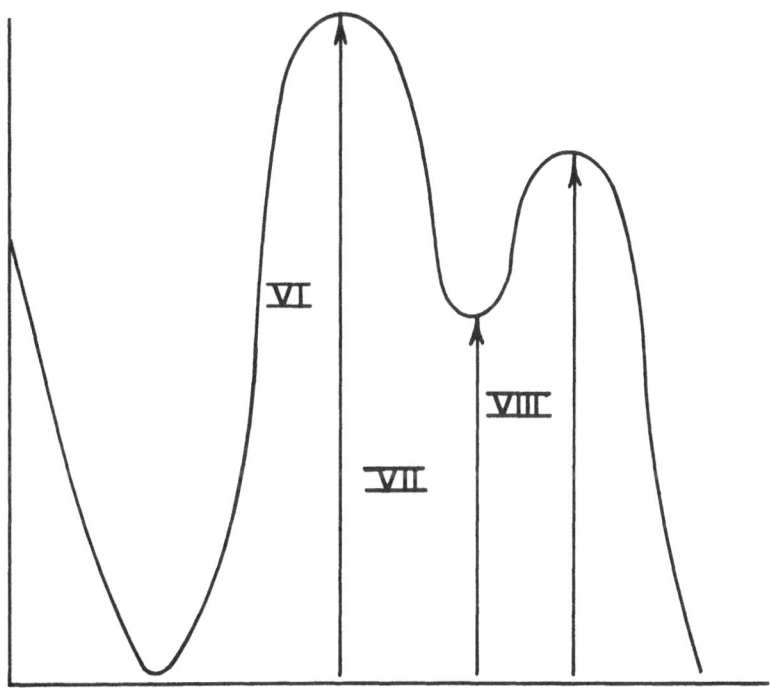

Fig. 2. Schematic form of double humped potential surface

The form of the potential surface is of great importance to the present work. So a brief review of the ideas which lead to the establishment of the double humped potential is appropriate.

Traditionally, the energy as a function of deformation was studied using the liquid drop model. This model was very useful in the study of fission, which is closely related to the present discussion. Also the model predicts very well the gross structure of binding energies. As early as 1935 Weizsäcker[35] suggested the semi-empirical mass formula for the ground state mass which is based on this model. However, the Weizsäker formula predicts the mass to be a smooth function of A and Z, and does not explain fluctuation due to shell effects. Also the liquid drop model suggests a spherical shape for the ground state; as is very well known this is not the case. Obviously the fluctuations which are related to the magic numbers are due to single particle effects. The single particle model is quantum mechanical model, the liquid drop model is classical. Myers and Swiatecki [27] used a liquid drop model with shell correction and came up with a much better semi-empirical mass formula.

It is clear, therefore, that in order to study the potential surface one has to use the liquid drop with some shell corrections. This program was undertaken by Strutinsky [33]. Strutinsky adds all single particle energies. The stable deformation is determined by the position of the gap of single particle energies as a function of deformation. Should there be more than one gap in the single particle energies there will be more than one minimum in the potential surface. Strutinsky's [33] work was followed by many calculations [26] of potential surfaces. The calculations have many features in common. First, a shape and single particle potential related to this shape are determined. Second, the single particle and pairing energies are determined for this potential. Finally, the energy surface is obtained by summing all single particle energies. Different groups parametrize the potentials in different ways. For example, the Greiner [26] group uses a two centered potential. Most groups introduced recently also use asymmetric deformations. It should be realized that in general more than one deformation coordinate is used. If one considers two degrees of freedom for deformation one gets topographical maps. For more degrees of freedom a graphical representation is difficult.

3. Statistical Models for Shape Isomers

One of the first theoretical evaluations of isomer ratios for shape isomers is due to Jungclaussen [20]. Jungclaussen employs the sta-

tistical theory of nuclear reactions. Hence, in his simplified picture
the isomer ratio for shape isomers is equal to the ratio between the
number of open channels for the transitions through the first barrier
to the number of evaporation channels. This is the same as the ratio
between the probability for transition through the first barrier left
to right to the probability for neutron evaporation times the probability for neutron evaporation after the transition. The number of open
channels is obtained by integrating the appropriate density of levels
by the corresponding energy interval.

A more serious approach to enlist the statistical model for the
study of isomer ratio for shape isomers is due to Jägare [18]. Jägare
[18] very sensibly starts his work with treating (n,γ) reactions.These
reactions are simpler as there is no competition from nuclear decay.
Also Jägare [18] realizes rightly that the isomer ratio depends on
the population of states as a function of time. He then derives coupled differential equations for the population of states in both wells.
He solves these equations in which he includes gamma decay between
wells and penetration between wells. Unfortunately Jägare [18] does not
include the energy dependence of this population. This is a crucial
point, since the relative probability between the transition between
well and within well is strongly energy dependent. Close to the top of
the first barrier transition between barrier predominates. Close to
the bottom of the wells gamma decay between states within the wells dominates. Therefore, one has to obtain equations for the population of
states as a function of time and energy. Such equations are derived later in this paper. Jägare [18] differential equations are later replaced by integro-diferential equations. The differentiation is with respect to time and the integration with respect to energy. For (n,γ) reactions spin consideration can be neglected. Jägare's [18] calculation
is consistent with the independent hypothesis.

Lynn [23] classified the states in double humped potential as states of class I and states of class II. States of class I have a large
amplitude in the first well and small amplitudes in the second well.
States of class II have a small amplitude in the first well and large
amplitude in the second well. Jägare [18] considers rightly the states
of class II as doorway states to fission, therefore, the fission excitation function depends on the states of class II. Finally, Jägare [18]
fits the barrier height and well depth parameters to obtain the right
isomer ratio. As can be seen from his paper, these parameters are not
uniquely determined and a wide rage of parameters reproduces the experimental results. More important, the potential surface is not speci-

fied only by the values of the minima and maxima. The position (in
terms of a deformation parameter) of the maxima and the curvature of
the potential surfaces are significant. These features are not discussed in Jägare's [18] paper. Despite its shortcomings the paper makes
many reasonable assumptions, namely that i) the nuclei in both wells
are originally highly excited to allow statistical consideration; ii)
the population of class II states can be neglected at time t=0; iii)
the statistical model is valid for both wells, and iv) angular momentum effects can be neglected and v) the density of levels is the same
in both wells. The last assumption can not be justified from first
principles. However, at present little is known about the level density of levels for strongly deformed nuclei. Therefore, one uses the
same form for the density of levels for both wells. However, one has
to keep in mind that the zero point, for the purpose of the calculation of densities of levels, is shifted to the minimum in the second
well.

An important recent contribution to the statistical properties of
shape isomers is due to Britt [5] et al. This group studied many shape
isomers experimentally. The raw data is converted to ratios between
isomer to prompt fission. This quantity is given as a function of the
difference between the excitation energy E* and the sum of binding
energies evaporated neutrons in the reaction for which the ratio is
measured. In this paper the group analysis also included reactions which
were studied by other groups. An attempt is made to develop a theory
for this ratio. This theory is now briefly reviewed. For simplicity
they discuss in detail a reaction in which two neutrons are evaporated.
They first treat the decay of the original compound nucleus with A + 2
nucleons decaying to a nucleus of A + 1 nucleons. They determine the
fraction of nuclei which decay by neutron evaporation as $<\Gamma_n/(\Gamma_n + \Gamma_f)>$
$>_{A+2}$. For the ratio Γ_n/Γ_f values suggested by Vandenbosch and Huizenga
[34]. The distribution of energies in the residual nucleus of A + 1
nucleons is considered to be Maxwellian. It is implied that the original nucleus decays only by prompt fission or neutron decay. This is
very reasonable. The nucleus containing A + 1 nucleons is assumed to
decay to the final nucleus by neutron emission. Allowance is made for
the neutron to evaporate leaving the final nucleus of A nucleons in
either the first or the second well. Also prompt fission is allowed.
The width Γ for barrier penetration is calculated using

$$\Gamma = \frac{1}{2\pi\rho} \quad N \qquad\qquad (1)$$

In eq. (1) ρ is the density of levels in the region from which the
transition is made and N is the number of open channels into which the
transition is made. Britt [5] et al. use level spacing rather than level densities. The number of open channels is obtained by integrating
the level density in the transition region over energy. The nuetron
width is calculated using a sharp cut-off   approximation. Whenever
gamma decay is included it is assumed to be pure dipole, with its
strength used as an adjustable parameter. For the density of levels the
form suggested by Gilbert and Cameron [13] is used. The same density
of levels is used in the following discussion. This is a density of
levels for a Fermi gas. The details of this density of levels is discussed in the appendix. However, a few words about its parametrization
are appropriate now. As all other Fermi-gas densities of levels it has
the general form

$$\rho(E) \propto \exp 2(aE)^{1/2} \tag{2}$$

Britt [5] et al. use the same value of "a" for neutron evaporation and
the penetration of the barrier between the wells. This value of "a" is
the one suggested by Gilbert and Cameron [13]. The value of the parameter "a" for fission "a_f" is determined to obtain the values of the
average value $<\Gamma_n/\Gamma_f>$ of Vandenbosch and Huizenga [34]. This yields a
value of $a_f/a=1.2$. This is consistent with the work of Sikkeland et al.
[30]. The height of the first barrier is determined from fission thres-
holds when available [6]. Whenever this information does not exist the
same height for isotopes of the same elements are used. Usually the
height of a barrier for one isotope is known. Next the frequencies of
the wells and barrier are determined. Britt [5] et al. estimate the
frequencies by the analysis of experimental half lives for spontaneous
fission from the isomeric and ground state. The calculated life times
by Cramer and Nix [8] using the same parameters as Britt [5] et al.
agree reasonably well with the results of Malikin [24]. However, there
is no justification for the use of the same frequencies for the barriers of all the isotopes. Britt's statistical model includes three adjustable parameters, i) the height of the secondary minimum, ii) the
difference between the height of the second barrier in a nucleus of
A + 1 nucleons and the second minimum for a nucleus of A nucleons and
iii) the difference between the height of the second barrier and the
second minimum is the final nucleus. These parameters are determined
so as to fit the experimental results. Britt [5] et al. show that the
parameters depend on the way the neutron evaporation is treated. For

example, there is a difference in values for the second minimum if neutron evaporation is approximated by the Jacson [13] Monte Carlo method or the sharp cut-off approximation. Neutron evaporation plays an important role. Therefore, it has to be treated as rigorously as possible. This is shown below. In Britt's [5] paper a detailed analysis of the systematics of the deduced parameters is found. The authors note that there are still some inconsistencies in the fitted parameters. Also the authors notice the discrepancy between the theoretical and calculated values for the density of levels. This discrepancy is particularly important in the second well. A comparison of the barrier with the measured values of Pauli [26] et al. and Nix [26] et al. shows that for ^{240}Pu the extracted values of the height of the second minimum, the first and second maxima with respect to the first minimum are very reasonable.

4. Time Dependent Statistical Approach

Now a different method for the calculation of i) the ratio between cross section for isomeric fission to the cross section for prompt fission, ii) the ratio for the evaporation of a few particles and cross section for prompt fission and iii) isomer ratios for shape isomers is described. It is realized that all observable quantities depend on the population of states in all intermediate and final nuclei. In the present study only barrier penetration between wells, fission, neutron evaporation and gamma decay are considered.

The present work is motivated by previous work on the calculation of isomer ratios for spin isomers. Methods for the evaluation of isomer ratios for spin isomers have been studied by many groups (cf [16], [19], [22], [25], [28], [29], [32]). It has been shown [32], [22] that isomer ratios for spin isomers depends on the population of states at all times at all energies and with all possible spins. Integro-differential equations for the populations of states have been derived and solved. For spin isomers it was found that the easiest reactions to study, as far as first principles are concerned are (n,γ) reactions. In this case the knowledge of the population of states in only one nucleus is required and there are no coupled equations. For shape isomers, from a theoretical point of view the easiest reactions are also (n,γ) reactions for an even even nuclei. There is a distinct difference between the study of isomer ratios for spin and shape isomers. For shape isomers the knowledge of population of states in two wells is required. Therefore, even for the simple case of (n,γ) reactions the population of states in both wells is required and coupled integral equations are obtained. As will be seen below, the barrier penetrabilities depend on the form of the potential surfaces and inertial parameters as a function of deformation. Such parameters have been first calculated in detail for even even nuclei. Therefore, only even even nuclei can be treated so that all parameters are obtained from theoretical considerations. Only little information is available concerning (n,γ) reactions for nuclei for which compound nucleus is even even . The present theory is developed from first principles. No parameters are adjusted. All parameters are obtained from other sources. As far as possible all parameters are obtained from theoretical consideretions only, without extrapolation.

First the differentio-integral equations for the population of states are derived. A state in each well can decay in many ways. For example, a state in the second well can decay by fission, barrier penetration to the first well, gamma decay to lower states in the same well and neutron emission if energetically allowed. Let $P_1^i(E,t)$ and $P_2^i(E,t)$ be the population of states after the emission of i neutrons in the first and second well respectively. The populations of states satisfy the following equations.

$$\frac{\partial P_1^i(E,t)}{\partial t} = (1 - \delta_{i,o}) \int P^{i-1}(E',t) S_n(E',E) dE'$$

$$- (1 - \delta_{i,x}) P_1^i(E,t) \int S_n(E,E') dE'$$

$$+ \int P_1^i(E',t) S_\gamma(E',E) dE'$$

$$- P_1^i(E,t) \int S_\gamma(E,E') dE$$

$$- \frac{P_1^i(E) \Gamma_{12}(E)}{\hbar} + \frac{P_2^i(E,t) \Gamma_{21}(E)}{\hbar} \qquad (3a)$$

$$\frac{\partial P_2^i(E,t)}{\partial t} = (1 - \delta_{i,o}) \int P^{i-1}(E',t) S_n(E',E) dE'$$

$$- (1 - \delta_{i,x}) P_2^i(E,t) \int S_n(E,E') dE'$$

$$+ \int P_2^i(E',t) S_\gamma(E',E) dE'$$

$$- P_2^i(E,t) \int S_\gamma(E,E') dE'$$

$$+ \frac{P_2^i(E,t) \Gamma_{12}(E)}{\hbar} - \frac{P_2^i(E,t)\ \Gamma_{23}(E)\ +\ \Gamma_{21}(E)}{\hbar} \qquad (3b)$$

In eqs. (3a) and (3b) x is the maximum number of evaporated neutrons, $S_n(E',E)$ and $S_\gamma(E',E)$ are the neutron and gamma decay rates respectively. In the same equations $\Gamma_{12}(E)$, $\Gamma_{23}(E)$ are the penetration widths of the first and second barriers from left to right whereas $\Gamma_{21}(E)$ is the penetration width of the first barrier from right to left. The integrals in eqs. (3a) and (3b) are finite. For convenience the limits of integration are not spelled out. However, the limits depend on the number of evaporated particles and whether the population of states is in the first, or second well. The forms for decay rates $S(E,E')$ and widths $\Gamma(E)$ are discussed in detail below. To demonstrate how the above equations are solved, a particular simple case is chosen, namely the case when no neutrons are evaporated. In this case only two functions, representing the density of states in both wells are sought after $P_1^o(E,t)$ and $P_2^o(E,t)$. The integro-differential equations for these functions become

$$\frac{\partial P_1^o(E,t)}{\partial t} = - \frac{P_1^o(E,t)\Gamma_{12}(E)}{\hbar} + \frac{P_2^o(E-\Delta E,t)\Gamma_{21}(E-\Delta t)}{\hbar}$$

$$+ \int_{E^o}^{E} P_1^o(E',t) S_\gamma(E',E) dE$$

$$- P_1^o(E,t) \int_o^E S_\gamma(E,E') dE' \qquad (4a)$$

$$\frac{\partial P_2^o(E-\Delta E,t)}{\partial t} = - \frac{P_2^o(E-\Delta E,t)\Gamma_{21}(E)}{\hbar}$$

$$- \frac{P_2^o(E-\Delta E)\Gamma_{23}(E-\Delta E)}{\hbar} + \frac{P_1^o(E,t)\Gamma_{12}(E)}{\hbar}$$

$$+ \int_{E-\Delta E}^{E_o} P_2^o(E',t) S_\gamma(E',E) dE'$$

$$- P_1^o(E-\Delta E,t) \int_{\Delta E}^{E-E} S_\gamma(E-\Delta E, E') dE' \tag{4b}$$

The energies appearing in the equation correspond to the energies as measured from the ground state in each well. In eqs. (4) ΔE represents the difference in energy between the two minima. Therefore a state at an energy E above the first minimum corresponds to a state at an energy $E-\Delta E$ above the second minimum. It will be seen below that for the purpose of evaluating observable quantities the knowledge of time integrated population of states is sufficient. Let $z_1^o(E)$ and $z_2^o(E)$ be

$$z_1^o(E) = \int_o^\infty P_1^o(E,t) dt \tag{5a}$$

$$z_2^o(E) = \int_o^\infty P_2^o(E,t) dt \tag{5b}$$

Integrating eqs. (4a) and (4b) over time and using (5a) and (5b) one obtains the coupled integral equations

$$P_1^o(E,t=\infty) - P_1^o(E,t=0) = - \frac{z_1^o(E) \Gamma_{12}(E)}{\hbar}$$

$$+ \frac{z_2(E-\Delta E) \Gamma_{12}(E-\Delta E)}{\hbar} - z_1(E) \Gamma_\gamma(E)$$

$$+ \int_{E^o}^E z_1(E') S_\gamma(E',E) dE' \tag{6a}$$

$$P_2(E-\Delta E,t=\infty) - P_2(E-\Delta E,t=0) = \frac{z_2(E-\Delta E) \Gamma_{21}(E)}{\hbar} - \frac{z_2(E-\Delta E) \Gamma_{23}(E-\Delta E)}{\hbar}$$

$$+ \frac{z_1(E) \Gamma_{12}(E)}{\hbar} - \frac{z_2(E-\Delta E) \Gamma_\gamma(E-\Delta E)}{\hbar}$$

$$+ \int_{E-\Delta E}^{E_o} z_2(E') S_\gamma(E',E) dE' \tag{6b}$$

The above equations can be simplified using the boundary conditions satisfied by the functions representing the population of states.

$$P_1^O(E,t=0) = C\delta(E_o,E) \tag{7a}$$

$$P_1^O(E>0,t=\infty) = 0 \tag{7b}$$

$$P_2^O(E,t=0) = 0 \tag{7c}$$

$$P_2(E>0,t=\infty) = 0 \tag{7d}$$

The integral equations for the function $z_1^O(E)$ and $z_2^O(E)$ become

$$z_1^O(E) = \frac{C\hbar\,\delta(E,E_o)}{\Gamma_{12}(E) + \Gamma_\gamma(E)} + \int_E^{E_o} K_1(E',E) z_1^O(E') dE'$$

$$+ M_{21}(E) z_2(E-\Delta E) \tag{8a}$$

$$z_2^O(E-\Delta E) = \int_{E-\Delta E}^{E_o} K_2(E',E-\Delta E) z_2(E') dE'$$

$$+ M_{12}(E) z_1^O(E) \tag{8b}$$

Here

$$K_1(E',E) = \frac{\hbar S_\gamma(E',E)}{\Gamma_{12}(E) + \Gamma_\gamma(E)} \tag{9a}$$

$$K_2(E',E-\Delta E) = \frac{\hbar S_\gamma(E',E)}{\Gamma_{21}(E-\Delta E) + \Gamma_{23}(E-\Delta E) + \Gamma_\gamma(E-\Delta E)} \tag{9b}$$

The integral equations (8a) and (8b) are coupled. However, the fact that $z_2^O \ll z_1^O$ allows one to decouple the equations thus one obtains

$$z_1^O(E) = z_{1,0}^O(E) + \int_E^{E_o} K_1'(E',E) z_1^O(E') dE' \tag{10a}$$

$$z_2^O(E-\Delta E) = z_{2,0}^O(E) + \int_{E-\Delta E}^{E_o} K_2(E',E-\Delta E) z_2^O(E') dE' \tag{10b}$$

Equation (10b) can be rewritten as

$$z_2^o(E) = z_{2,\theta}^o(E) + \int_o^{E_o - E} K_2(E,E') z_2^o(E') \qquad (11)$$

In eqs. (10a) and (10b)

$$z_{1,0}^o(E) = \frac{C\hbar\delta(E_o,E)}{[\Gamma_{12}(E) + \Gamma_\gamma(E)][1 + M_{12}(E)M_{21}(E)]} \qquad (12a)$$

$$z_{2,0}^o = M_{12}(E) z_1(E)$$

$$K_1'(E',E) = \frac{K_1(E',E)}{1 + M_{12}(E)M_{21}(E)} \qquad (12b)$$

The decoupled integral equations (10a) and (10b) are solved by the method of successive approximations. First the equation for z_1^o is solved, this enables the solution for z_2^o. The solutions for the two functions depend on the constant C, however, the observable quantities are independent of this constant.

Now the decay rates appearing in the integral equtions are discussed. The kernels depend on the gamma decay rates and on the width for the penetration of various barriers.

First the gamma decay rate $S_\gamma(E,E')$ is discussed. The nuclei under consideration are deformed. In such nuclei collective gamma decay predominates. In the present discussion only one degree of freedom, namely quadrupole deformation, which finally leads to fission is considered. The collective states in both states are vibrational. The decay rate for the transition between such states is given by Bohr and Mottelson [3] as

$$S_\gamma(E,E') = \frac{h60\alpha r_o^2}{A^{1/3}M} (\frac{E-E'}{hc})^4 \rho(E') \qquad (13)$$

In eq. (13) α is the fine structure constant, r_o is taken as 1.25 fermi, M is mass of a nucleon and $\rho(E')$ is the level density.

Now expressions for the width of barrier penetration are given. For example, expressions for penetrating the first barrier from left to right are given. Different expressions are used for different energy ranges. Let the energy E be measured from the bottom of the first well then

$$\Gamma_{12}(E) = \hbar\omega_1 \frac{1}{1 + \exp\frac{2\pi}{\hbar\omega_{12}}(B_1 - E)} \qquad (14a)$$

$$\Gamma_{12}(E) = \frac{1}{2\pi\rho_I} \frac{1}{1 + \exp\frac{2\pi}{\hbar\omega_{12}}(B_1 - E)} \qquad (14b)$$

$$\Gamma_{12}(E) = \frac{1}{2\pi\rho_I} \int_0^{E-B} \frac{\rho^*(E-B_1-\varepsilon)d\varepsilon}{1 + \exp\frac{2\pi}{\hbar\omega_{12}}(B_1 - E + \varepsilon)} \qquad (14c)$$

$$\Gamma_{12}(E) = \frac{1}{2\pi\rho} \int_0^{E-B} \rho^*(E-B_1-\varepsilon)d\varepsilon \qquad (14d)$$

Equation (14a) is valid for very low excitation just above the ground state, eq. (14b) is valid for energies about 1 MeV in the first barrier above the ground state and B_1 the height of the barrier, eq. (14c) is valid just above the first barrier, for higher excitation the expression (14c) can be approximated by (14d). In eq. (14a) ω_1 is the frequency in the first well, ω_{12} corresponds to the frequency of the parabolic well separating the two wells. In eqs. (14) ρ_I is the density of levels in the first well ρ^* is the level density of transition region above the saddle point. In eq. (14a) the expression for the width is the expression for fission width from the ground state. The factor $[1 + \exp 2\pi/h\omega_{12}(B_1 - E)]^{-1}$ is the penetrability as calculated for a parabolic barrier as suggested by Hill and Wheeler. Equation (14b) is based on the relation [2]

$$\Gamma = T \frac{D}{2\pi} \qquad (15)$$

In eq. (15) T is the penetrability and D is the level spacing. Equation (14d) is the standard expression for fission width above the barrier as derived by Bohr and Wheeler [4]. Expression (14c) is more appropriate just above the barrier. A similar expression was recently suggested by Britt et al. [5].

Now it is shown how the frequencies are determined. For example, it is shown how the frequency in the first well is determined. Let $\bar{\varepsilon}$ be the value of the deformation parameter ε for which the potential assumes its first minimum than the potential V near this deformation can be approximated by

$$V = \frac{1}{2}K (\epsilon - \bar{\epsilon})^2 \tag{16}$$

The constant K is determined for even even nuclei from theoretical calculations. The frequency ω_1 is given by

$$\omega_1 = \sqrt{K/B(\omega)} \tag{17}$$

In eq. (17) $B(\bar{\epsilon})$ is the inertial parameter corresponding to the deformation $\bar{\epsilon}$. The above indicated method for the evaluation of the frequency is possible for only a limited class of nuclei. Whenever a determination like this is not possible the frequency is taken from Britt et al. [5] who determined them from the analysis of life times.

Now the forms for the decay rates for neutron decay are discussed. Instead of determining $<\Gamma_n>/<\Gamma_f>$ from old work the neutron decay rate is obtained from reciprocity applying the principle of reciprocity to each channel with specified angular momentum separately [31]. Accordingly the neutron decay rate $S_n(E,E')$ is

$$S_n(E,E') = \sum_{JJ'j\ell} S_n(E,J;E',J';\ell,j) \tag{18}$$

In eq. (18) $S_n(E,J;E',J';\ell,j)$ is the decay rate from a state with energy E and spin J to state with energy E' and spin J' by emitting a neutron with orbital angular momentum ℓ and total angular momentum j. $S_n(E,J;E',J';\ell,j)$ is given by [31]

$$S_n(E,J;E',J';\ell,j) = \frac{(2j+1)\pi\lambda^2 T_{\ell j}(\epsilon)}{8h\pi^4 R^2}$$

$$\times \{[1 - (\frac{\ell_o - \frac{1}{2}}{\ell_o})^2]^{\frac{1}{2}} - [1 - (\frac{\ell_o + \frac{1}{2}}{\ell_o})^2]^{\frac{1}{2}}\} \frac{\rho(E',J')}{\rho(E,J)} \tag{19}$$

The critical angular momentum ℓ_o is given by

$$\ell_o = \frac{R\sqrt{2M\epsilon}}{\hbar} \tag{20}$$

In eqs. (19) and (20) ε is the kinetic energy of the emitted neutron $T_{\ell j}(\varepsilon)$ is the transmission coefficient determined from the optical model, and R is the nuclear radius.

The cross section for prompt fission $\sigma_{p,f}$ is given by

$$\sigma_{p,f} = \frac{\sigma_c}{\hbar} \int P_2^i(E',t)\Gamma_{23}(E')dtdE'$$

$$= \frac{\sigma_c}{\hbar} \int z_2^i(E')\Gamma_{23}(E')dE' \tag{21}$$

The cross section for isomer fission $\sigma_{is,f}$ is given by

$$\sigma_{is,f} = \sigma_c \frac{\Gamma_{23}(o)}{\Gamma_{23}(o) + \Gamma_{21}(o)} \int_o^\infty \int P_2^x(E',t)S_\gamma(E',0)dtdE'$$

$$= \sigma_c \frac{\Gamma_{23}(o)}{\Gamma_{23} + \Gamma_{21}(o)} \int z_2^x(E')S_\gamma(E',0)dE' \tag{22}$$

The cross section for the emission of i neutrons $\sigma_{i,n}$ is given by

$$\sigma_{i,n} = \prod_{k=1}^{i-1} \frac{<\Gamma_n^k>}{<\Gamma_n^k> + <\Gamma_f^k>} \frac{\Gamma_n^i(E)}{\Gamma_n^i(E) + \Gamma_f^i(E)} \tag{23}$$

In eq. (23) $<\Gamma>$ is the calculated average width.

$$<\Gamma> = \frac{1}{E} \int \Gamma(E')dE' \tag{24}$$

In eqs. (21, (22) and (23), σ_c is the cross section for the formation of the compound nucleus, the value of which is not required for the calculation of ratios between cross sections.

The isomer ratio for (n,γ) reactions IR is the ratio between the yield Y_2 for populating the isomeric state and the yield for populating the ground state Y_1

$$IR = Y_2/Y_1 \tag{25}$$

The rates $R_1(t)$ and $R_2(t)$ for populating the isomeric and ground states are

$$R_2(t) = \int_0^{E_0-E} P_2^0(E',t)S_\gamma(E',0)\,dE' \tag{26a}$$

$$R_1(t) = \int_0^{E_0} P_1^0(E',t)S_\gamma(E',0)\,dE' \tag{26b}$$

From the rates $R_2(t)$ and $R_1(t)$ the yields Y_2 and Y_1 can be found.

$$Y_2 = \int_0^\infty R_2(t)\,dt = \int_0^\infty \int_0^{E_0-\Delta E} P_2^0(E',t)S_\gamma(E',0)\,dt\,dE'$$

$$= \int_0^{E_0-\Delta E} Z_2^0(E')S_\gamma(E',0)\,dE' \tag{27a}$$

$$Y_1 = \int_0^\infty R_1(t)\,dt = \int_0^\infty \int_0^{E_0} P_1^0(E',t)S_\gamma(E',0)\,dt\,dE'$$

$$= \int_0^{E_0} Z_1^0(E')S_\gamma(E',0)\,dE' \tag{27b}$$

From eqs. (21) - (24) it can be seen that the knowledge of the function $Z_1^i(E)$ and $Z_2^i(E)$ is indeed sufficient for the purpose of calculation of observable quantities.

5. Comparison with Experiment

The predictions of the theory are now compared with experiment. In table 1 the calculated and experimental values of the isomer ratio for the reactions $^{233}U(n,\gamma)^{234}U$, $^{235}U(n,\gamma)^{236}U$, $^{241}Am(n,\gamma)^{242}Am$ and $^{239}Pu(n,\gamma)^{240}Pu$ are found. For uranium and plutonium nuclei the frequencies are determined using eq. (16). The potential surfaces are taken from the work of Nilsson et al. [26] and the inertial parameters are obtained from the work of Sobiczewski [31] et al. The calculated values for the isomer ratio for the reaction $^{241}Am(n,\gamma)^{242}Am$ are performed using the parameters suggested by Lynn [23]. The experimental

results for the isomer ratio for uranium and plutonium isotopes are
taken from the work of Elwyn and Ferguson [9]. The experimental value
of the isomer ratio for americium is taken from the work of Gangrsky
et al. [12]. There is some doubt as to the validity of the experimental data of Elwyn and Ferguson [9]. These doubts and the interpretation
of the results in table 1 are discussed later.

Table 1

Comparison between experimental values of the isomer ratio in IRE in
units of 10^{-4} and calculated values of the isomer ratio IRC in units
of 10^{-4}

Energy	Target Nucleus	Final Nucleus	IRE	IRC
0.0	^{241}Am	^{242}Am	2×10^{-2}	5.6×10^{-1}
1.0	^{241}Am	^{242}Am	5.0	5.0
2.2	^{239}Pu	^{240}Pu	80 - 160	56
2.2	^{235}U	^{236}U	60 - 120	60
2.2	^{233}U	^{234}U	100 - 200	64
0.5	^{233}U	^{234}U	150 - 300	59

The comparison of the calculated and measured values of the ratio
of isomer to prompt fission for some typical reactions is found in figs.
3-5. The ratios are given as a function of the excitation energy minus
the binding energy of the emitted neutrons. The solid lines present the
results of the present calculation where no parameters are adjusted.
The parameters describing the potential surface are obtained from the
work of Hilsson [26] et al. and Sobiczewski [31] et al. and the use of
eq. (16). For nuclei for which such information is not available the
parameters are obtained directly from experiment or by extrapolation.
The dashed curves represent the results of Britt [5] et al. Flerov [10]
et al., S. Bjørnholm [1] and Lark [21] et al. Finally, for the reaction
^{235}U(α, xn)Pu the competition between the emission of three neutrons and
the prompt fission following the emission of two neutrons is given in
fig. 6. The experimental points are due to Britt [7]. Many more reactions have been studied, the results of this analysis is available.

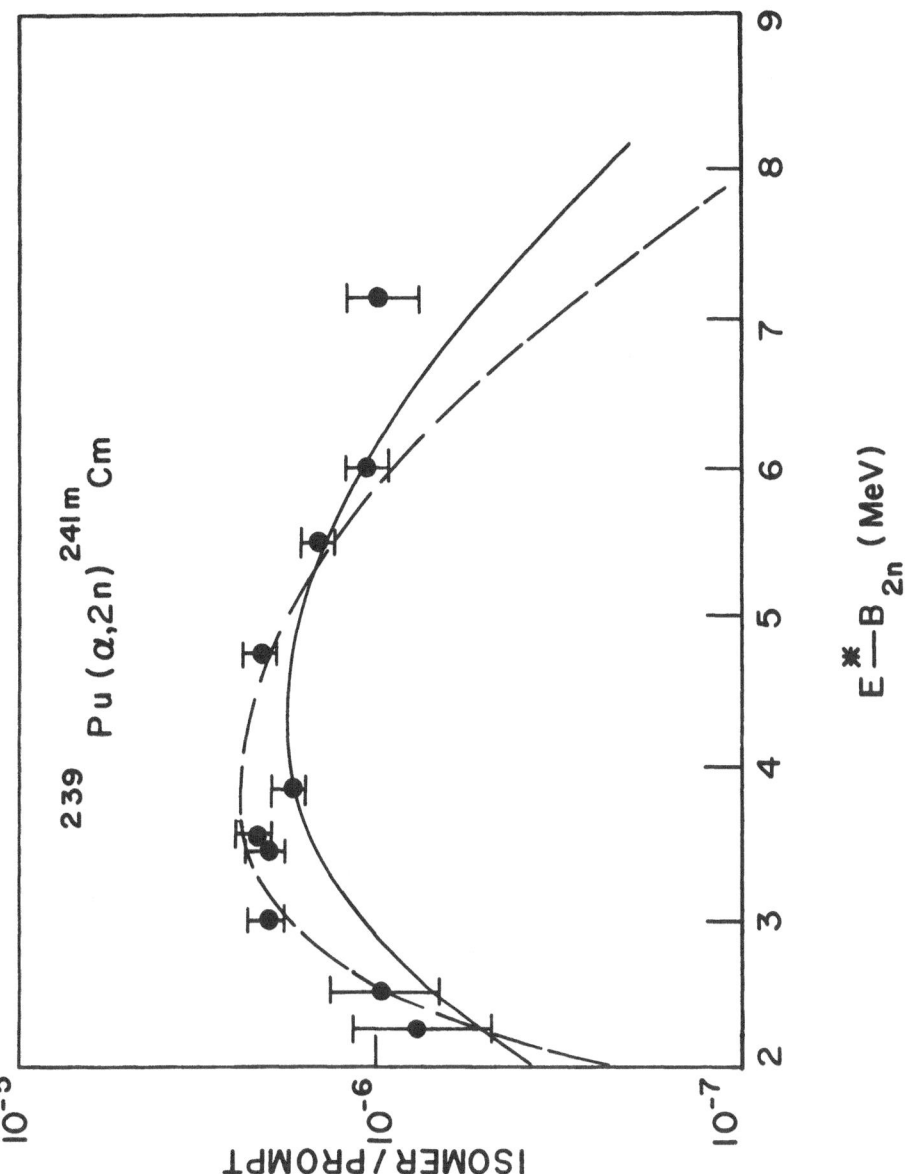

Fig. 3. A comparison between experimental and theoretical values of the ratio of isomeric to prompt fission for the reaction $^{239}Pu(\alpha,2n)^{241m}Cm$. Here E* is the excitation energy of the compound nucleus and B_{2n} is the sum of the neutron binding energies of the two evaporated neutrons. The experimental points are from Britt et al. (ref. [5]). The dashed line is taken from Britt et al. (ref.[5]) and the solid curve is from the present work

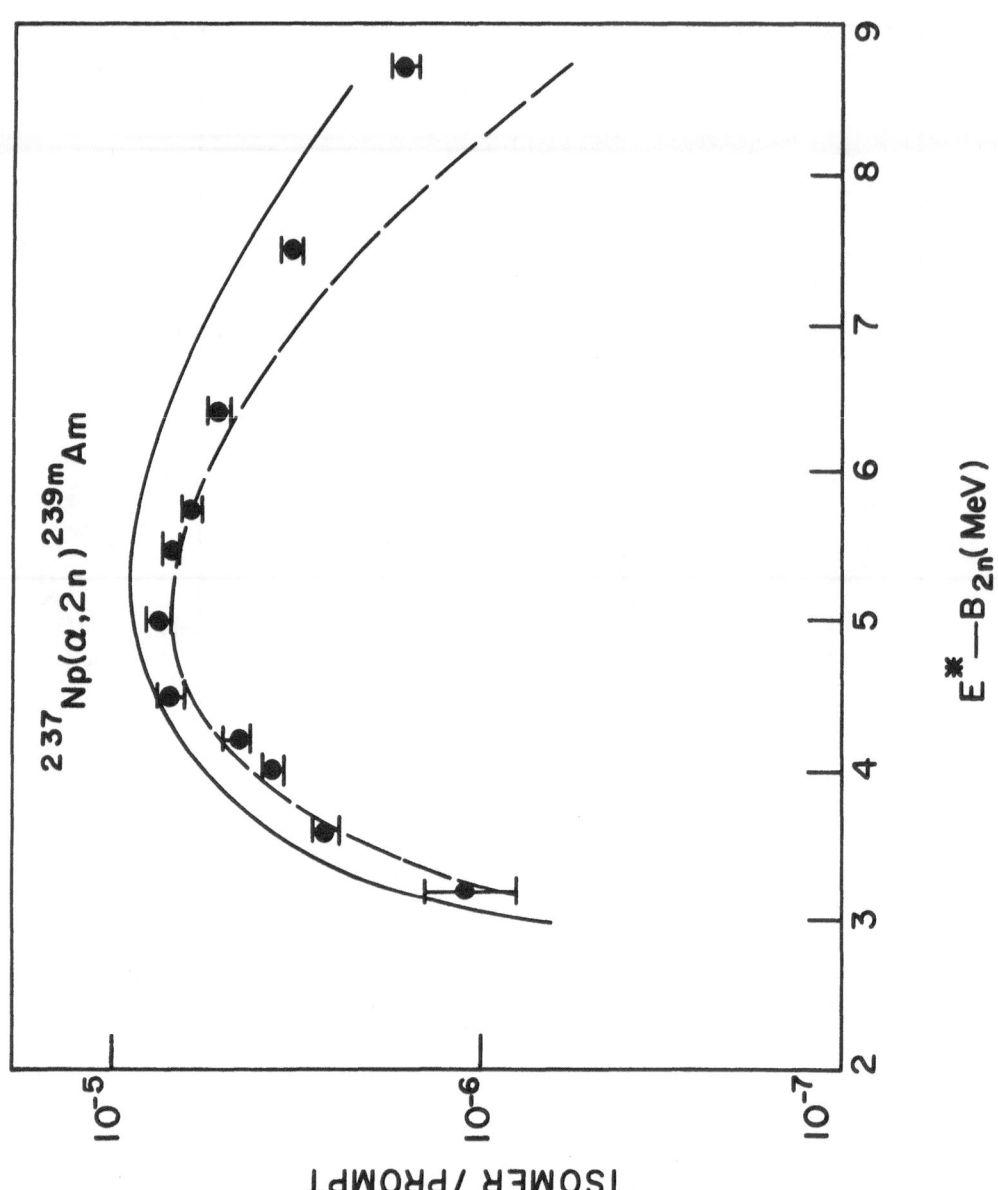

Fig. 4. Same for the reaction $^{237}Np(\alpha,2n)^{239m}Am$. Experimental points from Britt et al. (ref. [5])

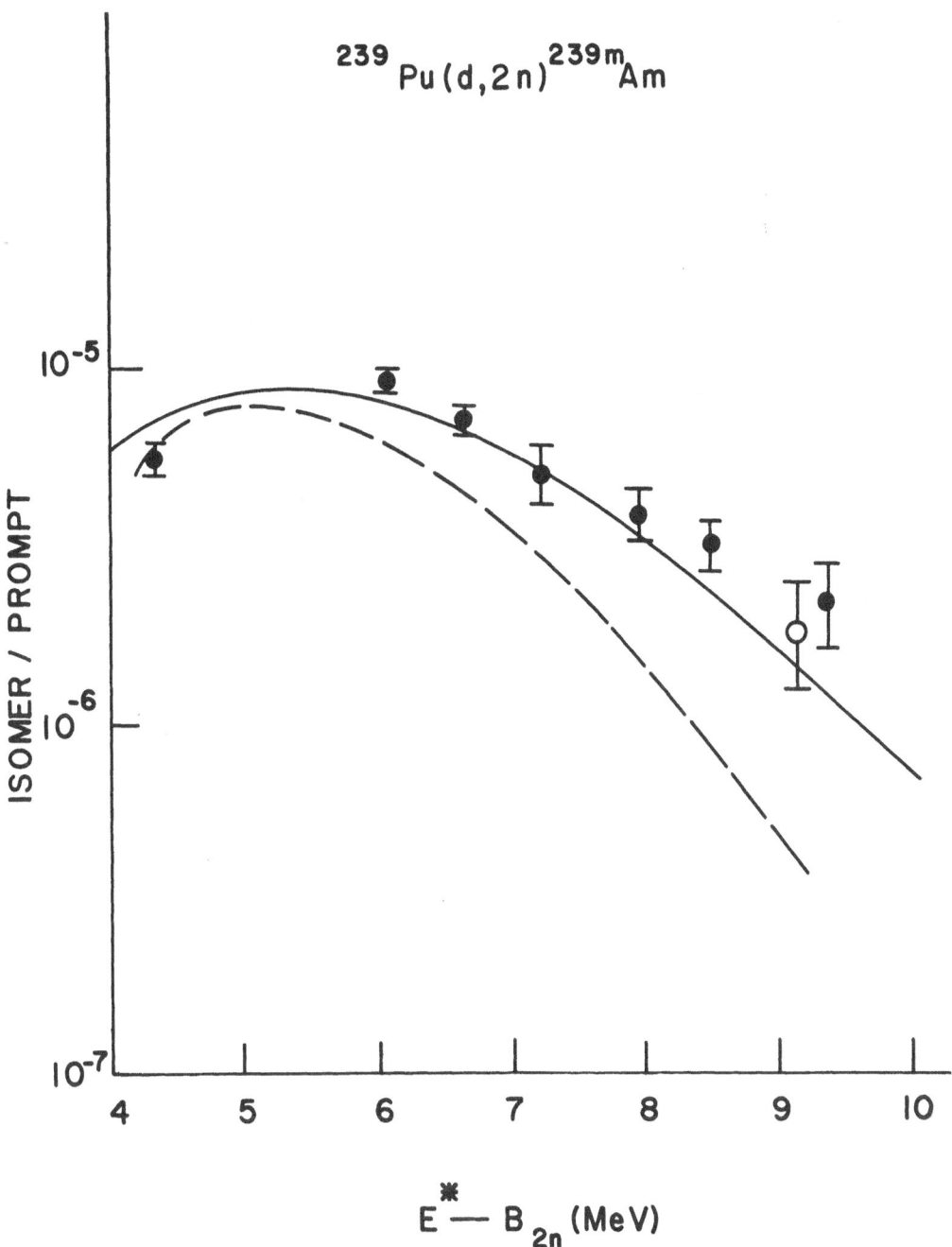

Fig. 5. Same for the reaction $^{239}\mathrm{Pu}(d,2n)^{239m}\mathrm{Am}$. Experimental points from Britt et al. (ref. [5])

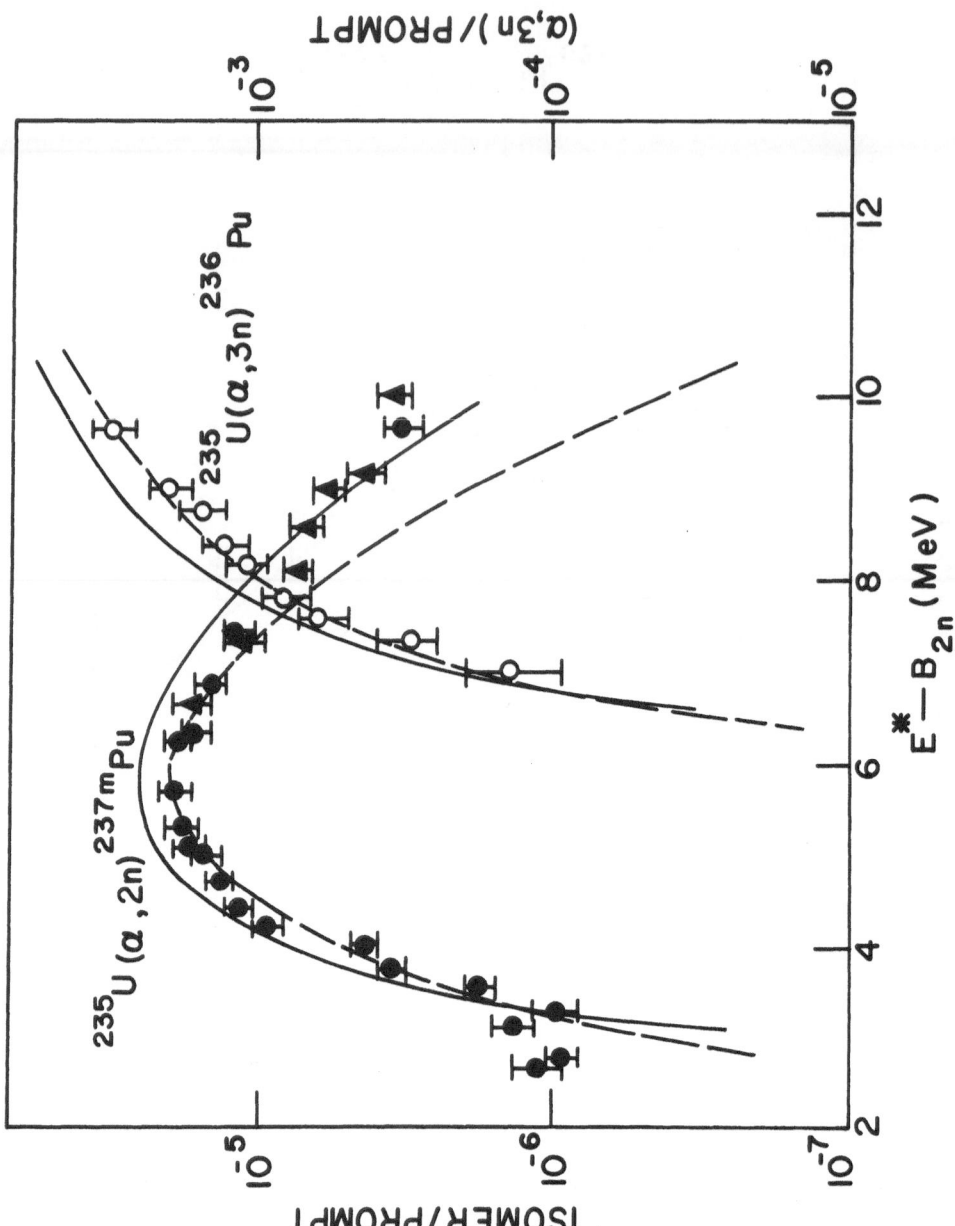

Fig. 6. A comparison between experimental and theoretical values of the ratio of isomeric to prompt fission for the reaction ^{235}U $(\alpha,2n)^{237m}$Pu and the ratio between the cross section for evaporation of three neutrons and prompt fission. Experimental points from Britt et al. (ref. [5])

6. Discussion

As can be seen from table 1 and figs. 3-6 good agreement between
theory and experiment is obtained. However, the limitation of the theory has to be recognized. First the fission barriers are changed by
asymmetric deformation [26] and should be included. Second, dipole radiation may play a significant role. Third, in the present work the
dependence of the density of levels on deformation is not included in
a rigorous way. Finally, there may be some doubt concerning the results of Elwyn and Ferguson [9]. In particular, the lighter uranium
isotopes do not have shape isomers. This fact is supported by more modern calculations of potential surfaces. Nevertheless a comparison between his theory is presented. This is mainly due to the fact that
earlier barrier calculations do not exclude the existence of these
isomers.

There is no doubt that the inclusion of asymmetric deformation,
dipole radiation and better shape dependent densities of levels will
yield more realistic results. However, at present such information
exists in preliminary form. The formalism discussed in this paper can
very well be adopted to calculate more realistic ratios once the required information becomes available.

Recently Greiner [14] suggested that the zero energy has to be
considered for the determination of the position of the ground state
in both wells. Calculations considering this effect for the evaluation
of the ratio between isomer to prompt fission have been performed. The
calculations indicate that the ratio is not very sensitive to the position of the ground state. However, in some cases calculations with
shifted ground states yield better agreement between theory and experiment. A typical case can be found in fig. 5.

The present theory agrees asymptotically with the results of
Britt [5] et al. on one hand and the results of Jägare [18] respectively. In particular the present results suggest that the more phenomenological approach of Britt [5] et al. is a very good approximation to
the rigorous method. Admittedly, the calculated values differ slightly from the experimental values. Of course , one can adjust parameters
to represent the experimental points. In fact only a slight change in
one of the parameters reproduces the experimental points exactly.
However, this is not the purpose of this study.

140

References

1. BJØRNHOLM, S., BORGGREEN, J., WESTGAARD, L. and KARNAUKOV, V.A., Nucl. Phys. A95, 513 (1967).

2. BLATT, J.M. and WEISSKOPF, V.F., Theoretical Nuclear Physics (John Wiley & Sons, New York, 1952) p. 389.

3. BOHR, A. and MOTTELSON, B.R., in Beta and Gammy Spectroscopy, Edited by K. Sieagbahn (North Holland Publishing House, Amsterdam, 1955) p. 448.

4. BOHR, N. and WHEELER, J.A., Phys. Rev. 56, 426 (1939).

5. BRITT, H.C., BURNETT, S.C., ERKKILA, B.H., LYNN, J.E. and STEIN, W.E., Phys. Rev. C4, 1444 (1971).

6. BRITT, H.C. and CRANIER, J.D., Phys. Rev. C2, 1758 (1970).

7. BRITT, H.C., private communication.

8. CRAMER, D. and NIX, J.R., Phys. Rev. C2, 1048 (1970).

9. ELWYN, A.J. and FERGUSON, T.G., Nucl. Phys. A148, 337 (1970).

10. FLEROV, G.N., PLEVE, A.A., POLIKANOW, S.M., TRETYAKOVA, S.P., MONTALOGU, N., POENARU, D., SEZON, M., VILCOV, I. and VILCOV, N., Nucl. Phys. A97, 444 (1967).

11. FLEROV, G.N. and POLIKANOV, S.M., Comptes Rendus du Congress International de Physique Nucleaire, edited by P. Gugenberger (Edition du Centre National de la Researche Scientifique, Paris, 1964). p. 407.

12. GANGRANSKY, Yu., GAVINKOV, K.A., MARKOV, B.N., KAHN, N.K. and POLIKANOV, S.M., Yad. Fiz. 10, 65 (1969) (English translation: Soviet Journal of Nucl. Phys. 10, 38 (1970)).

13. GILBERT, A. and CAMERON, A.G.W., Cand. J. Phys. 43, 1446 (1965).

14. GREINER, W., private communication.

15. HILL, D.L. and WHEELER, J.A., Phys. Rev. 89, 1102 (1953).

16. HUIZENGA, J.R. and VANDENBOSCH, R., Phys. Rev. 120, 1305 (1960) and VANDENBOSCH, R. and HUIZENGA, J.R., Phys. Rev. 120, 1313 (1960).

17. JACKSON, J.D., Cand. J. Phys. 34, 767 (1956).

18. JAGARE, S., Nucl. Phys. A137, 241 (1969).

19. JAGARE, S., Nucl. Phys. A103, 241 (1967).

20. JUNGCLAUSSEN, H., Yad. Fiz. 7, 88 (1968)(English translation Soviet J. of Nucl. Phys. 7, 60 (1968)).

21. LASK, N.L., SLELTEN, G.S., PEDERSON, J. and BJØRNHOLM, S., Nucl. Phys. A139, 481 (1969).

22. LIGGETT, G. and SPERBER, D., Phys. Rev. C2, 447 (1971).

23. LYNN, J.E., in Theory of Neutron Resonance Reactions (Clarendon Press, Oxford 1968); LYNN, E., in Physics and Chemistry of Fission,

Procedings of a Symposium, Vienna, July 28 - Aug. 1, 1969 (IAEA, Vienna 1969) p. 249.

24. MALKIN, J.D., At. Energ. (USSR) 15, 158 (1963) (trans. Soviet J. Atomic Energ. 15, 851 (1964)).

25. MANDLER, J.W. and SPERBER, D., Nucl. Phys. A113, 689 (1968).

26. Examples for the calculation of potential surfaces and shell corrections are given in this list of references, which is by no means complete.

MOSEL, U., Phys. Rev. Let. 25, 678 (1970)

BABA, H., Nucl. Phys. A-159, 625 (1970)

SCHARNWEBER, D., GREINER, W. and MOSEL, U., Nucl. Phys. A-164, 257 (1971)

GNEUSS, G. and GREINER, W., Nucl. Phys. A-171, 449 (1971)

PASHKEVICH, V.V., Nucl. Phys. A-169, 257 (1971)

GAEEV, F.A., IVANOVA, S.P., PASHKEVICH, V.V., Sov. Jour. Nucl. Phys. 11, 667 (1970)

ADEEV, G.D.,GAMALYA, I.A. and CHERDANTSEV, P.A., Sov. Jour. Nucl. Phys. 12, 148 (1971)

RAMAMURTHY, V.S., KAPOOR, S.S. and KATARIA, S.K., Phys. Rev. Let. 25, 386 (1970)

NILSSON, S.G., et al., Nucl. Phys. A-131, 1 (1969)

TSANG, C.F. and NILSSON, S.G., Nucl. Phys. A-140, 289 (1970)

BASSICHIS, W.H. and KERMAN, A.K., Phys. Rev. C-2, 1768 (1970)

LARSSON, S.E., RAGNARSSON, I. and NILSSON, S.G., Phys. Let. 38B, 269 (1972)

GOTZ, U, PAULI, H.G., ALDER, K., JUNKER, K., Phys. Let. 38B, 274 (1972)

BRACK, M., DAMGAARD, J.,JENSEN, A.S., PAULI, H.C., STRUTINSKII, V. M., WONG, C.Y., Rev. Mod. Phys. Vol. 44, 321 (1972)

HASSE, R.E., Nucl. Phys. A-128, 609 (1969)

NILSSON, B., Nucl. Phys. A-129, 445 (1969)

MOLLER, P. and NILLSON, B., Phys. Let. 31B, 171 (1970)

MOLLER, P. and NILSSON, S.G., Phys. Let. 31B, 283 (1970)

BJØRNHOLM, S. and STRUTINSKII, V.M., Nucl. Phys. A-136, 1 (1969)

MOLLER, P., Nucl. Phys. A-142, 1 (1970)

SCHARNWEBER, D., MOSEL, U. and GREINER, W., Phys. Rev. Let. 24, 601 (1970)

BUNATIAN, G.G., KOLOMIETZ, V.M. and STRUTINSKY, V.M., Nucl. Phys. A-188, 225 (1972)

ROSS, C.K. and BHADURI, R.K., Nucl. Phys. A-188, 566 (1972)

LEDERGERBER, T. and PAULI, H.C., Phys. Let. 39B, 307 (1972)

GOTZ, U., PAULI, H.C. and JUNKER, K., Phys. Let. 39B, 436 (1972)

ADEEV, G.D. and CHERDANTSEV, P.A., Phys.Let. 39B, 485 (1972)

FAESSLER, A., GOTZ, U., SLAVOV, B., LEDERGERBER, T., Phys. Let. 39B, 579 (1972)

MORETTO, L.G., Phys. Let. 38B, 393 (1972)

MORETTO, L.G., THOMPSON, S.G., ROUTTI, J. and GATTI, R.C., Phys. Let. 38B, 471 (1972)

DICKMANN, F., METAG, V. and REPNOW, R., Phys. Let. 38B, 207 (1972)

WILLIAMS, F.C., Jr., CHAM, G. and HUIZENGA, J.R., Nucl. Phys. A-187,225 (1972)

JOHANSSON, T.A., Nucl. Phys. A-183, 33 (1972)

MAHARRY, D.E. and DAVIDSON, J.P., Nucl. Phys. A-182, 371 (1972)

MORETTO, L.G., Nucl. Phys. A-182, 641 (1972)

BOLSTERLI,M., FISET, E.O., NIX, J.R. and NORTON, J.L., Phys. Rev. C5, 1050 (1972)

MOELLER, P., NILSSON, B. et al., Phys. Let. 26B, 418 (1968)

HASSE, R.W., Nucl. Phys. A-118, 577 (1968)

STRUTINSKY, V.M., Nucl. Phys. A-122, 1 (1968)

LAMMN, I., Nucl. Phys. A-125, 504 (1969)

NILSSON, S.G. et al., Phys. Let. 30B, 437 (1969)

SWIATECKI, W.J., Nucl. Phys. A-81, 1 (1966)

STRUTINSKY, V.M., Nucl. Phys. A-95, 420 (1967)

GEILIKMAN, B.T., Sov. Jour. Nucl. Phys. 9, 521 (1969)

PAULI, H.C. and LEDERGERBER, T., Nucl. Phys. A-175, 545 (1971)

SLAVOV, B., GALONSKA, J.E. and FAESSELER, A., Phys. Let. 37B, 483 (1971)

MOSEL, U., MARUHN, J. and GREINER, W., Phys. Let. 34B, 578 (1971)

ANDERSON, B.L., DICKMAN, F. and DIETRICH, K., Nucl. Phys. A-159, 337 (1970)

MOSEL, U. and SCHMITT, W., Nucl. Phys. A-165, 73 (1971).

27. MYERS, L. and SWIATECKI, W., Nucl. Phys. 81, 1 (1966).

28. POTNIZ, W.P, Z. Physik. 197, 262 (1966).

29. SARANTITES, D.G., Nucl. Phys. A93, 576 (1967).

30. SIKKELAND, T., CLARKSON, J.E., STEYER-SHAFRIR, N.H. and VIOLA, V. E., Phys. Rev. C3, 329 (1971)

31. SOBICZEWSKI, A., SZYMANSKI, Z., WYCECH, S., NILSSON, S.G., NIX, J. R., TSANG, C.F., GUSTAFSON, C., MÖLLER, P. and NILSSON, B., Nucl. Phys. A131, 67 (1969).

32. SPERBER, D., Nucl. Phys. A90, 665 (1967).

33. STRUTINSKY, V.M., Nucl. Phys. A95, 420 (1967).

34. VANDENBOSCH, R. and HUIZENGA, J.R., in Procedings of Second United

Nations International Conference on Peaceful Uses of Atomic Energy,
Geneva (1958)(United Nationa, Geneva, Switzerland), Vol. 15.

35. WEIZSÄCKER,C.F., Z. Physik <u>96</u>, 431 (1935).

NUCLEAR MOLECULAR STRUCTURE IN HEAVY ION SCATTERING *

W. SCHEID, H.J. FINK and H. MÜLLER

Institut für Theoretische Physik der Universität Frankfurt

Frankfurt Main, Germany

1. Introduction

Among the most studied heavy ion reactions we find the elastic
scattering of identical nuclei (cf [4], [13], [14], [35]). Typical examples for that are the elastic scattering of ^{12}C on ^{12}C or ^{16}O on
^{16}O. With such scattering experiments one tries to get insight into
the interaction between complex nuclear systems. The measured cross
sections serve to test the theoretically predicted nucleus-nucleus potentials, kinetic energy operators and the coupling potentials to inelastic channels.

In this article we discuss the possibility that nuclear molecules
can be formed during the scattering process. The concept of nuclear molecules was first introduced by Bromley et al. after the discovery of
structures in the ^{12}C - ^{12}C cross section [6]. In a molecular state
the two nuclei attract each other along their surfaces by the long
ranging part of the nuclear forces. They do not amalgamate, i.e. fuse,
if the system rotates and the centrifugal forces are in equilibrium
with the attractive nuclear force. Microscopically the binding of nuclear molecule is generated by a few nucleons which are orbiting around
both nuclear centers. This can be understood with the aid of a two-center shell model which is used in the microscopical description of the
nucleus-nucleus system [14] (see e.g. Fig. 6): the nucleons in the deeper shells are concentrated around the individual centers whereas the
loosest bound nucleons surround both centers, i.e. the binding is homeopolar in the language of chemistry.

Let us first briefly review the experimental situation [4], [35].
In fig. 1 the elestic 90°-cross sections for three different systems,
namely $^{16}O-^{16}O$, $^{14}N-^{14}N$ and $^{13}C-^{13}C$ are given (cf [6], [18], [32], [36]).
In these systems the differential cross sections are symmetric around
the 90°-direction in the center of mass system and, therefore, the 90°
cross section shows the most prominent structures arising from the nuclear forces. All cross sections follow the same pattern: at low bombarding energies the nuclei do not overcome their Coulomb barrier. In

*This work has been supported by the Bundesministerium für Bildung und
Wissenschaft and by the Gesellschaft für Schwerionenforschung.

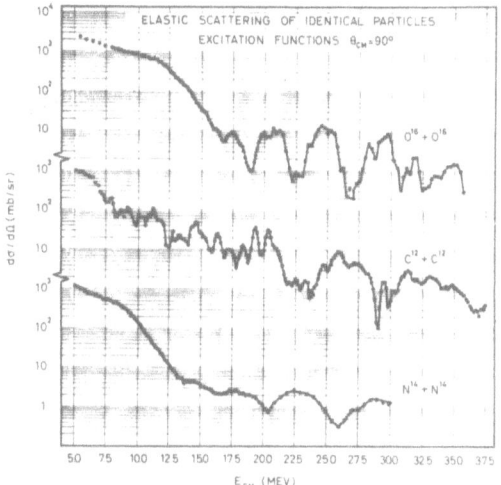

Fig. 1. The experimental 90°-differential cross sections for the elastic scattering of ^{16}O-^{16}O, ^{14}N-^{14}N and ^{12}C-^{12}C (from ref. [4])

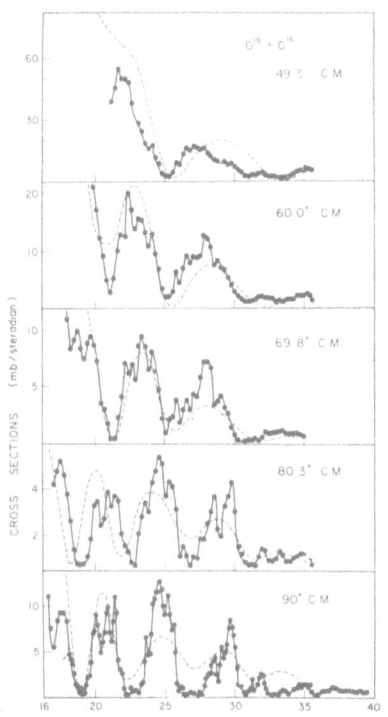

Fig. 2. Differential cross sections for the elastic scattering of ^{16}O on ^{16}O (from ref. [36])

this energy range the cross section follows the Mott cross section
with a $1/E^2$-dependence.Above the Coulomb barrier the nuclei penetrate
into each other and a large number of inelastic reactions can happen:
e.g. compound nucleus formation, transfer of nucleons and clusters, and
the inelastic excitation of the individual nuclei. All these reactions
absorb flux from the elestic channel so that the cross section drops
by a factor 50 to 100.

At higher bombarding energies these cross sections show structures. They can be subdivided into gross structures with a width of about
2 MeV and intermediate structures with widths of about 0.2 MeV. In fig.
2 the differential cross section for ^{16}O-^{16}O clearly reveals the difference between the two structures. The dashed curve represents a fit of
the Yale group [36] in which an optical potential was used. Therefore,
the gross structure can be explained by potential scattering.

The purpose of this paper is to show that intermediate structures
arise if molecular states are excited [37] . This does not exclude the
possibility that also compound elastic processes can produce intermediate structures[2], [20] , [21] , [40] .

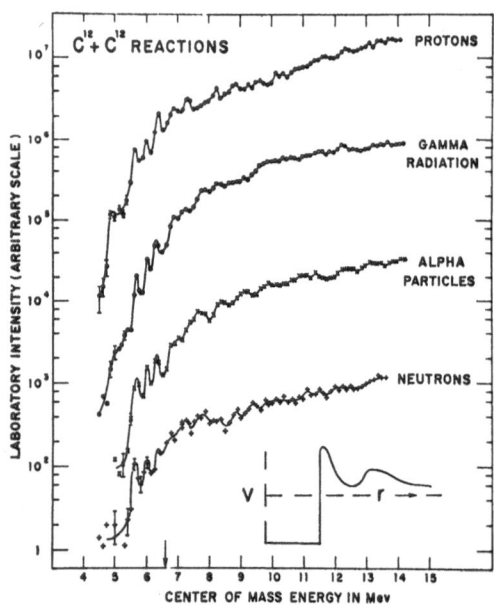

Fig. 3. Reaction cross sections for the scattering of ^{12}C
on ^{12}C near the Coulomb barrier (from ref. [1])

A clear example for intermediate structures which proceed through molecular states are the resonances at 6 MeV CM-energy in the $^{12}C-^{12}C$ system [1]. Fig. 3 shows the different reaction cross sections for proton, α-particle, neutron and γ-ray-emission in the $^{12}C-^{12}C$ scattering. The resonances were explained by Vogt and McManus [41] and Davis [9] as states in a quasi-molecular potential. Imanishi [17] proposed an indirect excitation mechanism which will be presented in a later section. Michaud and Vogt [22] gave a qualitative explanation in the framework of an α-cluster model.

In the following we discuss first the nucleus-nucleus potential. Then the molecular states and their excitation will be considered. At last a coupled channel calculation for the $^{12}C-^{12}C$ system is presented which shows that intermediate structures can be generated by the indirect excitation of quasibound molecular states.

2. The Nucleus-Nucleus Interaction

To answer the question about the existence of the molecular states we have first to ask for the nucleus-nucleus interaction. Since it is impossible to solve the many-body problem we have to restrict the degrees of freedom of the nucleus-nucleus system to all those which we want to treat explicitly. These are the relative motion, inelastic excitation of the individual nuclei and perhaps transfer of α-clusters. Therefore, the system will be described only by a small number of open channels. Under such restrictions the exact Hamiltonian changes to an effective Hamiltonian [10], [29]. In the effective Hamiltonian the effects of all neglected channels are mainly reproduced by an imaginary potential [29].

Because we neglect so many degrees of freedom we have to state with the help of physical assumptions how the nuclei penetrate into each other, i.e. we have to define the elastic channel in the penetration region of the nuclei. This leads us to discriminate between the sudden and adiabatic approximation used in the calculation of the scattering potentials.

2.1. Sudden and Adiabatic Approximations

In the sudden approximation one assumes that the scattering process is so fast that the densities of the individual nuclei overlap and a local compression of nuclear matter happens in the contact region. In that case the potential rises steeply if the nuclei overlap (fig. 4). For the adiabatic approximation one supposes that the scattering proceeds slowly. The energy of the nucleus-nucleus system is minimized with respect to the shape parameters of the nuclear density

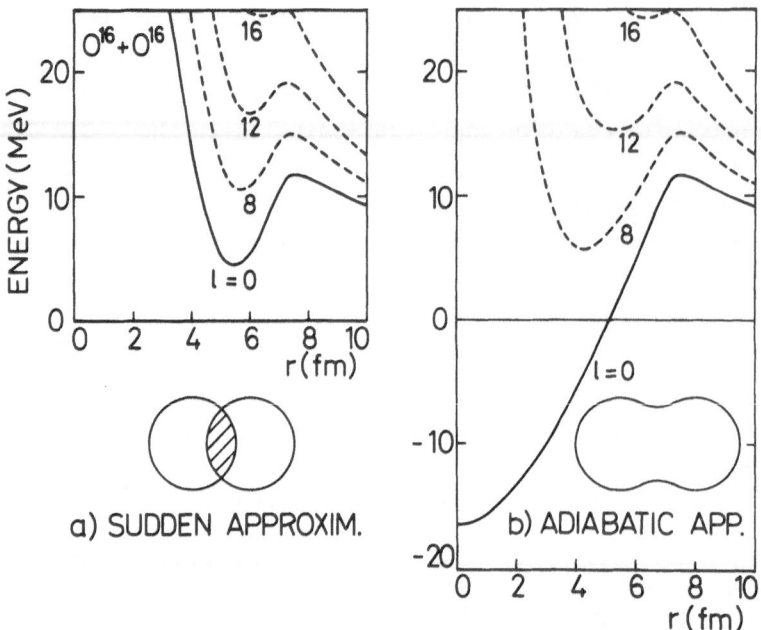

Fig. 4. The real part of the nucleus-nucleus potential in
the sudden and adiabatic approximations for ^{16}O-^{16}O.
The centrifugal potentials for various angular momenta have been added

distribution for each distance of the nuclear centers. The potential
decreases to the ground state of the compound system (fig. 4). Both
approximations produce similar potentials in the region where the nuclear surfaces come into contact and where no nuclear matter is compressed (fig. 4).

The question arises which of the two approximations is more realistic. For that we compare the velocity of the nuclei which is of the
order v=c/10 with the velocity of sound in nuclear matter which is
also of the same order of magnitude (see eq.(9)). With the latter velocity local compressions are removed. Therefore, we conclude that during
the scattering process not all the compressed regions can be expanded.
A further estimation supports this assumption. The scattering time is
of the same magnitude as the rearrangement time of the nuclear shells.
Both times are of the order of $5\cdot10^{-22}$ sec [34]. Therefore, the shells
of the two nuclei cannot completely rearrange into the shells of the
compound system. That favours also the idea that the scattering process
develops in between the two extreme cases of the sudden and adiabatic

approxiamtions.

2.2. The Calculation of the Real Potential

The real part of the nucleus-nucleus potential can be approxima-
tively obtained from the expectation value of the nuclear forces with
appropriate wave functions.

$$V(\vec{r}) = <\psi | \sum_i T_i + \sum_{i<j} V_{ij} - T_{CM} | \psi> \tag{1}$$

Several assumptions about the spatial distributions of the nucleons in
the wave functions have been made: one can construct the many-body wave functions ψ by the single particle solutions of a two-center oscillator potential [31]. Another possibility is to use one-center functions as single particle states [12], [42].

A different procedure to calculate potentials is the application
of the Strutinsky method [38]. This method is used widely for fission
potentials [24]. Also heavy ion potentials can be calculated according
to this method as done by Prüß [30], Mosel et al. [25] and Morović
[26]. In the Strutinsky method one calculates first the general behaviour of the nucleus-nucleus potential with the liquid drop model and
adds then the effects of the shell structures.

Since we want to investigate also compression effects in the potential, an extended liquid drop model is used [34],[7]. In this model
the binding energy is a functional of the nuclear density distributions
given by

$$E[\rho] = W_o A + \frac{C}{2\rho_o} \int (\rho - \rho_o)^2 d\tau$$

$$+ \frac{V}{8\pi} \int \rho(\vec{r}_1) \frac{e^{-|\vec{r}_1 - \vec{r}_2|/u}}{|\vec{r}_1 - \vec{r}_2|} (\rho(\vec{r}_2) - \rho(\vec{r}_1)) d\tau_1 d\tau_2$$

$$+ \frac{1}{2} (\frac{eZ}{A})^2 \int \rho(\vec{r}_1) \frac{1}{|\vec{r}_1 - \vec{r}_2|} \rho(\vec{r}_2) d\tau_1 d\tau_2$$

$$+ \frac{G}{2\rho_o} (\frac{2Z}{A} - 1)^2 \int \rho^2 d\tau \tag{2}$$

The expression consists of a term proportional to the nucleon number A,
a compression term, a Yukawa-term which simulates a surface tension,
the Coulomb-term and the symmetry energy. The functional is able to

predict the experimental binding energies of the nuclei. Further details are given in [34].

The sudden approximation within this model is defined by the superposition of the unperturbed densities of the two nuclei (see fig.
5).

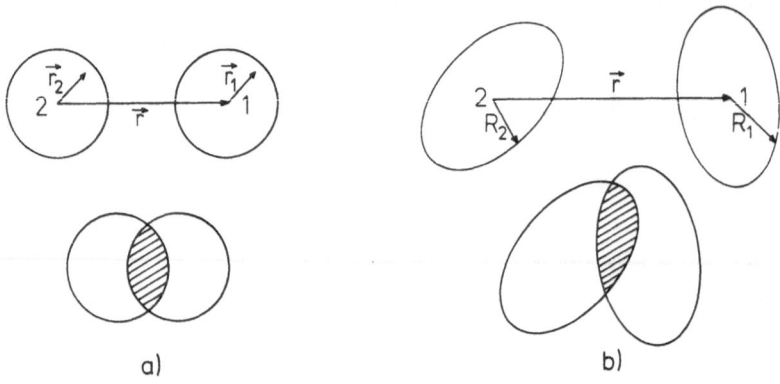

a)   b)

Fig. 5. Coordinates for the calculation of the real potential in the sudden approximation for: a) spherical
b) deformed nuclei

$$\rho = \rho_1(\vec{r}_1) + \rho_2(\vec{r}_2) \qquad (3)$$

This leads to the sudden nucleus-nucleus potential as a function of
the relative coordinate

$$V_{SD}(\vec{r}) = E[\rho_1(\vec{r}_1) + \rho_2(\vec{r}_2)] - E[\rho_1] - E[\rho_2] , \qquad (4)$$

where the energy functional E is given by eq. (2).

In the adiabatic approximation the density is uncompressed, i.e.
the nuclear volume is conserved. Then the compression term does not
contribute. The resulting potentials are already shown in fig. 4 for
the $^{16}O-^{16}O$ system.

To determine the shell effects we apply the two-center shell model [16]. In that model the single particle Hamiltonian is given by
(simplest version, no l^2- and $\vec{l}\vec{s}$ -terms):

$$h = \frac{p^2}{2M} + \frac{M\omega^2}{2} (x^2 + y^2 + (|z|-z_o)^2) . \tag{5}$$

The distance between the potential centers is $r=2z_o$. The two-center oscillator and the wave functions for the two lowest eigenstates in z-direction are drawn in fig. 6. Fig. 7. shows the single particle energies of the two center oscillator. For the distance $z_o=0$ we recognize the single particle states of the compound system, for $z_o \rightarrow \infty$ the single particle states of the two individual nuclei.

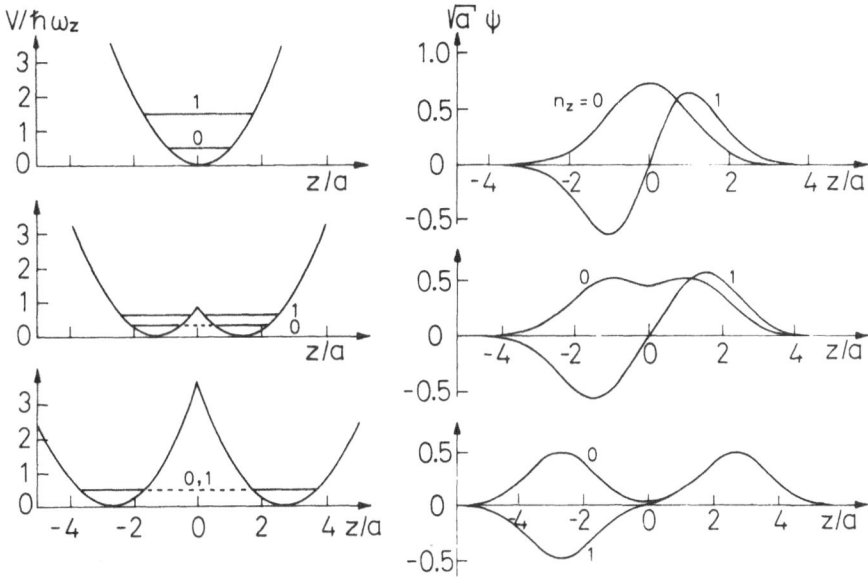

Fig. 6. The two-center oscillator potential for different
two-center distances and the corresponding wave
functions of the two lowest eigenstates

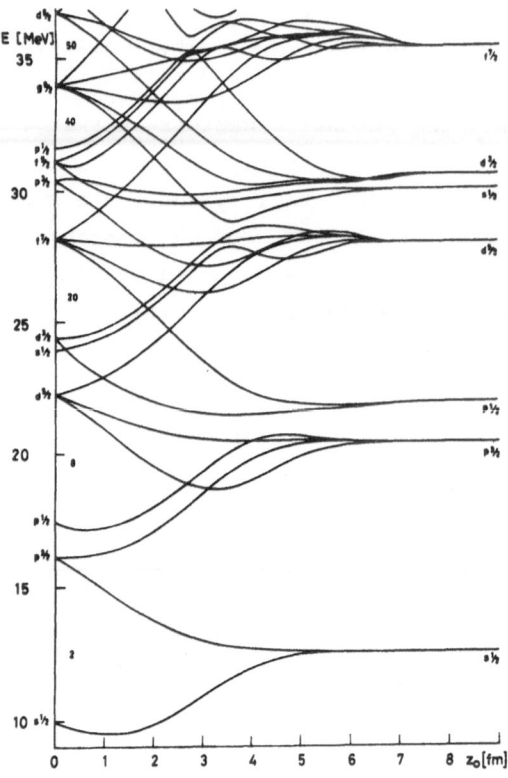

Fig. 7. The single particle levels for the symmetric
break-up in the two-center shell model

The single particle energies depend not only on the distance z_o
but also on the frequency ω of the oscillator which plays the role of
a radius parameter. Thus we have the possibility to describe the adiabatic and sudden process. If we let the frequency ω constant during
the collision ($\omega = \omega_o$) we get obviously compression effects. In the adiabatic approximation the frequency ω has to be changed so that for every
distance the volume enclosed in an equipotential surface remains constant. In fig. 8 the single particle energies are calculated in the
sudden and adiabatic approximations. For $r \to 0$ most of the single particle states rise steeply in energy in the sudden approximation.

Filling the lowest states with the appropriate number of particles
we get the instantaneous internal energy for the elastic configuration

$$U(r) = \sum_{i=1}^{A} \varepsilon_i(r) \qquad (6)$$

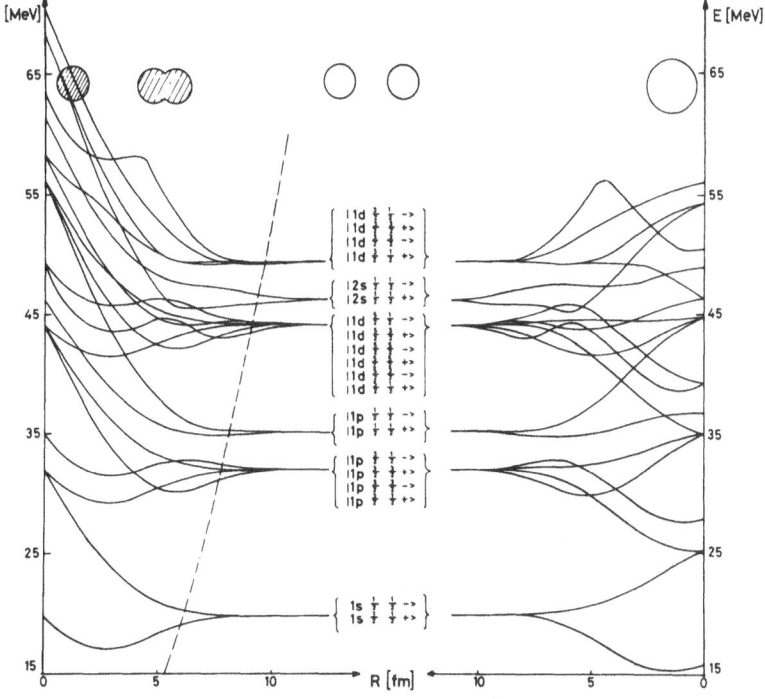

Fig. 8. The single particle levels for the ^{16}O-^{16}O collision with $\hbar\omega_\infty=13.22$ MeV and spin orbit strength
$\kappa=0.08$ as function of the relative distance. The
sudden and adiabatic cases are depicted on the
left and right side, respectively. The shape of
the nucleus-nucleus systems is shown on the top
of the figure (from ref. [30])

In fig. 9 the sum of the single particle energies is shown for ω=const
and for the volume-conserving frequency ω in the case of the systems
^{16}O + ^{16}O and ^{12}C + ^{12}C. The sudden approximation is marked by the
compression effect. The dip in the region where the nuclei come into
contact is the binding effect arising from nucleons near the Fermi
surface. These nucleons surround already both centers.

Following the method described by Strutinsky one calculates the
shell effects from the single particle energies

$$\delta U(r) = U(r) - \bar{U}(r) \qquad (7)$$

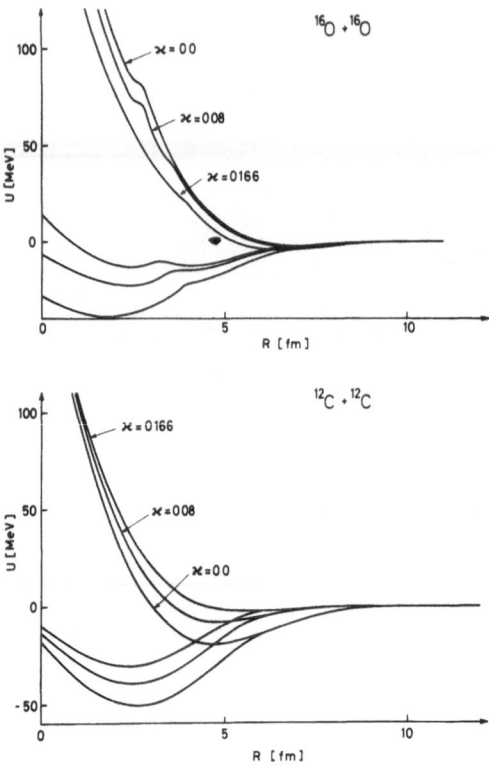

Fig. 9. The sum of the single particle energies (eq.(6)) for
the $^{16}O-^{16}O$ and $^{12}C-^{12}C$ systems in the adiabatic and
sudden approximations (different spin orbit strengths)

In the second term $\bar{U}$ one averages over the distributions of the single
particle levels. Thereby, all contributions independent from shell
effects are subtracted out of U. The shell effects are added to the
liquid drop potential to get the final potential

$$V(r) = V_{LD}(r) + \delta U(r) \qquad (8)$$

In fig. 10 the $^{16}O-^{16}O$ and $^{12}C-^{12}C$ - potentials are shown for the
adiabatic case using the described method. The Coulomb contribution is
disregarded.

2.3. Models for Energy-Dependent Potentials

To get more insight into the connection between the adiabatic and
sudden approximation we have considered two models for the scattering
of identical nuclei. The models are treated by time dependent methods
of classical mechanics [27]. In both models the density distribution
of the nucleus-nucleus system is defined by two parameters: by the re-

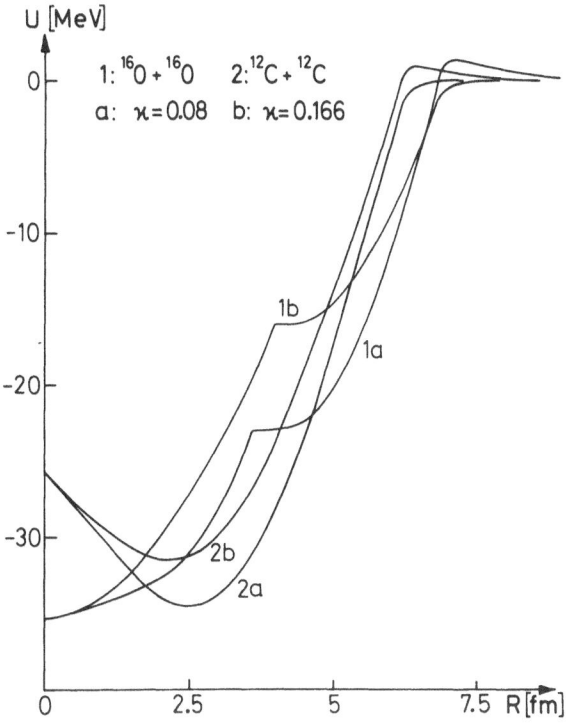

Fig. 10. The real part of the nucleus-nucleus potential in
the adiabatic approximation including shell effects
for different spin orbit strengths (without Coulomb
energy). Further details are given in ref. [30]

lative distance and the radius of the nuclei. In the first model the
nuclear density is homogeneously distributed over the volume shown in
fig. 11 . In the second model we assume that the densities of the nuclei overlap additively (fig. 12). In both models the system can carry
out two movements: the relative motion of the centers and compression
vibrations, i.e. oscillations of the radius coordinate R.

The adiabatic potential is obtained when the binding energy is
minimized at each two-center-distance with respect to the radius R. If
we choose the ^{16}O-^{16}O scattering as an example, the radius R increases
along the adiabatic curve from the ^{16}O-radius to the ^{32}S-radius. The
sudden curve is defined by the prescription that the radius R stays
constant during the penetration of the nuclei (see fig. 11-12).

Solving the dynamical problem we find the time-dependence of the
process: the compressed density tries to reach the adiabatic distribu-

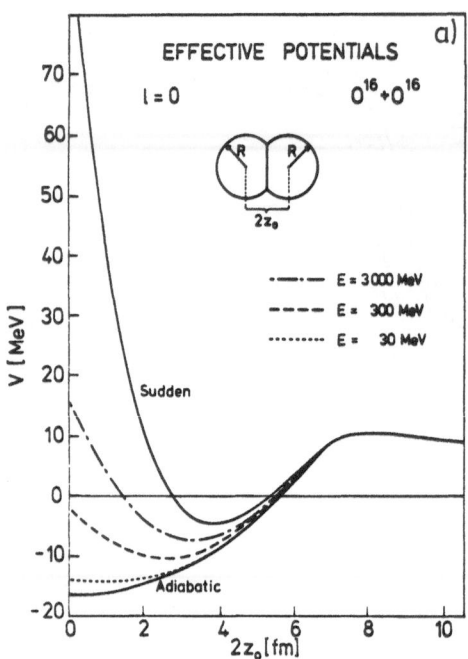
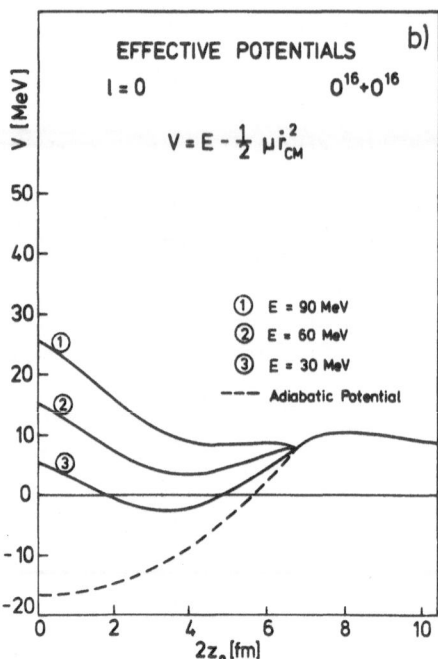

Fig. 11. The compression effect on the potential as function
of the bombarding energy for the model in which the
nuclear density is homogeneously distributed over
the volume shown in fig. a).
a) For various bombarding energies the potential
energy of the compression mode is added to the adiabatic potential. The adiabatic and the sudden potentials arethe limiting curves for slow and fast processes.
b) The kinetic and potential energies of the compression
mode are added to the adiabatic potential for different bombarding energies

157

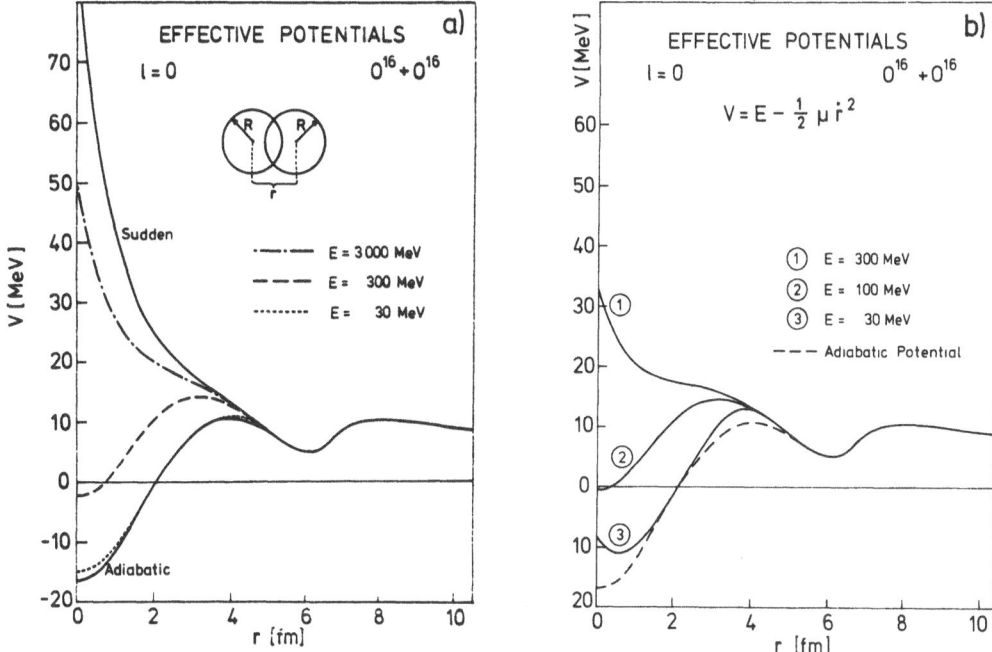

Fig. 12. The compression effect on the potential calculated
for the model in which the densities are additively
superposed as shown in fig. a). The figures a) and
b) give the same information as in fig. 11

tion. In the intrinsic motion, i.e. compression mode, kinetic and potential energies are stored. Both kinds of energies have to be added to the adiabatic potential in order to obtain the effective potential for the relative motion.

In figs. 11a and 12 a the compression potential is added to the adiabatic potenital for various bombarding energies. The energy dependence is too small, because the models are not fully able to describe the expansion of locally compressed matter towards uncompressed regions.

We expect that nuclear matter is locally compressed because the sound velocity in nuclear matter is of the same order of magnitude as the velocity of the nuclei. The ratio between these velocities is given for the scattering of two equal nuclei with mass number A and relative energy E by

$$V_{REL.}/V_{SOUND} = 6 \cdot \sqrt{E/K \cdot A} \qquad (9)$$

where K = 150-200 MeV is the compression constant [28]. For E = 30 MeV, K = 150 MeV and A = 16 we find already 0.7 from (9). A necessary extension of the second model (fig. 12) would be to allow the compressed region to expand <u>independently</u> from the motion of the uncompressed region. For that one has to introduce more shape parameters.

If we want to describe the system only by the relative coordinate, we are allowed to include the intrinsic kinetic energy in the energy of the relative motion in two ways: (a) We can add this energy to the kinetic energy of the relative motion and obtain an effective mass which depends on the relative coordinate and is larger than the reduced mass. (b) Or we add the intrinsic kinetic energy on the potential energy of the relative motion given in figs. 11a and 12a. In that case potentials result which are shown in figs. 11b and 12b and which reveal a strong energy-dependence.

3. Molecular States

In the following we discuss the states in the quasi-molecular potential. We add the centrifugal potentials to the adiabatic or sudden potential. According to classical mechanics the system can rotate in the potential wells (fig. 4). Quantum-mechanically the nuclei vibrate against each other, i.e. they lead out the zero point motion. Fig. 13 gives the position of the states found by a phase shift analysis for the $^{12}C-^{12}C$ system. The applied potential is calculated in the sudden approximation according to eq. (4) [11]. In this example states up to

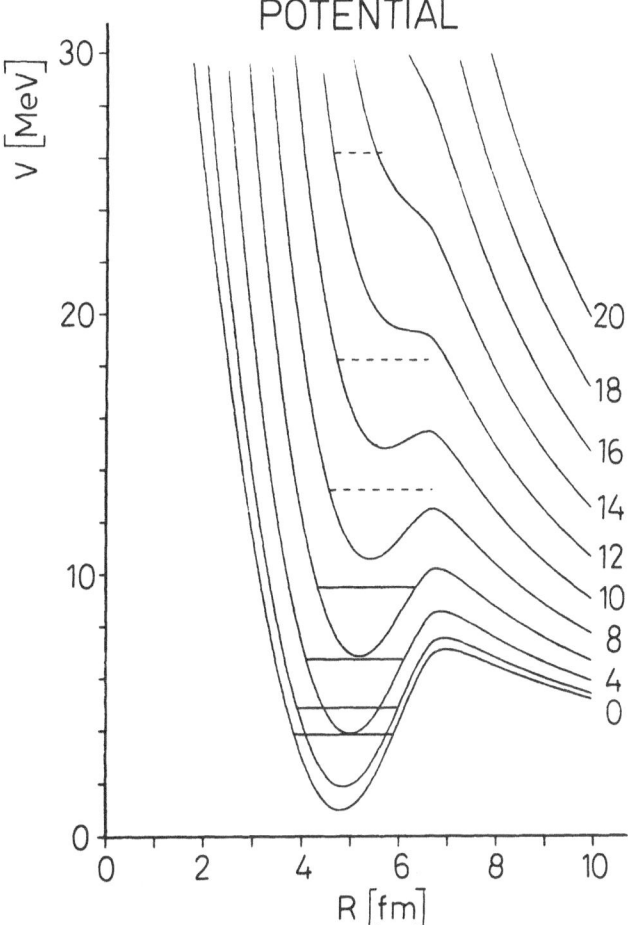

Fig. 13. The real potential for $^{12}C-^{12}C$ scattering in the
sudden approximation for various angular momenta.
The virtual states are indicated by dashed hori-
zontal lines, the quasibound states by full lines.
The position of these states has been chosen such
that nuclear phase shift has the value $\delta_1 = \pi/2$ at
these energies (see fig. 14)

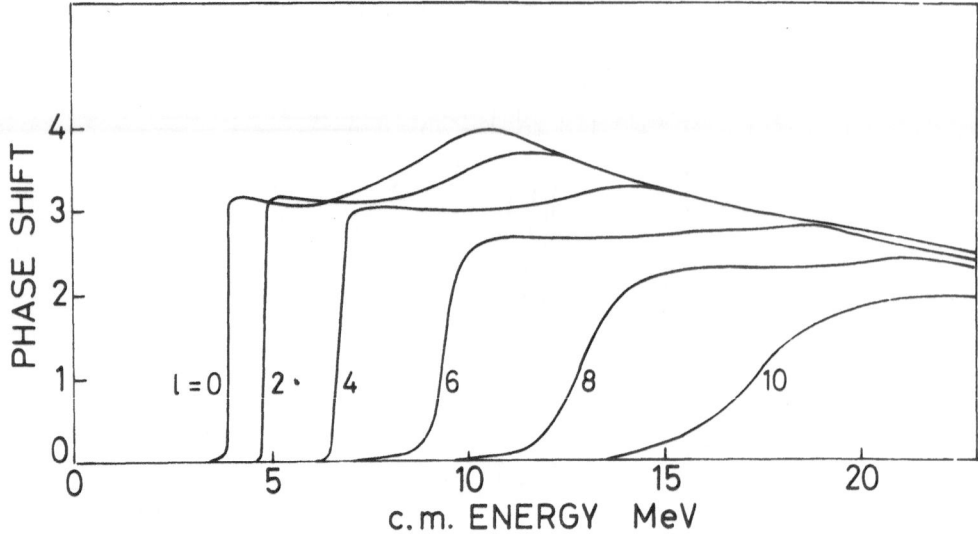

Fig. 14. The nuclear phase shifts for the elastic $^{12}C-^{12}C$
scattering calculated with the potential of fig.
13 and without an imaginary potential

an angular momentum $\ell=6$ are possible which lie in the potential well.
We call these states quasibound, since the system can decay through
the Coulomb barrier and into the compound nucleus.

Above the quasibound states we find resonance states denoted as
virtual states which do not lie in the potential well. In the next
section we prove that the virtual states are responsible for the gross
structure.

3.1. The Gross Structure

The virtual states have a larger width than the quasi-bound states. This is evident from the energy dependence of the nuclear phases
for the various partial waves (fig. 14). Over a quasibound state the
phase changes very rapidly, whereas the virtual states show a width
of about 2-3 MeV.

The existence of the virtual states has the following consequences
for the cross section: with increasing energy the partial waves successively overcome their corresponding Coulomb and centrifugal barriers,
penetrate into the overlap region and resonate with their virtual state. Then gross structures appear in the cross section. The interference
mechanism with the Coulomb waves has been extensively studied by Gobbi
[13].

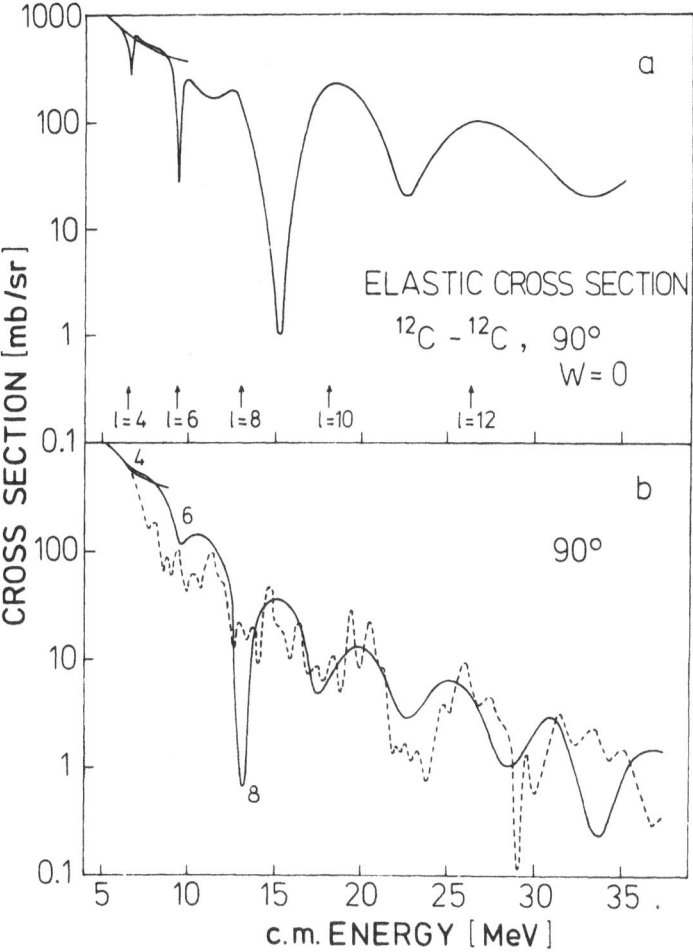

Fig. 15. The elastic 90°-excitation function for ¹²C on ¹²C.
a) In the upper half the theoretical curve is calculated with no imaginary potential included. The
angular momenta of the mainly contrubuting partial
waves are indicated.
b) In the lower half the imaginary part for the theoretical cross section (full line) is taken from ref.
[11] (see fig. 16). The dashed curve represents the
experimental data [4]

In fig. 15a the cross section is calculated without an imaginary potential. At the maxima the angular momenta of the partial waves are indicated, which resonate in this energy range. In fig. 15b an absorptive potential is included which will be discussed in the next section. By that the maxima of the cross section are lowered to the experimental values. This can be explained as follows: for a fixed bombarding energy all partial waves which overcome the Coulomb barrier are nearly completely absorbed except one, namely the partial wave with the highest angular momentum which penetrates nearly undamped into the overlap region. This partial wave can resonate with its corresponding virtual resonance state and produces the gross structures. We have the unusual situation that the resonating partial wave finds no or only a few open channels with the same high angular momentum through which flux can be carried out of the elastic channel.

So we conclude from the experiment that the gross structure is connected with the virtual resonances. This is proved for the systems $^{12}C-^{12}C$ and $^{16}O-^{16}O$ (cf [4], [14],[3]).

3.2. The Imaginary Potential

The gross structure is sensitive to an energy and angular momentum dependent imaginary potential [8]. Such an imaginary potential can be derived from the assumption that in first order the transition probability from the elastic channel is given by [15]:

$$\frac{\Gamma_{el.\rightarrow comp.}}{\hbar} = \frac{2\pi}{\hbar}\ \rho(E^*,I) \cdot |<el.|V|comp.>|^2 \tag{10}$$

The density of compound states ρ depends on the excitation energy E^* and angular momentum I. The average transition matrix element out of the elastic channel is abbreviated by $|<el.|V|comp.>|^2$. If we assume that there are enough open channels, i.e. that the compound elastic contributions are small (never-come-back-approximation), the imaginary potential is given by

$$W = -\ \frac{i}{2}\ \Gamma_{el.\rightarrow comp.} \tag{11}$$

In the actual calcualtion we have set the square of the transition matrix element proportional to the nucleon number in the overlap region [15]. The density of compound levels can be taken from a statistical formula [39]:

163

$$\rho(E^*,I) = \omega(E^*) \cdot \frac{2I+1}{2\sqrt{2\pi}\ \sigma^3}\ \exp\left[-(I+1/2)^2/2\sigma^2\right] \tag{12}$$

The parameter σ cuts off all states with higher angular momenta. It is connected with the excitation energy and the moment of inertia by

$$\sigma^2 = \frac{\Theta}{\hbar^2}\ \sqrt{E^*/\beta} \tag{13}$$

where the constant β is adjusted to the experiment. The excitation energy E^* is measured from the adiabatic potential

$$E^* = E - V_{ad} \tag{14}$$

With these assumptions the following imaginary potential results depending on two free parameters α and β:

$$W(r,E,I) = \alpha \cdot N(r)\ \frac{2I+1}{\sigma^3}\ \exp\left[2\sqrt{aE^*} - (I+1/2)^2/2\sigma^2\right]$$

with

$$N(r) = \int_{overlap} \rho\, d\tau \tag{15}$$

We have applied such an imaginary potential in the case of the elastic scattering of ^{12}C on ^{12}C [11]. In fig. 16 the imaginary potential for a bombarding energy of 15 MeV is shown.

The radial dependence is proportional to the number of nucleons enclosed in the overlap region. In fig. 15b the experimental ^{12}C-^{12}C-excitation function is compared with the theoretical one which reproduces the gross structure quite well.

For the existence of the gross structure it is necessary that the partial wave which resonates with a virtual state is not much absorbed. To clarify this point, we have drawn fig. 17. Fig. 17 gives an impression of the absorptive strength out of the elastic channel drawn over the energy - angular momentum plane.

Below the Yrast-line which is extrapolated from the groundstate

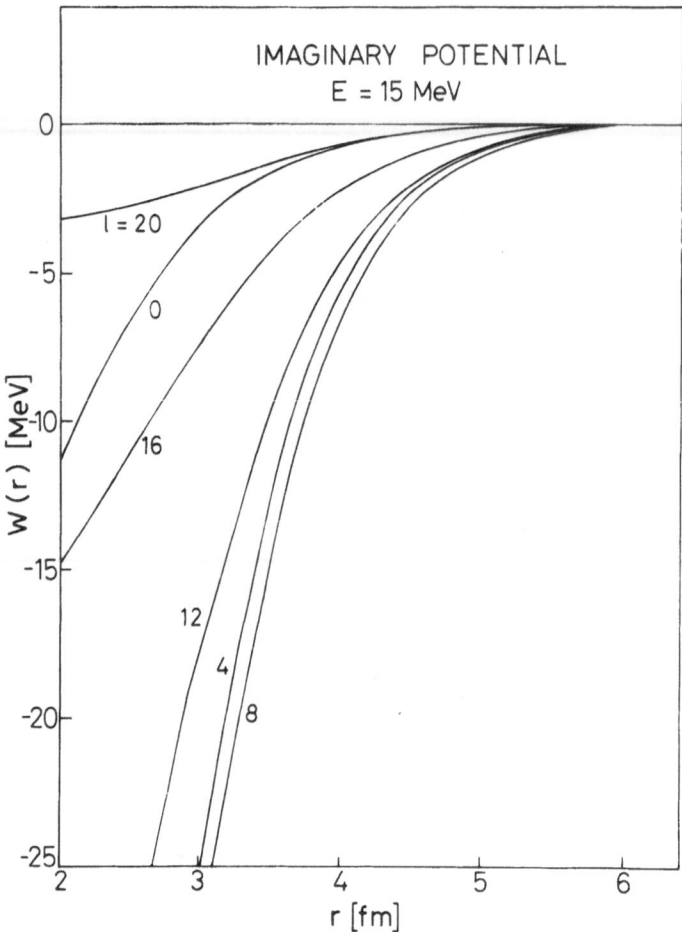

Fig. 16. The imaginary potential for $^{12}C-^{12}C$ and $E_{CM}=$
=15 MeV for various angular momenta according
to eq. (15)

band of ^{24}Mg the system has no compund states with such an high angular momentum. There the imaginary potential is exactly zero. Another
curve in the E-I-plane is the line of the maximum density of states.
Since the imaginary potential is proportional to the density of states
the absorption will also be largest along this line. This is strictly
true only if the transition matrix elements are independent of E and
I. The area between the two curves is the area of small and nearly
vanishing imaginary potential. Nuclear states in this area are of molecular nature. The figure shows the quasimolecular states of the sudden potential. Further such molecular configurations are for example

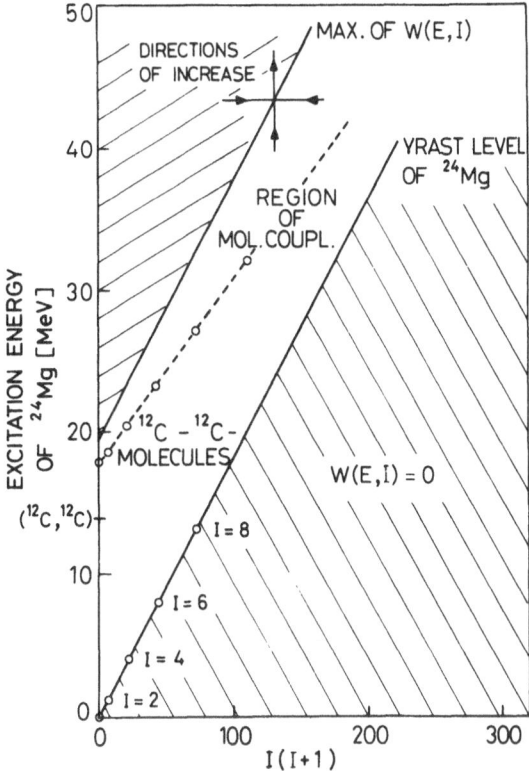

Fig. 17. The imaginary potential and its dependence on
energy and angular momentum. The Yrast line is
obtained by extrapolating the experimental rotational ground state band of ^{24}Mg. Between the
Yrast line and the line of maximum imaginary potential the ^{12}C-^{12}C molecular states are situa-
ted. This part of the (E,I)-plane is the region
of molecular coupling and small imaginary potential due to compound damping

^{12}C-^{12}C*, ^{12}C*-^{12}C*, ^{20}Ne-α, ^{16}O-^{8}Be etc.

Such configurations have only a small (or no) imaginary potential
in the unshadowed area of fig. 17. The damping of these states arises
essentially from their coupling between themselves (area of molecular
coupling).

3.3. The Excitation of the Quasibound States

Quasibound states connot be excited directly in the elastic scattering because of the inpenetrability of the potential barrier. Only
via an indirect, i.e. inelastic excitation mechanism, it is possible
to circumvent the Coulomb barrier and to excite quasibound states
with sufficient strength.

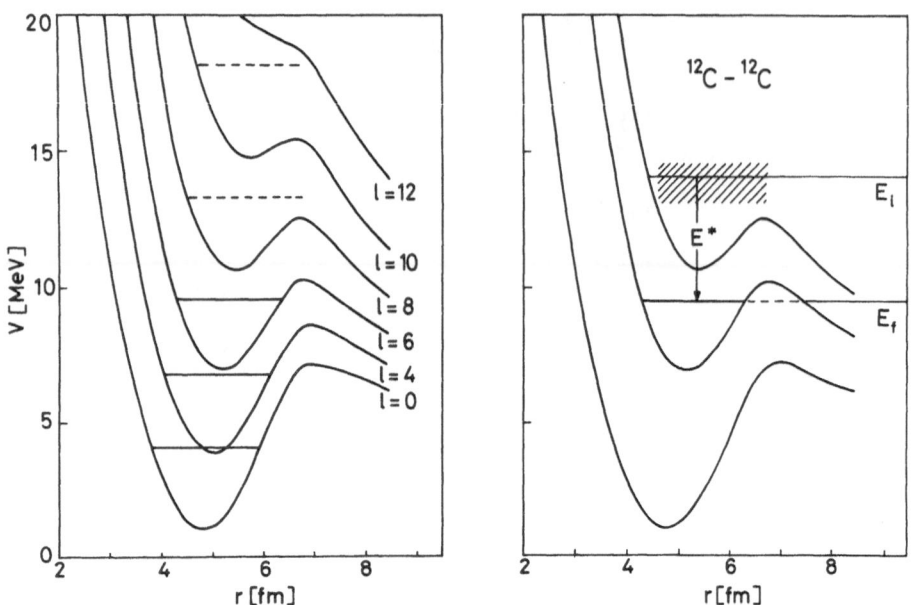

Fig. 18. Quasimolecular $^{12}C-^{12}C$ potential and the excitation mechanism. On the right-hand side the deexcitation of an elastic channel (at energy E_i)
by an energy $E*$ to an inelastic channel (at energy
E_f of the relative motion) is indicated

The following mechanism is suggested (fig. 18): after the nuclei
have crossed the potential barrier they lose kinetic energy by inelastic excitation of low energy levels in one or both of the nuclei.
Thereby they drop into the potential well and are able to form a quasi-
molecule. For that their new relative energy and angular momentum has
to coincide with the energy and angular momentum of a quasibound state.

This idea was first applied by Imanishi [17] who found that the resonances seen in the ^{12}C-^{12}C reaction below the Coulomb barrier can be
interpreted as molecule-like compund states of two ^{12}C-nuclei, one being in the groundstate and the other in the 2^+-state at 4.43 MeV.

Since the elastic and inelastic channels are coupled intermediate
structures arise in the elastic channel. Large effects can be expected
if the following conditions are fulfilled:

(a) The elastic and inelastic channels must be coupled strongly
enough to each other. This is true for the excitation of collective
surface vibration states, because the nuclei have to change their
shapes during the penetration. Therefore, one expects a larger inter-
mediate structure in the ^{12}C-^{12}C case than for ^{16}O-^{16}O since the ^{12}C-
nuclei have a relative soft shell structure. Also α-particle transfer
channels like ^{16}O+^{16}O $\rightarrow$ ^{20}Ne+^{12}C, are strongly coupled to the elastic
channel.

(b) The partial wave of the entrance channel by which the quasi-
bound state is excited has to resonate with a virtual state. A resonating partial wave has an enhanced amplitude inside the attractive potential region and can induce inelastic processes with sufficient
strength. In each energy range the partial wave which essentially con-
tributes to the gross structure fulfills this requirement. Therefore,
we are led to a double resonance mechanism (see fig. 18): two partial
waves are simultaneously resonating, namely with a virtual state in
the elastic channel and with a quasibound state in the inelastic channel. In order that this happens, the energy and angular momentum difference between the virtual and quasibound state has to be matched by
the inelastic excitation of one or both nuclei. This condition restricts
the number of the possible intermediate structures very much. That is
in agreement with the experimental excitation function.

(c) The imaginary potential for the quasibound states has to be
sufficiently weak. As shown, the imaginary potential depends on the
angular momentum and excitation energy of the total system. Because the
coupled channels have the same total angular momentum as the elastic
one, they should feel the same imaginary potential as the elastic chan-
nel.But since the inelastic channels are excited over an undamped partial wave in the elastic channel, the inelastic channels have also a
very small imaginary potential.

In the following we apply the outlined ideas to explain the intermediate structure of the ^{12}C-^{12}C scattering.

4. Intermediate Structure in $^{12}C-^{12}C$

First we discuss the Hamiltonian of the model and then the results.

4.1. The Model

The Hamiltonian which describes the scattering of the two ^{12}C-nuclei is composed of the kinetic energy of the relative motion, of the intrinsic Hamiltonians of the two ^{12}C-nuclei and of their interaction energy. The interaction energy depends on the relative distance of the nuclei and on their intrinsic coordinates.

$$H = T_r(\vec{r}) + H_{12_C}(1) + H_{12_C}(2) + W_{12_C-12_C}(\vec{r},1,2) \tag{16}$$

Since we are interested in the excitation of collective modes we choose surface deformation parameters as intrinsic coordinates. Therefore, we expand the radii of the two nuclei in a multipole expansion (fig. 5b):

$$R^{(1,2)} = R(1 + \sum_{\ell,m} \alpha_{\ell m}^{(1,2)} Y_{\ell m}^*(\Omega_{1,2})). \tag{17}$$

Then the nuclear density distributions have deformed shapes in accordance with eq. (17). To calculate the interaction W in the sudden approximation, we insert the deformed density distributions into eq. (3) and (4) and expand the interaction in the surface coordinates α:

$$W_{12_C-12_C} = U(r) + \sum_{\ell,m} Q_{\ell m} Y_{\ell m}^*(\Omega) \tag{18}$$

with

$$Q_{\ell m} = I_\ell(r) \cdot ((-)^\ell \alpha_{\ell m}^{(1)} + \alpha_{\ell m}^{(2)}).$$

The potential $U(r) = V+iW$, which is independent of the intrinsic coordinates $\alpha_{\ell m}$, is the average optical potential between the nuclei shown in figs. 13 and 16. The multipole potentials $Q_{\ell m} Y_{\ell m}^*$ couple the relative and the intrinsic motions together. The radial parts I_ℓ depend on the same parameters as the real part of U, namely on the parameters in the expression (2) for the nuclear binding energy. The functions I_ℓ are drawn if fig. 19 for the quadrupole and octupole transitions.

The wave functions for H can be expressed as follows:

$$\psi = \sum_{\ell,J,\lambda,I} R_{\ell J \lambda I}(r) [i^\ell Y_\ell \otimes \phi_{J\lambda}(1,2)]^I$$

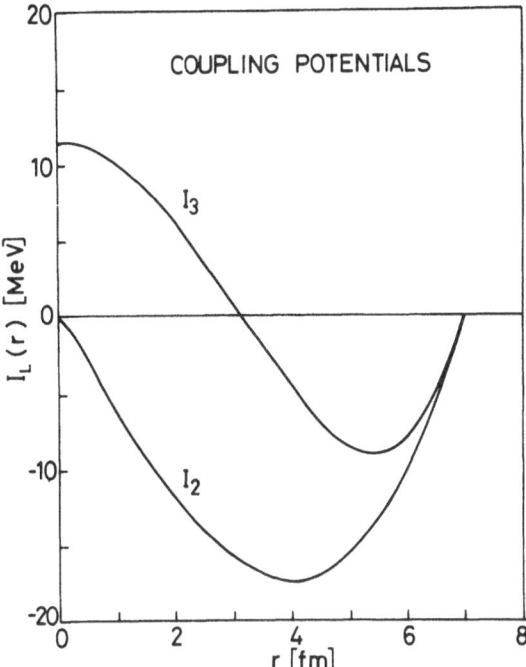

Fig. 19. The radial dependence of the quadrupole and octupole coupling potentials for $^{12}C-^{12}C$

with the channel function ($\lambda = (n_1, n_2)$):

$$\phi_{J\lambda} = \frac{1}{\sqrt{2(1+\delta_{n_1 n_2})}} \left[\chi_{n_1}^{(1)} \otimes \chi_{n_2}^{(2)} + (-)^{\ell} \chi_{n_1}^{(2)} \chi_{n_2}^{(1)} \right]^J \tag{19}$$

The states χ_{n_i} are the eigenstates of the intrinsic Hamiltonian H_{12C},
i.e. the $^{12}C-$states. We restrict ourselves to the following C^{12}-sta-
tes: ground state, $2^+(4.43 \text{ MeV})$, $3^-(9.64 \text{ MeV})$.

Under this restriction we derive a system of coupled equations
for the relative motion [11] as follows:

$$< i^{\ell} \left[Y_{\ell} \otimes \phi_{J\lambda} \right]^I | H\psi > = E < i^{\ell} \left[Y_{\ell} \otimes \phi_{J\lambda} \right]^I | \psi > \tag{20}$$

In these equations the total information about the ^{12}C-states is con-
tained in the reduced matrix elements: $<3^-||\alpha_3||0^+_{g.s}>$ and $<2^+||\alpha_2||0^+_{g.s}>$.
They can be related to the electromagnetic transition rates assuming
proportianality between charge and mass density. Beside these two ma-

trix elements all other constants in the real and imaginary part of the potential are already fixed by the elastic cross section. Therefore, the coupled channel calculations are carried out with no further adjusted parameters. All details of the calculation can be taken from ref. [11].

4.2. The Results

In fig. 20a we compare the 90°-excitation function, where no inelastic channel is coupled in, with the case, that one of the ^{12}C-nuclei is excited to the first 2^+-state. In fig. 20b we notice that the coupling of the 3^--state changes only the peak-to-valley ratio but causes no additional structures. More important are the effects arising from the simultaneous excitation of both fragments. In that case also energetically deeper bound states are reached (fig. 20c).

The next figures compare the calculated elastic and inelastic cross sections with the experimental data. As shown in fig. 21 the intermediate structure of the elastic excitation function can be explained up to $E_{CM} = 18$ MeV by the excitation of the 2^+-state in one or both of the scattered nuclei. In this region the intermediate structure is reproduced as well with regard to position as width. We expect that the remaining intermediate structure at higher energies can be explained by the excitation of states which have a larger excitation energy and higher angular momentum. Also the coupling of other molecular channels like ^{20}Ne-α and ^{16}O-^{8}B is important and should be considered in future calculations.

The corresponding inelastic cross sections (fig. 22) are of the correct order of magnitude and show a similar but displaced structure compared to the experiment [5]. Both facts are supporting our assumptions about the coupling strength and the imaginary potential in the inelastic channels. Since the energies of the quasibound states are sensitively fixed by the real potential, one has to change the real potential for the inelastic channels in order to get a quantitative agreement with the experimental inelastic cross sections. This is not done yet.

5. Summary and Conclusions

We have shown that nuclear molecules can be formed for very short times in the elastic heavy ion scattering. In a nuclear molecule the two nuclei attract each other on their surfaces. Microscopically the binding is produced by a few nucleons which orbit around both nuclei. This is a homeopolar binding.

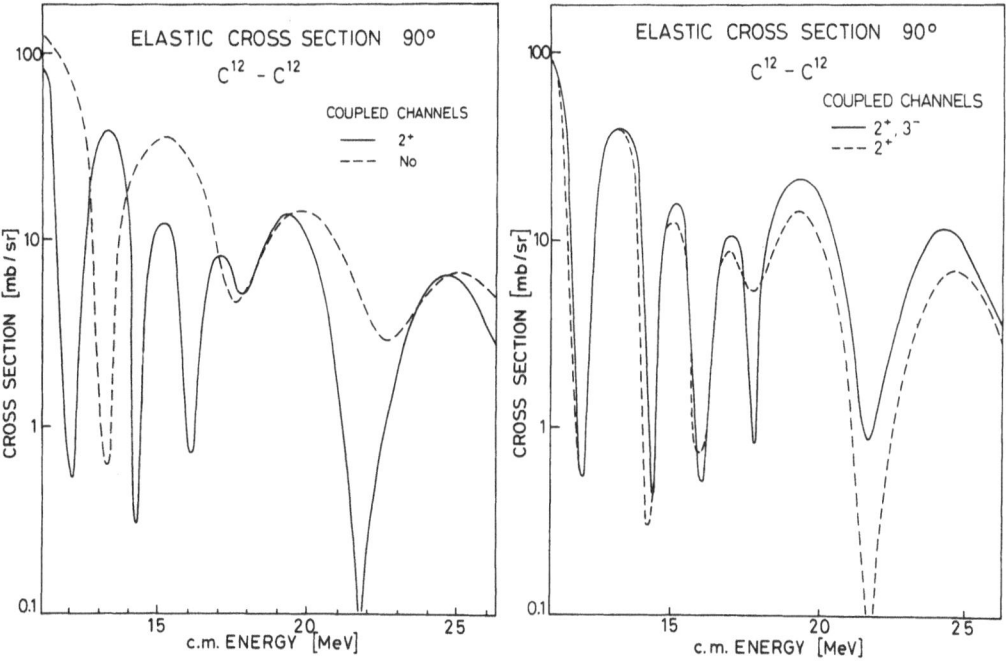

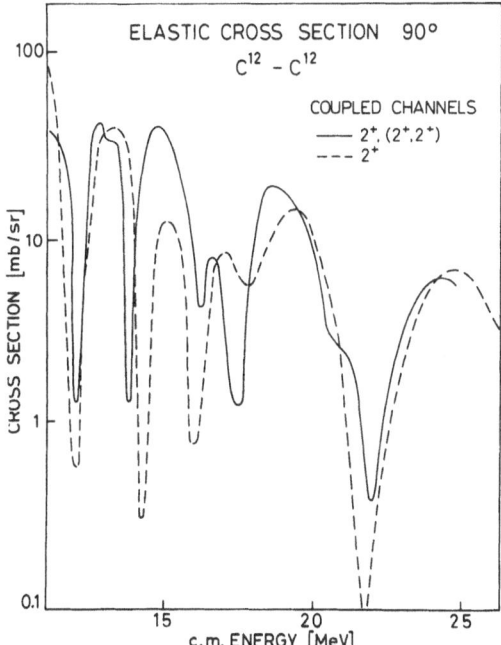

Fig. 20. The 90°-excitation for the elastic ^{12}C-^{12}C scattering.
a) No coupling and the coupling of the excitation of the 2$^+$-state in ^{12}C.
b) Comparison of the coupling of the 2$^+$- and the 3$^-$-state.
c) The simultaneous excitation of both ^{12}C nuclei to a 2$^+$-state

172

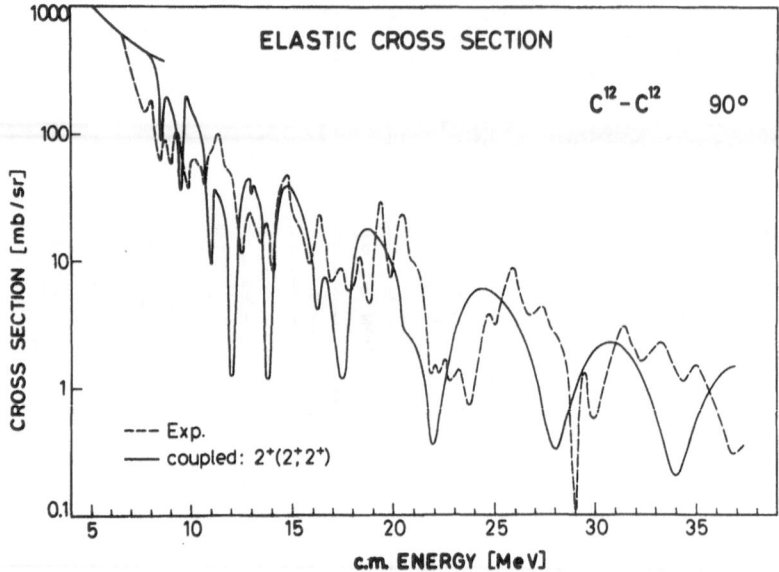

Fig. 21. Comparison of the experimental 90°-excitation
function (from ref. [4]) with the coupled channel
calculation of fig. 20c

The nuclear molecule potentials contain virtual and quasibound
states. The former states lie above, the latter in the potential wells.
The virtual states are directly excited in the elastic scattering and
generate gross structures in the elastic excitation function. The quasi
bound states can be excited over the double resonance mechanism. By
that intermediate structures are produced.

Measurements by Rossner et al. [33] show that only two kinds of
reaction channels are prominent in the ^{16}O-^{16}O scattering. These are
the channels for the inelastic excitation of one or both ^{16}O-nuclei
and the transfer channel to ^{12}C-^{20}Ne. So in future calculations one
should include also the α-particle transfer channels in the coupled
channel formalism. These channels give rise to additional intermediate
structures in the elastic channel.

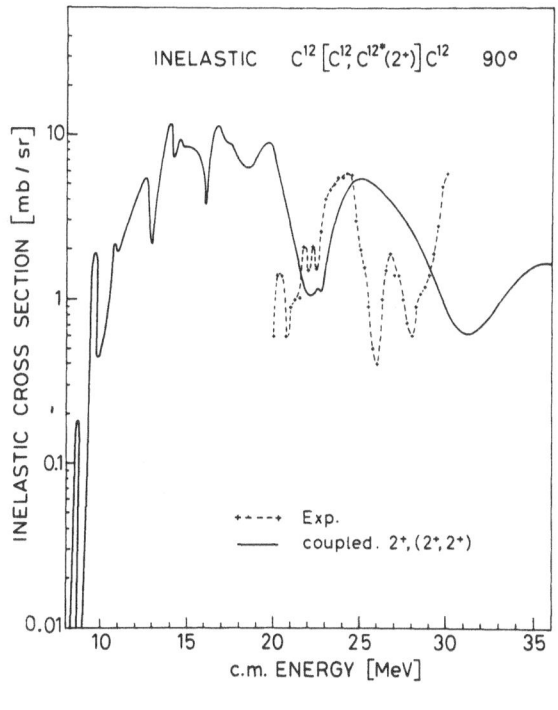

a) $^{12}C(^{12}C,\ ^{12}C(2^+))^{12}C$

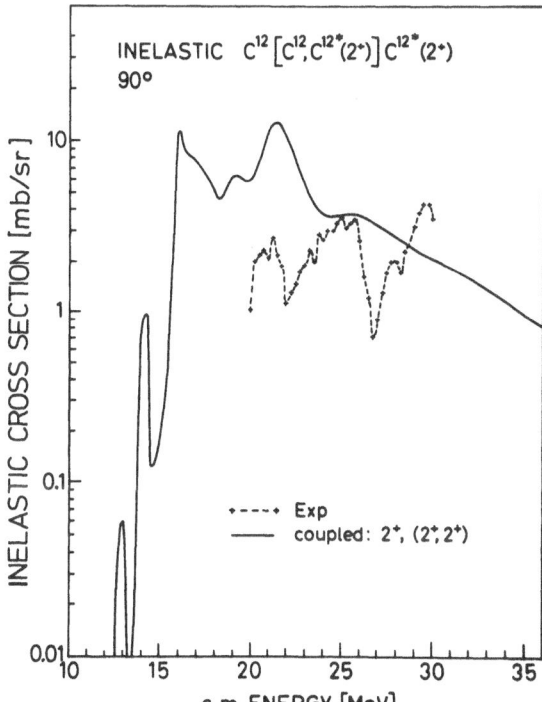

b) $^{12}C(^{12}C,\ ^{12}C(2^+))^{12}C(2^+)$

Fig. 22. Inelastic cross sections. The experiments are taken from ref. [5]

The question whether the real scattering potential has a core or not, cannot be answered by measuring the elastic cross section only. The reason is that partial waves with small angular momenta which can penetrate into the potential well are absorbed. Only waves which have passed the potential barrier at higher energy and which are lowered in their angular momenta and relative energy by an inelastic excitation of the nuclei, are sensitive to the core. Therefore, by analyzing the inelastic cross section one possibly can decide whether the potential is sudden or adiabatic.

A many-body description of heavy ion scattering involves a lot of new problems. One of these problems we want to mention: one calculates real nucleus-nucleus potentials with wave functions depending on a distance parameter, e.g. on the two-center distance $r=2z_o$ if two-center wave functions are applied (see fig. 6). To obtain the kinetic energy of the relative motion the mass has to be known which is moved when the collective distance parameter is changed. Asymptotically the motion of the potential centers is identical with the motion of the mass centers and, therefore, the mass of the relative motion is the reduced mass. But for smaller two-center distances, i.e. in the overlap region the mass of the relative motion varies appreciably with the relative distance of the potential centers. This is shown in fig. 23 for the symmetric fission of $^{236}_{92}U$. The curves are calculated in the framework of the cranking model with two-center wave functions [19] (see also [23]).

Acknowledgement

We thank Prof. W. Greiner for his collaboration in this work.

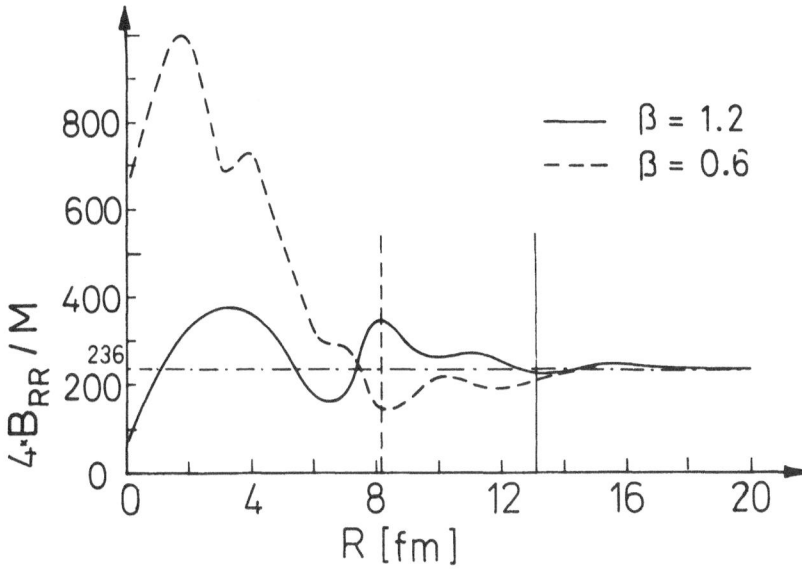

Fig. 23. Dependence of the mass on the two-center distance
$r=2z_0$ for the symmetric fission of ^{236}U. The parameter $\beta=a/b$ measures the deformation of the fission products, which have axes of length a and b.
Further details are given in ref. [19]

References

1. ALMQVIST, E., BROMLEY, D.A. and KUEHNER, J.A., Phys. Rev. Lett. $\underline{4}$, 515 (1960).

2. ALMQVIST, E., KUEHNER, J.A., McPHERSON, D. and VOGT, E.W., Phys. Rev. $\underline{136B}$, 84 (1964);
VOGT, E.W., McPHERSON, D., KUEHNER, J.A. and ALMQVIST, E., Phys. Rev. $\underline{136B}$, 99 (1964).

3. ARIMA, A., SCHARFF-GOLDHABER, G., McVOY, K.W., Phys. Lett. $\underline{40B}$, 7 (1972).

4. BROMLEY, D.A., in Proceedings of the International Conference on Nuclear Reactions Induced by Heavy Ions, Heidelberg 1969, edited by R.Bock and W.R. Hering (North-Holland, Amsterdam 1970) p. 27.

5. BROMLEY, D.A., GOBBI, A., private communication.

6. BROMLEY, D.A., KUEHNER, J.A. and ALMQVIST, E., Phys. Rev. Lett. $\underline{4}$, 365 (1960).

7. BRUECKNER, K.A., BUCHLER, J.R., JORNA, S. and LOMBARD, R.J., Phys. Rev. $\underline{171}$, 1188 (1968).
BRUECKNER, K.A., BUCHLER, J.R. and KELLY, M.M., Phys. Rev. $\underline{173}$, 944 (1968); PEREZ, J.D., Nucl. Phys. $\underline{A191}$, 19 (1972).

8. CHATWIN, R.A., ECK, J.S., ROBSON, D. and RICHTER, A., Phys. Rev. $\underline{C1}$, 795 (1970).

9. DAVIS, R.H., Phys. Rev. Lett. $\underline{4}$, 521 (1960).

10. FESHBACH, H., Ann. Phys. (N.Y) $\underline{5}$, 357 (1958), Ann. Phys. (n.Y.) $\underline{19}$, 287 (1962).

11. FINK, H.J., SCHEID, W. and GREINER, W., Nucl. Phys. $\underline{A188}$, 259 (1972.

12. FLIEβBACH, Z. f. Physik $\underline{238}$, 329 (1970); $\underline{242}$, 287 (1971); $\underline{247}$, 117 (1971), Nucl. Phys. $\underline{A194}$, 625 (1972).

13. GOBBI, A., in Proceedings of the Symposium on Heavy Ion Scattering, Argonne National Laboratory, 1971, Report No. ANL-7837, p.63.

14. GREINER, W. and SCHEID, W., J. de Physique $\underline{32}$,C6-91 (1971).

15. HELLING, G., SCHEID, W. and GREINER, W., Phys. Lett. $\underline{36B}$, 64 (1971).

16. HOLZER, P., MOSEL, U. and GREINER, W., Nucl. Phys. $\underline{A138}$, 241 (1969);
SCHARNWEBER, D., MOSEL, U. and GREINER, W., Phys. Rev. Lett. $\underline{24}$, 601 (1970); SCHARNWEBER, D., GREINER, W. and MOSEL, U., Nucl. Phys. $\underline{A164}$, 257 (1971).
MOSEL, U., MARUHN, J. and GREINER, W., Phys. Lett. $\underline{34B}$, 587 (1971);
MARUHN, J. and GREINER, W., Z.f. Physik $\underline{251}$, 431 (1972).

17. IMANISHI, B., Phys. Lett. $\underline{27B}$, 267 (1968); Nucl. Phys. $\underline{A125}$, 33 (1969).

18. JACOBSON, L.A., Phys. Rev. 188, 1509 (1969).

19. LICHTNER, P., DRECHSEL, D., MARUHN, J. and GREINER, W., Phys. Rev. Lett. 28, 829 (1972).

20. LOW, K.S. and TAMURA, T., Phys. Lett. 40B, 32 (1972).

21. MAHER, J.V., Thesis, Yale University 1969.

22. MICHAUD, G.J. and VOGT, E.W., Phys. Rev. C5 , 350 (1972).

23. MOSEL, U., Particles and Nuclei, June 1972.

24. MOSEL, U. and SCHMITT, H.W., Phys. Rev. C4, 2185 (1971); Nucl. Phys. A165, 73 (1971).

25. MOSEL, U., THOMAS, T.D. and RIESENFELDT, P., Phys. Lett. 33B, 565 (1970).

26. MOROVIĆ, T., MARUHN , J. and GREINER, W., to be published.

27. MÜLLER, H., diploma thesis, Frankfurt 1972.

28. PANDHARIPANDE, V.R., Phys. Lett. 31B, 635 (1970).

29. PRÜß, K., Nucl. Phys. A170, 336 (1971) and Thesis, University of Frankfurt, 1972.

30. PRÜß, K. and GREINER, W., Phys. Lett. 33B, 197 (1970).
    PRÜß, K., Thesis, University Frankfurt, 1972.

31. RAYET, M. and REIDEMEISTER, G., J. de Physique 32, C6-259 (1971).

32. REILLY, W., WIELAND, R., GOBBI, A., SACHS, M.W., MAHER, J.V., MINGAY, D., SIEMSSEN, R.H. and BROMLEY, D.A., in Proceedings of the International Conference on Nuclear Reactions, Heidelberg 1969, p. 80 (see ref. 4).
    REILLY, W., WIELAND, R., GOBBI, A., SACHS, M.W., BROMLEY, D.A., ibid. p.93.

33. ROSSNER, H.H., HINDERER, G.,WEIDINGER, A., EBERHARD, K.A., Proceedings of the European Conference on Nuclear Physics, Aix-en-Provence 1972, Vol. II, p. 42.

34. SCHEID, W., LIGENSA, R. and GREINER, W., Phys. Rev. Lett. 21, 1479 (1968).

35. SIEMSSEN, R.H., Elastic and Inelastic Scattering of Heavy Ions, Chapter IV, C, 1 of Nucl. Spectroscopy II, edited by J. Cerny (Academic Press, Inc., New York, in preparation).

36. SIEMSSEN, R.H., MAHER, J.V., WEIDINGER, A., BROMLEY, D.A., Phys. Rev. Lett. 19, 369 (1967), 20, 175 (1968).
    MAHER, J.V., SACHS, M.W., SIEMSSEN, R.H., WEIDINGER, A., BROMLEY, D.A., Phys. Rev. 188, 1665 (1969).

37. Reviews of further structures in the cross sections of heavy ion scattering are given by STOCKSTAD, R.G. and BOHLEN, H.G. in their contributions to this conference.

38. STRUTINSKY, V.M., Nucl. Phys. A95, 420 (1967); Nucl. Phys. A122, 1

(1968).

39. THOMAS, T.D., Ann. Rev. of Nucl. Sc. <u>18</u>, 343 (1968).

40. VANDENBOSCH, R., in Proceedings of the Symposium on Heavy Ion
    Scattering, Argonne 1971, p.103 (see ref. 13).

41. VOGT, E. and McMANUS, H., Phys. Rev. Lett. <u>4</u>, 518 (1960).

42. YUKAWA, T., Phys. Lett. <u>38B</u>, 1 (1972); Nucl. Phys. <u>A186</u>, 127
    (1972).

MOLECULAR RESONANCES AND INTERMEDIATE PROCESSES
IN HEAVY ION REACTIONS
A Review of Experimental Evidence*
ROBERT G. STOKSTAD
A.W. Wright Nuclear Structure Laboratory
Yale University, New Haven
Connecticut
and
The Niels Bohr Institute
University of Copenhagen
Denmark

1. Introduction

The concept of intermediate structure or doorway states [23] has
been most useful in the understanding of a wide range of phenomena in
nuclear reactions including isobaric analogue resonances, the giant
dipole resonance and neutron induced fission. In each of these examples
the primary characteristic of the doorway state is a relatively simple
structure which is easily excited in the entrance channel. The strength
of the coupling of this state to the many model of decay which may be
energetically allowed determines its width and the ease with which it
may be observed experimentally.

It is interesting to inquire whether processes analogous to the
formation of doorway states might play a role in heavy ion reactions.
In this case, the "simple structure" may refer to the spatial distribution of the nucleons which, in the entrance channel, consists initially of two large, well separeated groups of nucleons. To the extent
that such a spatial structure is a) approximately preserved during the
reaction and b) exists for a time which is long compared to the collision time, one may refer to the compound system as a "nuclear molecule". This elementary picture of molecular structure in nuclear reactions can be extended, if necessary, to include more complicated spatial configurations involving several clusters of nucleons.

The narrow resonance associated with the long-lived molecular
state envisaged here is one example of an intermediate process in heavy
ion reactions. There may exist other reaction mechanisms, however,
which produce structures with widths too large for the strict molecular

* Work supported in part by the U.S. Atomic Energy Commission under
contract AT(30-1) 3223.

interpretation above, but which nevertheless fall well within the spirit of this conference. In this case, the process is termed "intermediate" because it lies somewhere in between the extreme limits of compound nucleus formation (with coherence widths $\Gamma \lesssim 200$ keV) and the direct reaction ($\Gamma \gtrsim 3$ MeV).

This talk is a review of the experimental evidence for molecular structure and intermediate processes in heavy ion reacitons. However, experiments in which such effects have been sought but have not been observed will be discussed also, since such reactions can elucidate the nature of the intermediate process. Indeed, we know now that doorway states in heavy ion reacitons have another feature in common with those in analogue resonances, giant resonances and fission - to find them, you have to look in just the right place. Finally, this is an area of heavy ion physics which is growing rapidly and in which many physicists have a lively interest. As a consequence of this we will have occasion to discuss experimental results which have inspired a variety of different interpretations.

It is convenient to classify the experiments according to the energy in the entrance channel. Reactions with energies below and near the Coulomb barrier will be discussed in section 2 while those reactions proceeding well above the Coulomb barrier will be presented in part 3. A summary and conclusions are given in part 4.

2. Resonances Near the Coulomb Barrier

The reactions known to exhibit molecular-type resonances at bombarding energies below and up to the Coulomb barrier are $^{12}C + ^{12}C$, $^{12}C + ^{16}O$, and possibly $^{14}N + ^{14}N$. The $^{12}C + ^{12}C$ system was investigated initially at Chalk River some twelve years ago by Almqvist, Kuehner and Bromley [1]. Their observations, often referred to as a classical example of molecular resonances, are presented in fig. 1. At bombarding energies just below the Coulomb barrier at 6.5 MeV c.m., three prominent resonances are observed. They appear on a rapidly rising, smooth background coming form compound nucleus formation in a region of excitation where many levels overlap strongly. The elastic channel exhibits these resonances also but with a much smaller peak to valley ratio (about five percent) [2], [56] . Fig. 2, showing these resonances in the 90° elastic scattering cross section, is from the later work of Spinka [56] .

The main feature which distinguishes these resonances from those appearing further on above the Coulomb barrier [3,][12] , is that they are strongly cross-correlated in the various exit channels. This corre-

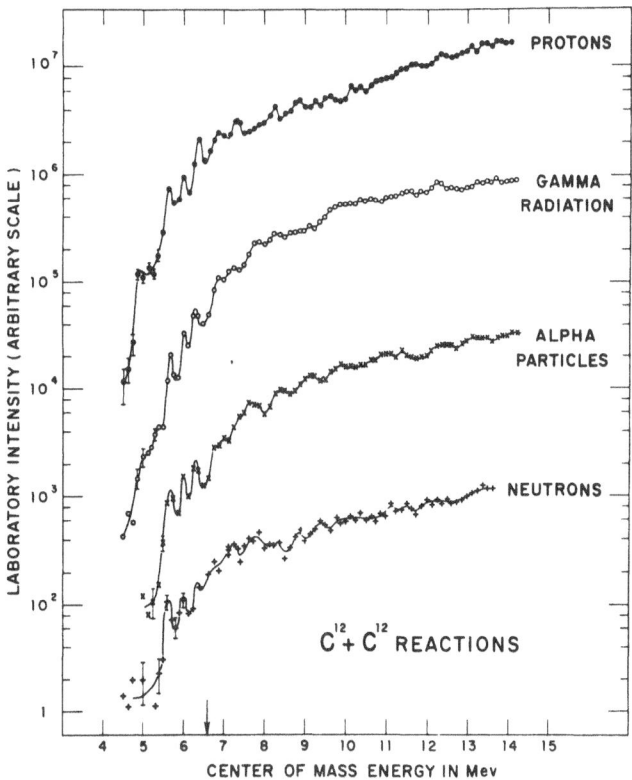

Fig. 1. Excitation functions for ^{12}C + ^{12}C induced reactions (ref.[1]). The various exit channel parti-
cles were observed at single angles. The arrow
indicates the Coulomb barrier, defined using a
radius parameter r_o=1.7 fm

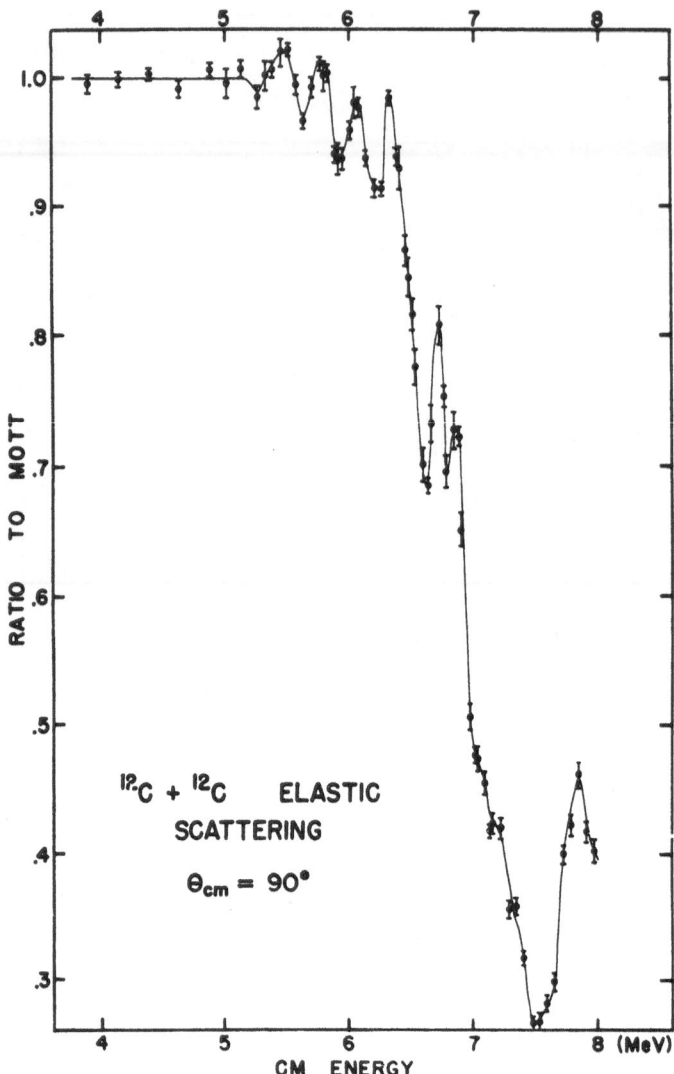

Fig. 2. ^{12}C + ^{12}C elastic scattering at 90° c.m. plotted
as the ratio to Mott scattering (ref. [56])

lation is one of the characteristics of an intermediate process in the reaction mechanism and allows such processes to be distinguished from the uncorrelated resonances or fluctuations arising from statistical compound nucleus formation [22].

Measuremements of the proton and alpha-particle yields from the $^{12}C + ^{12}C$ reaction have been extended down to an energy of 3 MeV c.m. by Patterson, Winkler and Zaidins [46] . By lowering their detection limits to less than a microbarn they obtained the results shown in fig. 3.

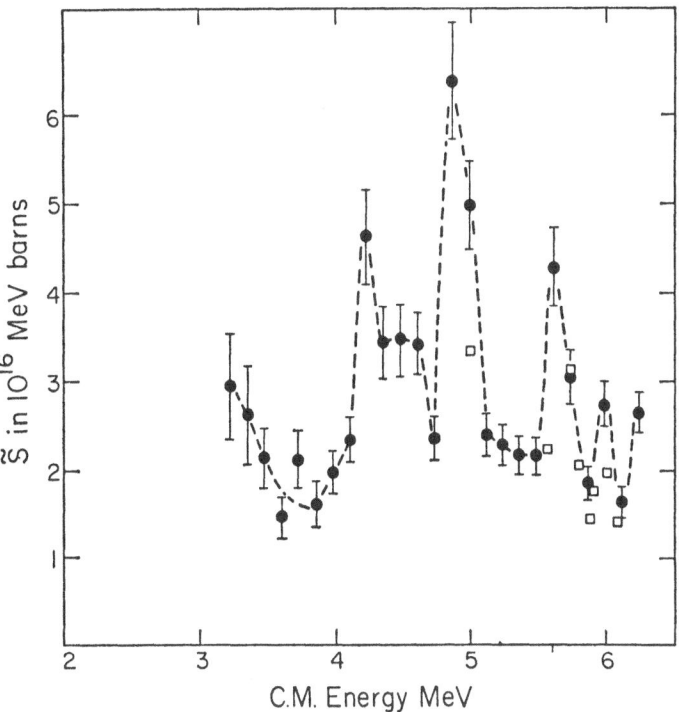

Fig. 3. The total reaction cross section for $^{12}C + ^{12}C$, extended to lower energies (ref. [46]). The three peaks at the highest energies are those also seen in fig. 1. The factor $\tilde{S}$ is defined by $\sigma_{tot} = \tilde{S} \; P/E$ where P is a penetration factor. σ_{tot} varies by seven orders of magnitude across the measured region

Here, the experimental cross section has been multiplied by the factor
E/P, where E is the center of mass energy and P is the Coulomb barrier
penetrabitility. The quantity Ŝ thus contains all the intrinsic nuclear
factors influencing the cross section. The presence of additional resonances at energies below those seen in figs. 1 and 2 is clearly indicated. The resonance widths are of the order of 100 to 200 keV.

Additional and important experimental information on the partial
widths of the resonances near 6 MeV has been obtained by Almqvist et
al. [56] and will be discussed later.

A qualitatively similar situation obtains for the $^{12}C + ^{16}O$ reaction at low energies. Fig. 4 gives the results of Pätterson et al.
[47] for the proton and alpha-particle exit cahnnels. Again, the angle-
integrated proton and alpha-particle yields are correlated. The widths,
not so clearly defined in this case, are about 400 keV. In fact, the
notable difference in the results in figs. 3 and 4 is that the resonances are narrower and hence more pronounced for $^{12}C + ^{12}C$ than for
$^{12}C + ^{16}O$.

The remaining system which has shown a sub-barrier resonance [4]
is $^{14}N + ^{14}N$. In this case only the gamma-ray yield, which indicates
a single resonance about 2 MeV below the Coulomb barrier, has been
measured. The results are shown in fig. 5.

The experimental results on the above three reactions take on a
special significance in the light of the large number of reactions in
which similar investigations have failed to reveal any notable resonant structure. These reactions are listed in table 1. The bombarding
energies used in all cases extended at least one MeV (c.m.) above and
below the Coulomb barrier. Typical results for a few of the reactions
listed in table 1 are presented in figs. 6-9.

The remarkable contrast found in the $^{12}C + ^{12}C$ and $^{16}O + ^{16}O$ reaction channels had already been noted by Almqvist, Bromley and Kuehner
[1]. Fig. 6 is taken from the work of Spinka which extends to lower
energies and which, by plotting the structure factor Ŝ on a linear
scale instead of the cross section on a logarithmic scale, displays
this contrast in a particularly striking manner. Fig. 7 is from the
early Chalk River [5] work and shows the rather structureless features
of the gamma-ray yields for a variety of systems. In searching for correlated resonant structure, it is important to sum over as many states
as possible in the residual nuclei in order to damp out any uncorrelated statistical fluctuations. In this respect, a measurement of the
gamma-ray yield is advantageous since gamma decay, to a large extent,
follows rather than precedes the initial particle decay of the compo-

und nucleus. Thus the gamma-ray yield automatically averages over all the particle yield except that leading to ground states of residual nuclei. The latter yield should be fairly small. The fact that the resonances seen in the particle yield measurements of Patterson et al. [47] on $^{12}C + ^{16}O$ are not immediately apparent in the gamma-ray yield shown in fig. 7 arises most likely from the lower statistical precision of the latter measurements. The early gammy-ray measurements were designed to detect the more prominent resonances as observed for $^{12}C + ^{12}C$.

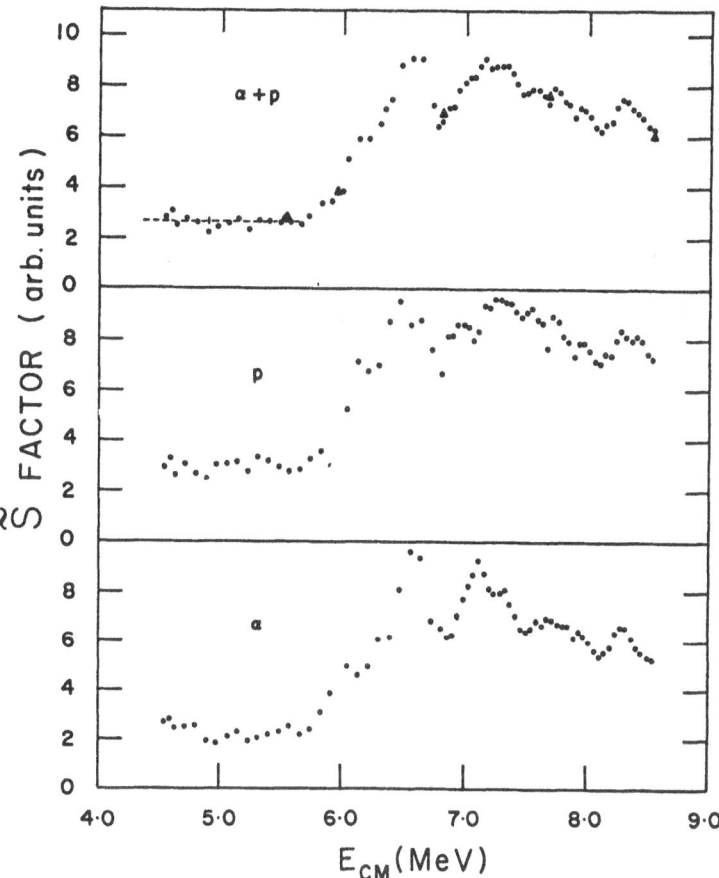

Fig. 4. The total reaction cross section for $^{12}C + ^{16}O$ (ref. [47]). The Coulomb barrier is at ~ 8.5 MeV

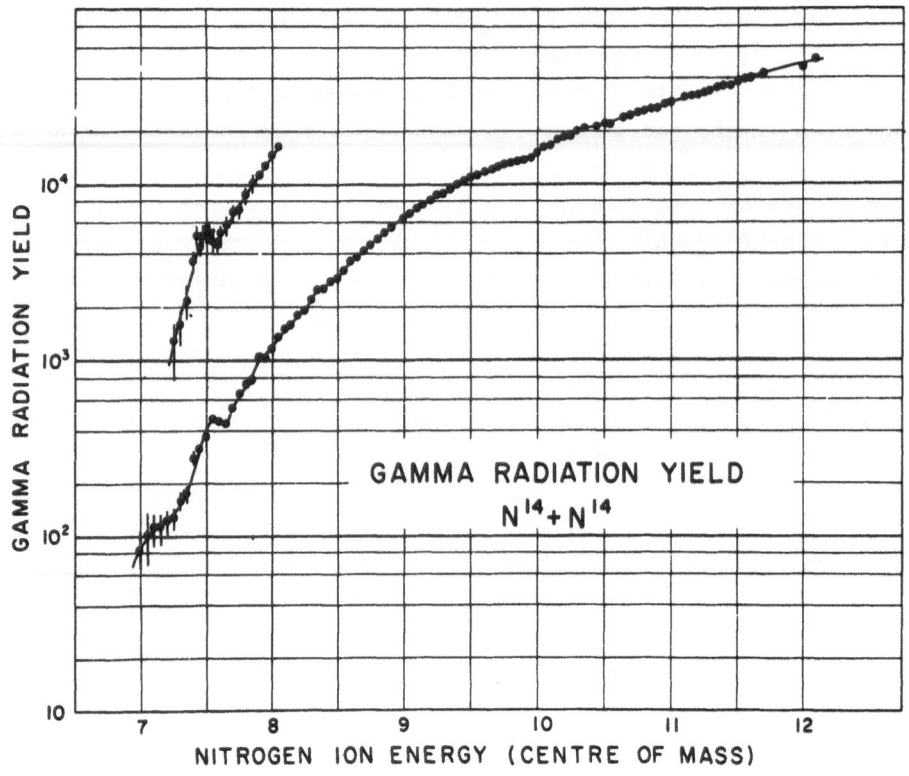

Fig. 5. Excitation function for gamma-ray yield in ^{14}N + ^{14}N (ref. [4]). The upper and lower curves corres- pond to an energy loss in the target of 50 and 100 keV, respectively

Table 1

Rections in which resonances similar to those in ^{12}C + ^{12}C have
not been observed at bombarding energies near the Coulomb barrier.

Compound System	Reaction beam, target	Exit Channels observed	Ref.
^{24}Mg	^{14}N + ^{10}B	α, ^{12}C	64
^{25}Mg	^{13}C + ^{12}C	p,α	64
		γ,α	32
^{26}Mg	^{13}C + ^{13}C	γ	32
^{26}Na	^{12}C + ^{14}N	γ	5
	^{14}N + ^{12}C	p,α	64
^{28}Si	^{12}C + ^{16}O	γ	5 *
^{31}P	^{12}C ^{19}F	γ	5
^{32}S	^{16}O + ^{16}O	γ,α	1
		γ,p,n,d,α	56
	^{12}C + ^{20}Ne	γ	5
^{36}Ar	^{16}O + ^{20}Ne	γ	5

* See text for comments on the lack of structure in this particular
measurement.

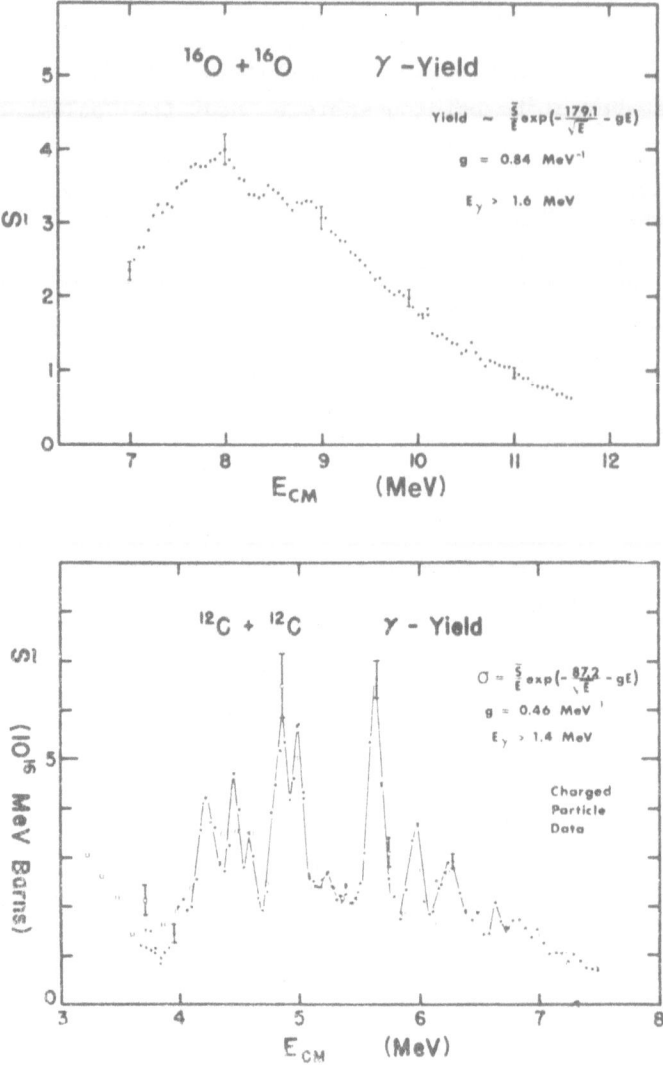

Fig. 6. Comparison of the gamma-ray yields for ^{16}O + ^{16}O
^{12}C + ^{12}C. The dependence of the cross section on
the penetration of the Coulomb barrier has been
removed. (ref. [56])

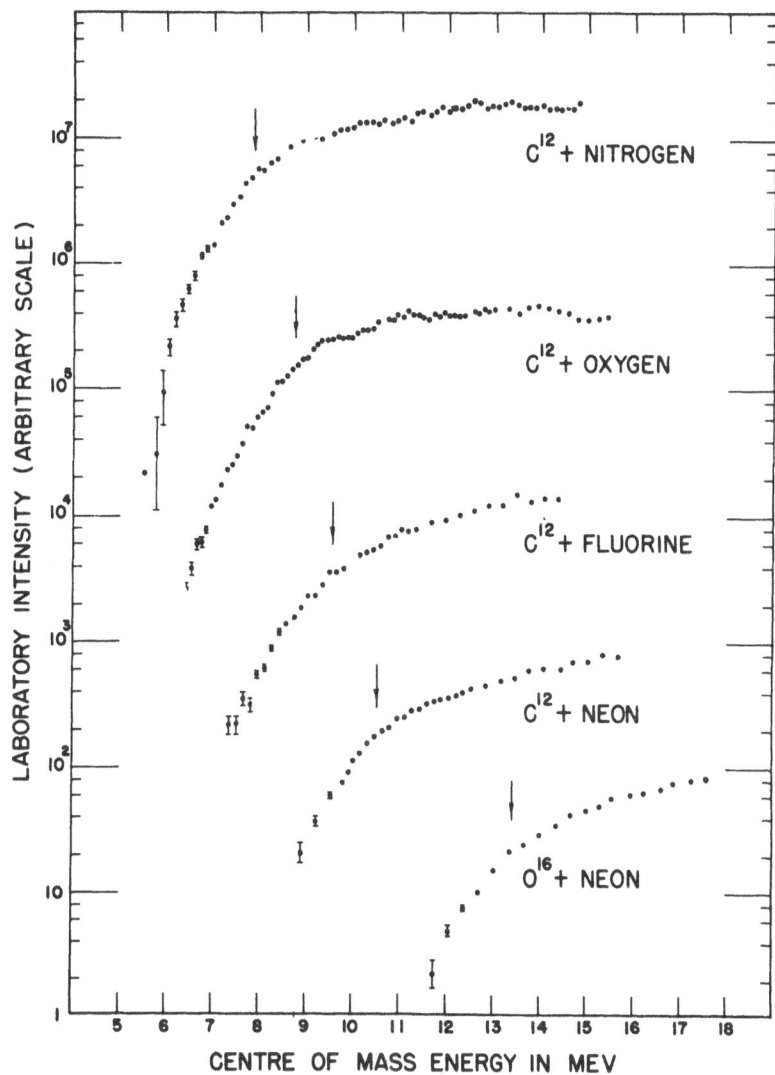

Fig. 7. Gamma-ray yields for a number of heavy ion reactions (ref. [5]). The arrow indicates the position of the Coulomb barrier

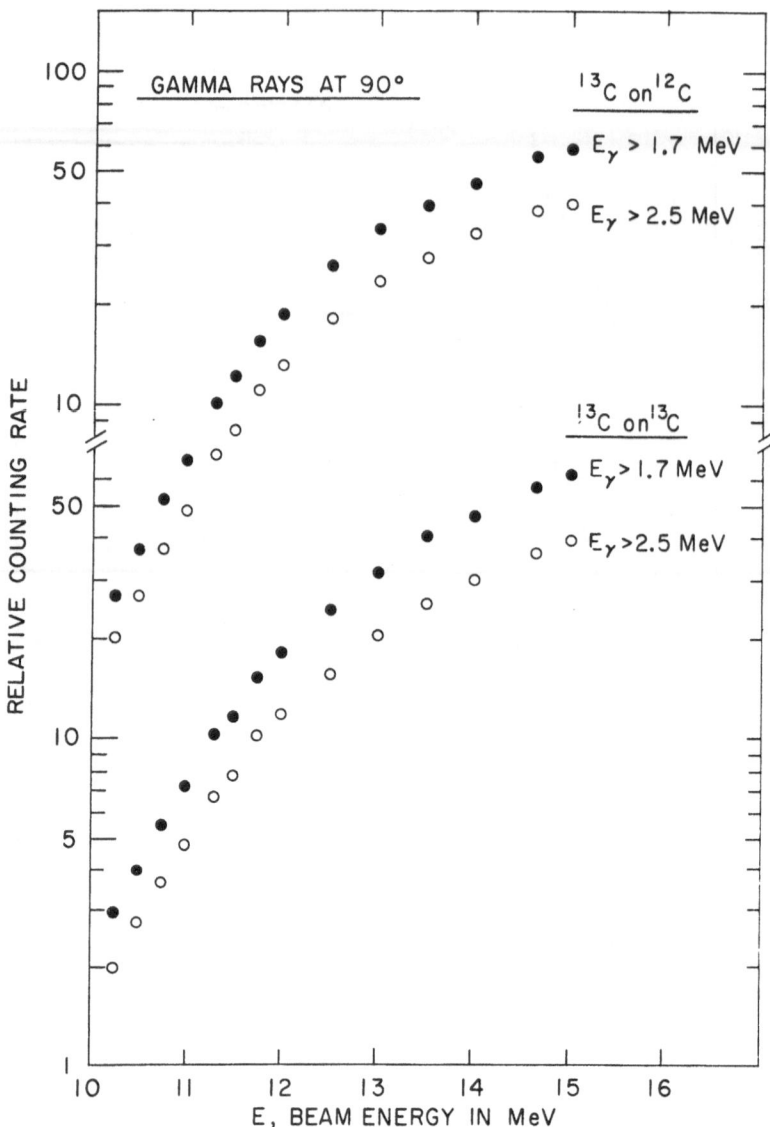

Fig. 8. Gamma-ray yields for $^{12}C + ^{13}C$ and $^{13}C + ^{13}C$
(ref. [32])

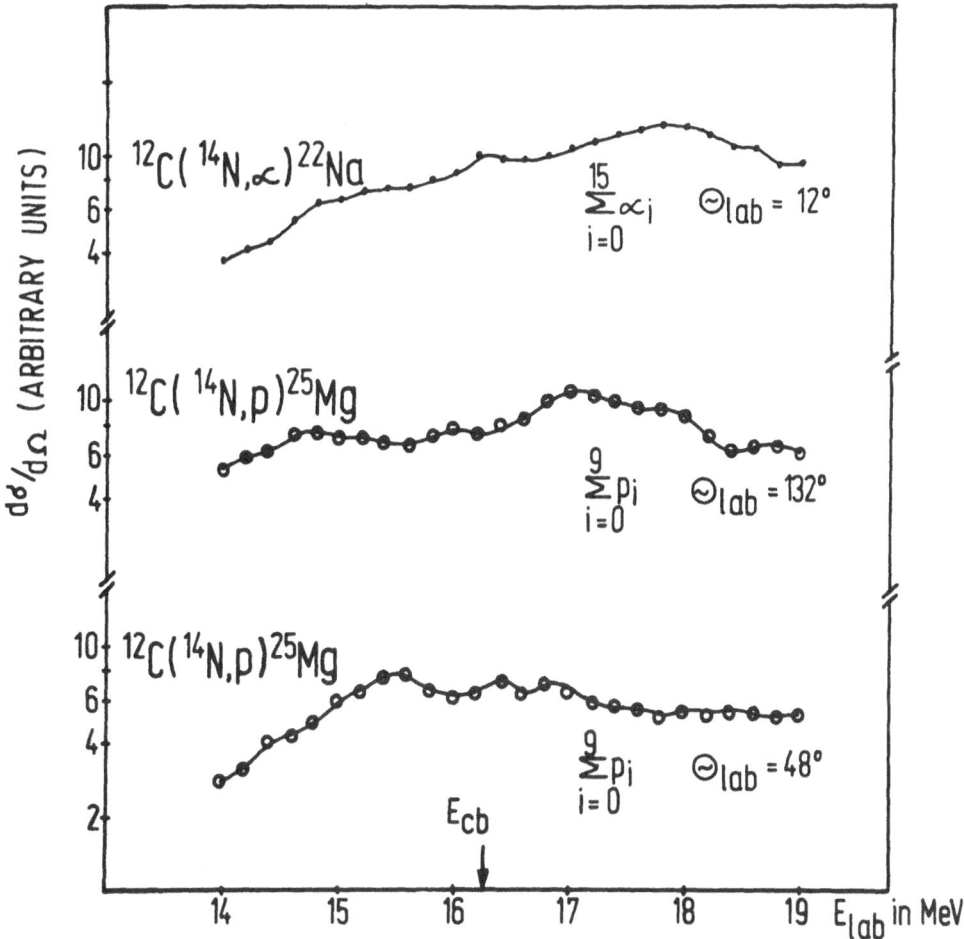

Fig. 9. Charged particle yields for $^{12}C + ^{14}N$ (ref. [64]).
The lack of any correlations in the structure indicates that this structure arises from statistical
fluctuations which would be strongly damped if a
sufficient number of states were included in the
summation

The gamma-ray yield for the $^{12}C + ^{13}C$ and $^{13}C + ^{13}C$ systems is presented in fig. 8, taken from the work of Halbert and Nagatani [32]. Voit and collaborators [64] at Erlangen have examined a number of reactions in a systematic search for correlated resonances. (and were the first to investigate the $^{12}C + ^{12}C$ system). Their results for the proton and alpha particle yield from the $^{12}C + ^{14}N$ reaction are presented in fig. 9. The slight structure present here is not correlated and would disappear if a sufficient number of final states were included in the summation.

The experimental situation may be summarized briefly as follows. The pronounced correlated resonances observed in the $^{12}C + ^{12}C$ system from the Coulomb barrier down to 3 MeV, are clear evidence for an intermediate process, i,e. they cannot arise from statistical fluctuations. Similar structure, although broader in width and less pronounced, is also clearly observed for $^{12}C + ^{16}O$. With the exception of the single bump in the case of $^{14}N + ^{14}N$ (fig. 5) none of the nine other reactions investigated to date, all with projectiles and targets of comparable masses, has shown any indication of a similar behavior. These are indeed remarkable features for which a physical interpretation must now be sought.

The Chalk River measurements on $^{12}C + ^{12}C$ have inspired a great amount of experimental and especially theoretical activity. This stems from the importance of this reaction both for nuclear structure and for astrophysics. New ideas on the nature of these resonances continue to appear. The first physical interpretation of the resonances centered on the formation of a quasi-bound, molecular state of the two ^{12}C ions, as described in the introduction. The additional experimental information, alluded to earlier, which has been largely responsible for the molecular interpretation of these resonances, comes from a careful analysis [2] of the strengths of the resonances in the various exit channels. These results are presented [13] in table 2. The very large reduced width for the decay of these long-lived ($\sim 10^{-20}$ sec) resonances into two ^{12}C ions suggests that the spatial localization of the nucleons in two distinct regions is maitained for the lifetime of the state. A stronger statement would be that the resonance consists of two ^{12}C ions in a single-particle type potential well. These suggestions [1] were developed further by Vogt and McManus [60] and by Davis [20], the former exploiting the deformability of the ^{12}C ions to justify a very large potential-well radius of 11 fm, and the latter noting that a standard optical model plus centrifugal potential can produce a resonant condition for the grazing partial wave. Of the two approaches,

Table 2

Characteristics of Quasimolecular Resonances in $^{12}C + {}^{12}C$ (ref. [2]).

Resonance Energy	J^π	Γ_{total} keV	Γ_c keV	$\Sigma\Gamma_\alpha$ keV	$\Sigma\Gamma_p$ keV	$\Sigma\Gamma_n$ %	γ_c^2 %	γ_α^2 %	γ_p^2 %	γ_n^2 %
5.6	2^+	130	20	78	22	4	14* (400)*	0.5	0.03	0.02
6.0	4^+	100	7.5	65	19	4	11* (1800)*	0.6	0.04	0.05

✶ Using $R\backsim 5F$ as in remaining γ^2 entries.

* Using $R\backsim 7F$.

that of Vogt and McManus enjoyed greater success because it explained
the lack of resonances in the $^{16}O + ^{16}O$ system (these nuclei do not
deform easily), and with the large radius could reproduce approximately
the close level springs and spins for the resonances in table 2.

Another of the new ideas to explain the Chalk River molecular resonances was proposed by Imanishi [34]. In this case, the deformation
of the ^{12}C nucleus is also important because it produces a strong
coupling of the ground state to the (therefore) rotational 2^+ state at
4.4 MeV. This 2^+ state is excited virtually in the collision of the
two nuclei and produces a sudden change in both the potential and the
available energy such that the nuclei become quasi-bound. This happens
only at certain bombarding energies for which an energy matching condition is satisfied, thus producing well-spaced resonances. Since Dr.
Scheid will be describing theoretical results based on a similar approach (but applied to scattering at much higher energies), I will not
go into further detail. Imanishi's calculation to date remains the
most quantitative treatment of the three Chalk River resonances, and
the agreement with experiment is excellent.

The emphasis in the above theoretical treatments [20], [34], [60]
has been naturally placed on finding mechanisms which will produce resonant structure in $^{12}C + ^{12}C$ and relatively little effort has been
devoted to an explanation of the lack of structure in the other systems,
many of which already had been studied experimentally at about the same time as $^{12}C + ^{12}C$. The more recent experiments [64], [32], especially
on $^{12}C + ^{13}C$ and $^{13}C + ^{13}C$, demonstrate that an explanation of the resonance phenomena cannot be based solely on the gross properties of
the heavy ions, e.g. the optical potential. Even the coupling to low
lying excited states is a feature which, although differing in detail
from nucleus to nucleus, should be common to many of the systems listed
in table 1, and which exhibit no structure. A difficulty with the above
models, therefore, is that they would apparently predict resonances
where none are observed.

Another crucial shortcoming of all of the above approaches is
that they are unable [39] to account for the subsequent discovery [46]
of additional resonances in $^{12}C + ^{12}C$ occuring at energies below 5.5
MeV c.m. As Michaud and Vogt [39] point out, this inability to reproduce the large number of closely spaced ($\sim$300 keV) resonances derives
from the use of a single-particle potential and the consequent assumption that the two ^{12}C nuclei retain their identity. They replace the
concept of a doorway state involving two ^{12}C nuclei with a mechanism
whereby one of the ^{12}C nuclei disassociates into three alpha particles.

Thus the doorway states now involve a ^{12}C nucleus and three alpha clusters. In other words, the molecular picture must be generalized from that of a simple diatomic molecule to include much more complex spatial configurations.

The proposal of Michaud and Vogt [39] is supported by a quantitative comparison to experiment in one case (the average ^{12}C + ^{12}C total reaction cross section at low energies) and by qualitative comparisons in other cases (anomalous ^{20}Ne + α branching ratios [3], [61] and the large number of molecular resonances below the Coulomb barrier [1], [46]). It is worth-while therefore to attempt a brief and very schematic description of their model. Their basic arguments are as follows: a) The 7.66 MeV state of ^{12}C is probably an intrinsic alpha-particle state. b) Since the average alpha-particle separation energy is about 7 MeV, one can expect some states at about 20 MeV in Mg^{24} to consist of a ^{12}C core and three loosely bound alpha particles. c) These latter states then correspond to the ∿100 keV wide resonances in the total cross section. They consider three equivalent Hamiltonians for ^{24}Mg, but each expanded using different subgroups of nucleons in the configuration basis - namely, twenty-four nucleons, six alpha particles and two ^{12}C nuclei. In practice, they argue, one can maintain twelve of the nucleons grouped together to form a ^{12}C core. Fig. 10 is a crude attempt to display the various interacitons encountered in the ^{12}C + ^{12}C reaction. Beginning with the compound nucleus ^{24}Mg, one has in fig. 10a the nucleon-nucleon interaciton (straight lines) between the 12 valence particles which defines the many states of this system. The 12 valence nucleons are then grouped in fig. 10b into three alpha-particles, with the α - C interaction (and α - α interacitons) defining the intermediate states. The wavey line denotes a nucleon-nucleon interaction* between nucleons of different alpha-particles. This can be thought of as a residual interaction which couples the intermediate state to the compound nucleus. The three alpha-particles are then grouped into a ^{12}C nucleus, (fig. 10c) where the residual interaction leading back to the intermediate system is now the α - C interaction. Fig. 10c corresponds to the single-particle state of two ^{12}C ions interacting through the optical potential. The interactions denoted by the wavey lines correspond to the imaginary parts of the interaction potentials and hence determine the lifetimes of the respective configurations. It is apparent from this figure that, if the intermediate state were

---

* More precisely, that part of the nucleon-nucleon interaction which is not already contained in α-α interaction

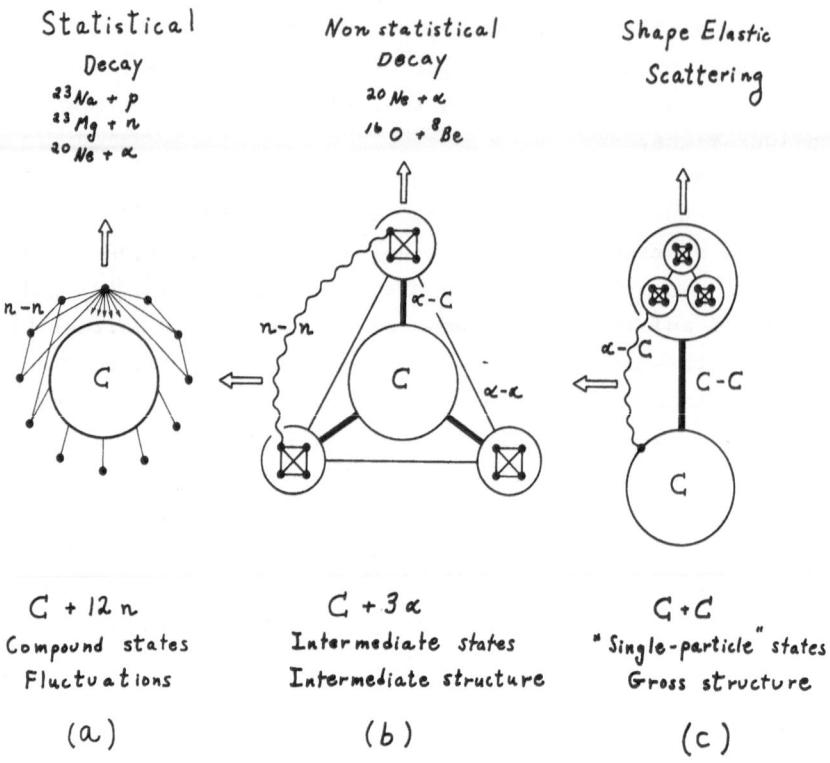

Fig. 10. Schematic representation of the alpha-particle
intermediate structure model. (ref. [39])

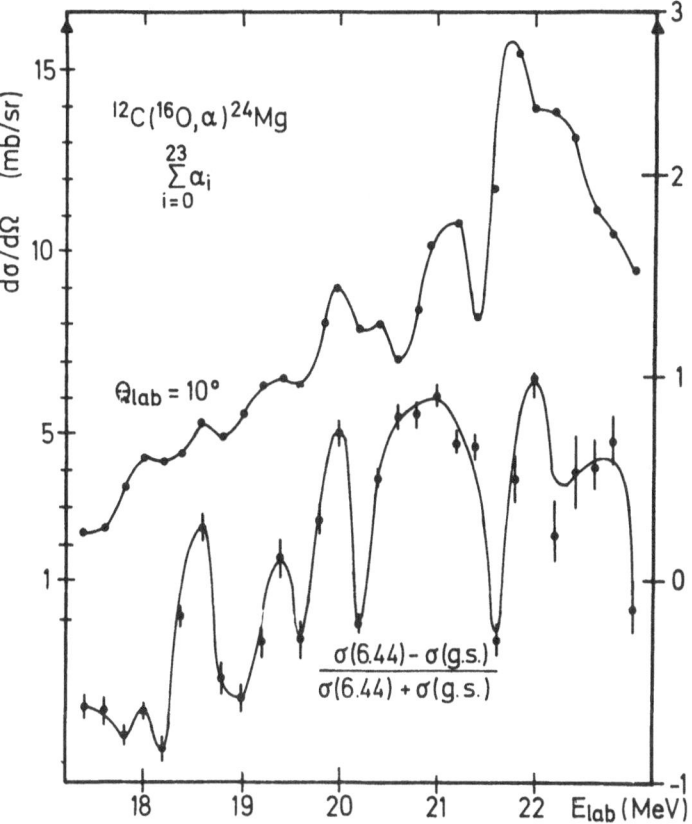

Fig. 11. Alpha-particle yields from the $^{12}C(^{16}O,\alpha)^{24}Mg$
reaction (ref. [65]). The yield to the 0^+ 6.44
MeV state is enhanced over that to the ground
state at energies for which the summed alpha-
particle and proton yields show resonances

to decay directly without forming the compound nucleus, then the ^{20}Ne + α channel would be favored over the ^{23}Na + p channel (i.e. favored relative to the statistical model prediction). This is observed experimentally [3], [61]. Furthermore, certain states in ^{20}Ne, namely those with a ^{12}C + 2α or ^{16}O + α parentage, would be favored over other states in ^{20}Ne. Direct reactions involving the transfer of one and two α-particles are also an obvious possibility.

These ideas are easily extrapolated to the ^{12}C + ^{16}O reaction where very recent data [65] can provide a test. In the absence of intermediate structure, the statistical model [12], [61], [62] predicts that the cross sections for any two excited states of the same spin and parity and with comparable excitation energies should fluctuate in a random way, though, on the average their cross sections should be comparable. At Erlangen, Voit et al. [65] have recently measured α-particle and proton yields to low-lying states in ^{24}Mg and ^{27}Aℓ in the ^{16}O + ^{12}C reaction at energies in the vicinity of the Coulomb barrier. They observe a number of correlated resonances and, in fig. 11, they compare the α-particle yields to the ground and first excited 0^+ (6.44 MeV) states of ^{24}Mg. Their results indicate that the 6.44 MeV state is preferentially populated whenever there is an increase in the overall α-particle yield. Furthermore, the 6.44 MeV state may be interpreted [6] as having a ^{20}Ne + α parentage, and the same interpretation [6] would predict states in the compound system ^{28}Si with ^{20}Ne + 2α parentage at about the right energy for these resonances. Thus, the recent work of Voit et al. [65] provides another qualitative confirmation of the existence of α-particle intermediate structure. They are presently conducting similar experiments on the ^{12}C(^{12}C,α)^{20}Ne reaction [66].

Thus, the intermediate structure mechanism proposed by Michaud and Vogt [39] qualitatively accounts for the resonances observed in ^{12}C + ^{12}C and ^{12}C + ^{16}O. However, quantitative predictions of level spacings and widths (especially the partial widths for ^{12}C + ^{12}C) are needed before any conclusive judgement on their model can be made. As Michaud and Vogt note, there may also be other explanations [45].

It is interesting to note the qualitative similarity between the α-particle model of Michaud and Vogt [39] and the pre-equilibrium theories of Griffin [31] and Blann [9] which arises from their common founding in the concept of intermediate structure.

As is the case with earlier authors [20], [34], [60], the important question of the <u>lack</u> of resonances in other systems also receives little attention in Michaud's and Vogt's articles to date. Should not $^{16}O + ^{16}O$ and $^{16}O + ^{20}Ne$ be excellent candidates for α-particle intermediate structure? Why should the addition of a neutron to one of the ^{12}C cores apparently destroy the effect of the α-particle intermediate sturcture and eliminate the associated resonances?

Lacking theoretical calculations of the strengths and widths of the intermediate resonances for any of these systems, the next approach in trying to answer these questions is to examine the other properties of the individual cases and note any correspondences to the observation or non-observation of structure. One such property which is easily considered is the number of open channels. This is done in figs. 12 and 13. Here it is seen that, at excitation energies in the compound nucleus corresponding to the Coulomb barrier, many more channels are available for the decay of ^{32}S than for ^{24}Mg. The number of channels open in the case of ^{28}Si, where broader and less pronounced resonances were observed, lies in between ^{24}Mg and ^{32}S. Thus, there is a correspondence between the number of open channels and the observation of intermediate structure. This correspondence extends to the other systems listed in table 1. The more relevant quantity for comparison may be the density of levels, ρ, in the compound nucleus which, in the statistical model, is related to the number of open channels by

$$\rho = \frac{1}{2 \pi \Gamma} \sum_{c,\ell} T_\ell(c)$$

where Γ is the average width of the levels and c and ℓ denote summations over the various exit channels and orbital angular momenta. T_ℓ is a transmission coefficient, usually taken from the optical model. Since Γ varies slowly with excitation energy and from nucleus to nucleus (it is typically 80-150 keV), and since only low angular momenta are involved at Coulomb barrier energies, figs. 12 and 13 can provide a qualitative indication of the density of levels in the compound nucleus. Because of the Q-values, non-identical particles, and higher density of levels in the odd-A nucleus Ne^{21}, evaluation of the above expression [26] yields a larger density of levels for ^{25}Mg. Similar conclusions can also be reached by using a Fermi-gas model to estimate the level density derectly. Rough estimates of the relative level densities are

200

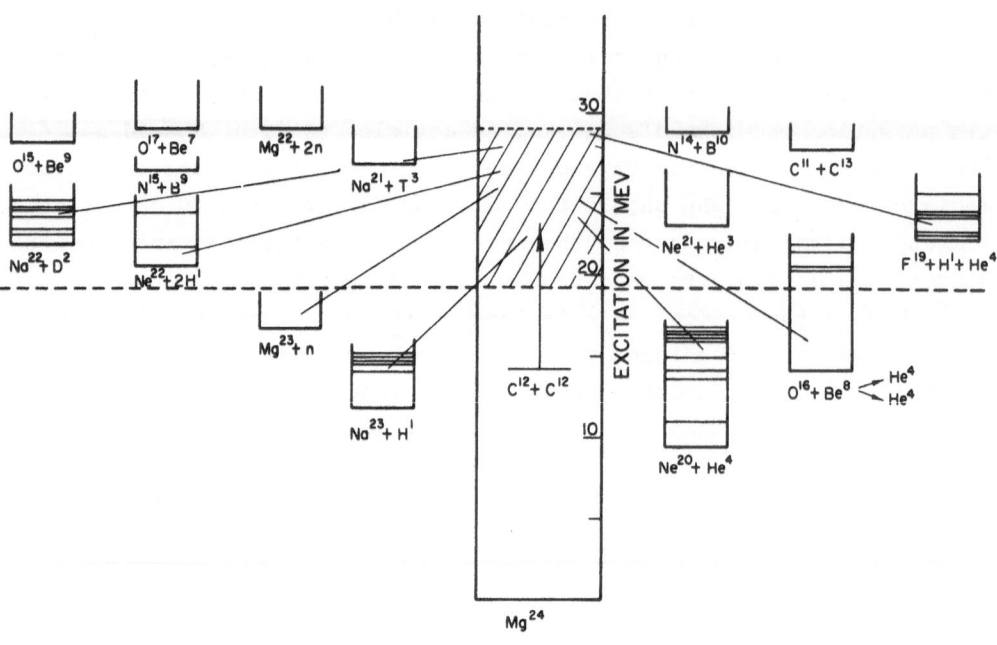

Fig. 12

Reaction channels for ^{12}C + ^{12}C. The lower dotted line indicates the excitation energy in ^{24}Mg when the ^{12}C + ^{12}C bombarding energy is at the Coulomb barrier. (ref. [5])

201

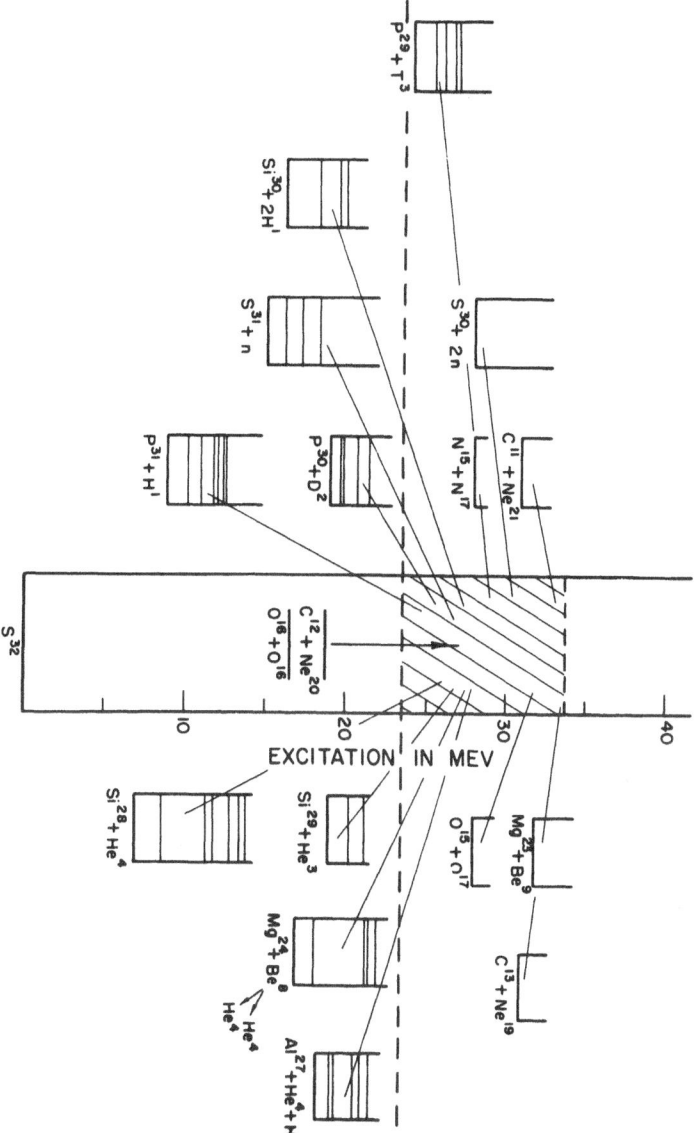

Fig. 13. Reaction channels for $^{16}O_9 + ^{16}O_9$ (ref. [5])

given in table 3. Thus, the empirical observation is, when there are many open channels and the level density in the compound nucleus is high (relative to that for $^{12}C + ^{12}C$), the intermediate structure is either less pronounced or absent.

Table 3

Estimates of the relative level density in the compound nucleus when the bombarding energy is at the Coulomb barrier.

A_1	A_2	A	E_x	ρ/ρ (A=24)
^{12}C	^{12}C	^{24}Mg	20.6	1
^{12}C	^{13}C	^{25}Mg	23.6	8
^{13}C	^{13}C	^{26}Mg	28.9	36
^{12}C	^{16}O	^{28}Si	25.3	20
^{14}N	^{14}N	^{28}Si	35.9	700
^{16}O	^{16}O	^{32}S	27.3	150

The values are obtained from the simple Bethe formula*

$$\rho(U) = \frac{\sqrt{\pi} \exp 2\sqrt{aU}}{12 \, a^{1/4} \, U^{5/4}} \quad \text{where } U = E_x - \Delta \quad \text{and} \quad \Delta = 3.6 \text{ MeV for even-even}$$

nuclei. The value of a is A/7 MeV^{-1} and a radius R = 1.7 $(A_1^{1/3})$ is used to obtain the Coulomb barrier. For $^{12}C + ^{12}C$, $\rho \sim 10^4$ levels/MeV. The estimates above are very approximate; the effects of symmetry in the entrance channel and angular momentum are neglected.

---

* A. Gilbert and A.G. Cameron, Can. J.P. __43__, 1446 (1965)

It was noted earlier that the width of the intermediate structure depends on the extent to which intermediate states are coupled to other possible modes of decay. If this coupling and hence the width of the intermediate structure were to vary as the density of levels in the compound nucleus or, equivalently, as the number of open channels, then this would provide a physical explanation of the above empirically observed correspondence*. In other words, the intermediate states would still be present in ^{12}C + ^{13}C, ^{16}O + ^{16}O, ^{16}O + ^{20}Ne, but so broadened that they are not observed. Another alternative possibility is that the density of the intrinsic intermediate states might vary rapidly from one system to another. Thus in the case of ^{16}O + ^{16}O, for example, intermediate structure might not be observed because the individual intermediate states overlap. This possibility seems unlikely, however, when one considers the spacings of the α-cluster or quartet states calculated by Arima et al. [6]. It is clear that these are interesting questions deserving a thorough theoretical investigation.

There remains, however, one experimental observation which is in striking exception to the foregoing ideas on α-particle intermediate structure and its observed correspondence with the number of open channels. This exception is the single resonance observed [4] in the γ-ray yield for ^{14}N + ^{14}N, fig. 5. Although one might speculate on the possibility of a strong ^{12}C + d parentage in ^{14}N producing an intermediate mechanism whereby two deuterons form an α-particle, the large Q-values for the reaction channels imply that any structure should be strongly damped (see table 3). The observed resonance thus indicates a particularly strong selection rule which decreases the spreading width of the intermediate state. One may note that, in contrast to ^{12}C and ^{16}O, ^{14}N has a ground state with spin 1 and a low lying T = 1 state. Clearly, this is a most interesting case for further experimantal study, and investigations of the individual exit channels are now being prepared.

The subject of intermediate structure near the Coulomb barrier is thus one in which there is still much to be discovered. It is also an area in which, despite remarkable and innovative theoretical progress, much of what already has been discovered still remains to be fully understood.

---

* It also follows from the statistical model that a similar correspondence between the intermediate structure and the intensity of compound elastic scattering will exist, since the latter also depends on the number of open channels.

3. Intermediate Structure above the Coulomb Barrier

3.1. Introduction

There are several features which distinguish reactions initiated by heavy ions at energies well above the Coulomb barrier from the reactions discussed in the preceeding section. 1) The level density in the compound nucleus for a given total angular momentum is much larger at higher bombarding energies. 2) Projectiles with small impact parameters, i.e. the lower partial waves, penetrate into the nuclear surface of the target. Thus, the collisions for small impact parameters are violent rather than "gentle" as before, and correlations or cluster structure in the colliding ions will be effectively destroyed. 3) The grazing partial waves, where such correlations may be preserved, have large angular momentum. Keeping these differences in mind, it will not be surprising that the experiments described in the following exhibit phenomena quite different from those we have just discussed. I will give a brief review in section 3.2. of the experimental results and interpretation for elastic scattering of selected systems and, in section 3.3., I will concentrate on what has been observed recently in the reaction channels.

3.2. Elastic Scattering

The elastic scattering of $^{12}C + ^{12}C$, $^{12}C + ^{16}O$, and $^{16}O + ^{16}O$ has been studied up to center of mass energies of $\sim$ 16 MeV by Bromley, Kuehner and Almqvist [14]; their by now long familiar results are shown in fig. 14. The main feature to be noted here is the remarkable decrease in the amount of the resonant structure* above the Coulomb barrier as the mass of the compound system increases. The newer tandem Van de Graff accelerators enabled the continuation of these measurements to higher bombarding energies - $^{12}C + ^{12}C$ (refs. [13], [48]) and $^{16}O + ^{16}O$ (refs. [53], [36]) at Yale, and $^{12}C + ^{16}O$ at Argonne (ref. [54]). Figs. 15 (logarithmic scale) and 16 (linear scale) present the data for carbon-carbon and oxygen-oxygen scattering, respectively. The strong resonant structure also appearing at lower energies is readily seen in fig. 15. A quite new phenomenon emerges at higher energies, however, in the form of gross structure with a width of several MeV. This gross

---

* In order to avoid confusion in the following discussion I will refer to the rapid variations observed in the experimental cross sections as resonant structure rather than "intermediate structure", as is often done in the literature. In this talk and in this conference in general, "intermediate structure" implies a statement about the physical origin of the resonant structure, e.g. a doorway state.

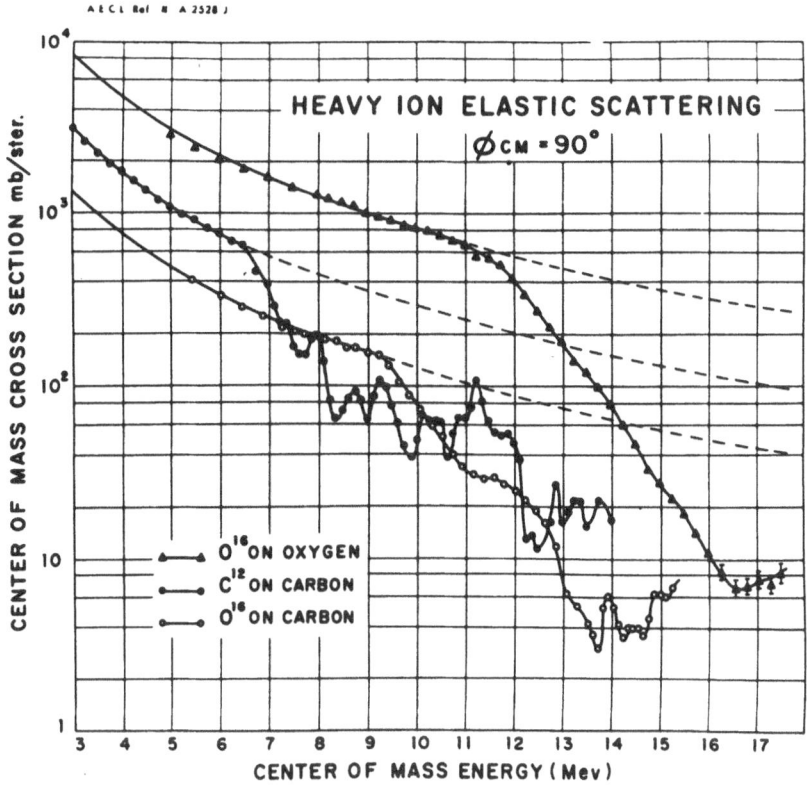

Fig. 14. Elastic scattering of ^{12}C + ^{12}C, ^{12}C + ^{16}O, and ^{16}O + ^{16}O (ref. [14])

206

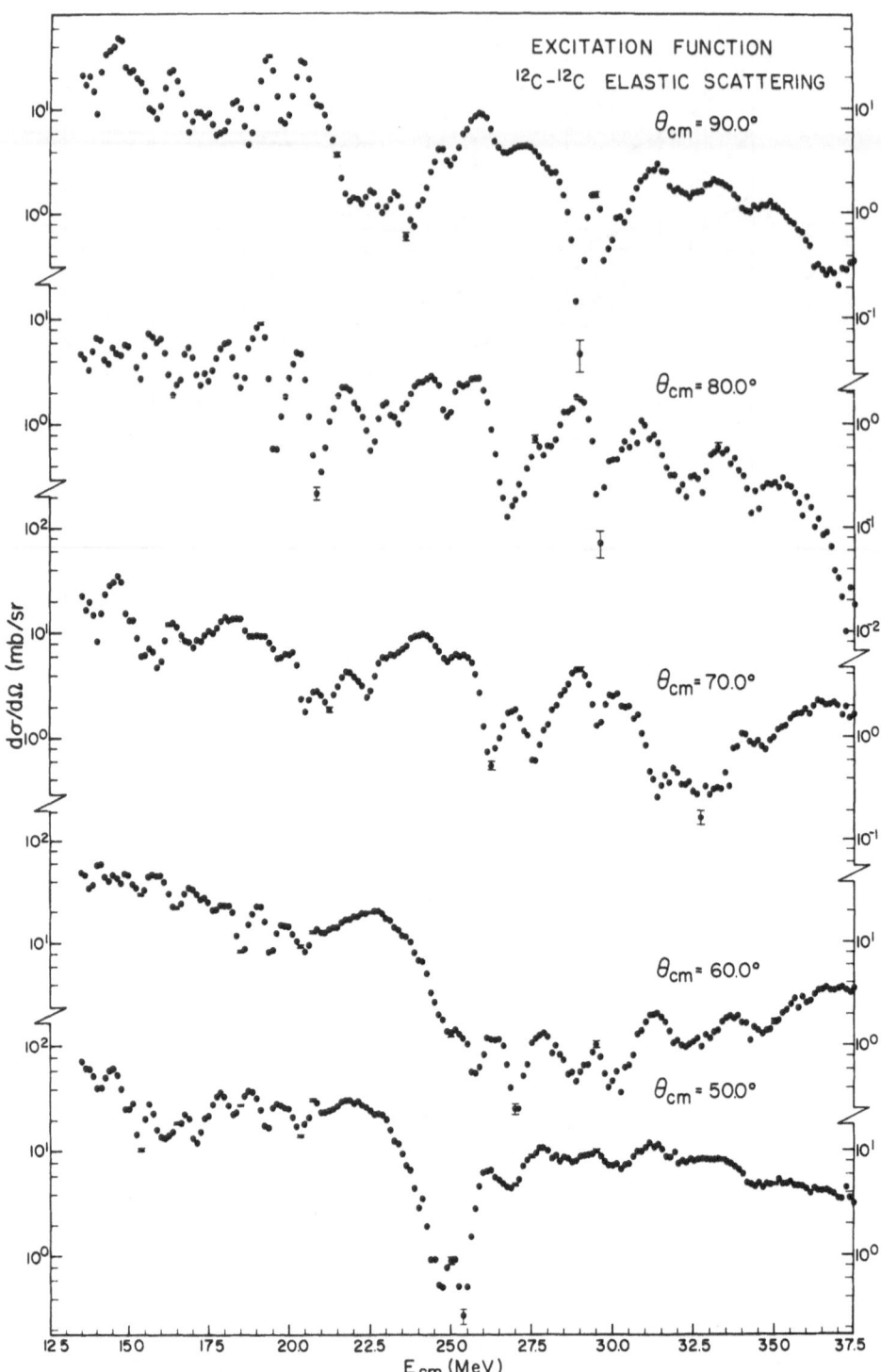

Fig. 15. Elastic scattering of ^{12}C + ^{12}C extended to higher
energies. (ref. [48]). Note the logarithmic scale

structure is much more apparent in fig. 16 for $^{16}O + {}^{16}O$ because the
relative intensity of the narrower resonant structure is much less in
this case. The $^{12}C + {}^{16}O$ scattering [54] exhibits a behavior lying in
between that shown in figs. 15 and 16.

The gross structure, observed in many scattering systems, is a
potential-scattering phenomenon for which the real potential is quite
shallow (∿17 MeV) and the absorption of the grazing partial waves is
generally weak. Within the optical model, each of the large peaks in
the 90° excitation function in fig. 16 can be associated with a single
partial wave [27]. An interesting comparison of the optical model to
the $^{16}O + {}^{16}O$ data has been made by Gobbi [27] and is shown in fig. 17.
An energy-angle surface is used to produce a relief map of the gross
structure; the agreement with the data is excellent in this case and,
for $^{12}C + {}^{12}C$, almost as good [27]. An extensive discussion of gross
structure and other related topics on heavy ion elastic scattering can
be found in the Proceedings of the Symposium on Heavy Ion Scattering
held at Argonne National Laboratory, March 1971.

A visual inspection of the data suggests [36], [48], [53] widths
of ∿0.6 - 1.2 MeV and 0.2 - 0.3 MeV for the resonant structure in
$^{12}C + {}^{12}C$ and $^{16}O + {}^{16}O$, respectively. Measurements with very thin targets and small steps over limited energy regions indicate a yet finer
structure of ≲100 keV in $^{16}O + {}^{16}O$ scattering [36], [52] and a lack of
finer structure for $^{12}C + {}^{12}C$ scattering [48]. Shaw et al. [52] measured the 90° excitation function for $^{16}O + {}^{16}O$ in fine steps over the
energy region 17.5-19.5 MeV c.m. An auto-correlation analysis yielded
a coherence width of 80 keV but gave no evidence [59] for any broader
structure of 200 - 300 keV such as reported in ref. [36]. This latter
negative result is perhaps not so surprising when one notes the energy
region over which the fluctuation analysis was performed. The region
17.5 to 19.5 MeV c.m. covers the minimum between the gross structure
peaks, and the ∿200 keV wide peaks appear predominently over the maxima
of the gross structure. However, an auto-correlation analysis by Maher
et al. [36] of excitation functions for the region 19 - 26 MeV c.m.
(including two gross structure peaks) yielded a coherence width of 50
- 100 keV, in agreement with ref. [52], but no conclusion on the existence of broader structure was drawn from this analysis. It is difficult in this particular case to establish or reject by an auto-correlation analysis the existence of intermediate widths in the excitation
function because of the difficulties in removing the effects of the
gross structure. The possibility therefore remains that the three
peaks superimposed on the gross structure in the region between 19 and

and 22 MeV c.m. at 80° and 90°, for example, are statistical fluctuations arising from overlapping levels in the compound nucleus with average widths of ~80 keV.

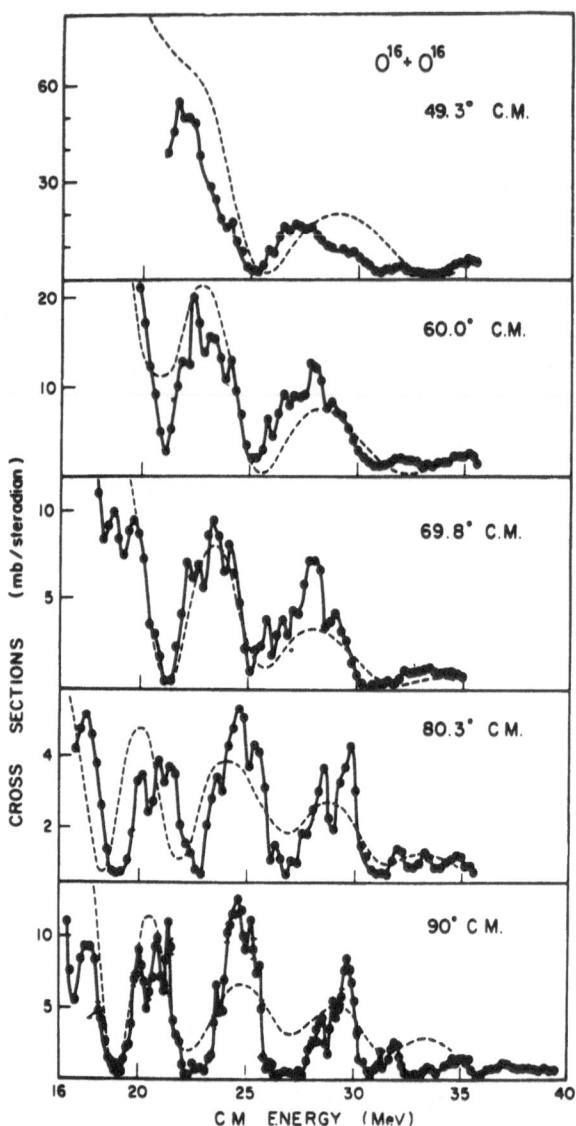

Fig. 16. Elastic scattering of ¹⁶O + ¹⁶O, extended to higher
energies (refs. [53], [36]). The dotted line is an
early optical model calculation. Note the linear scale

209

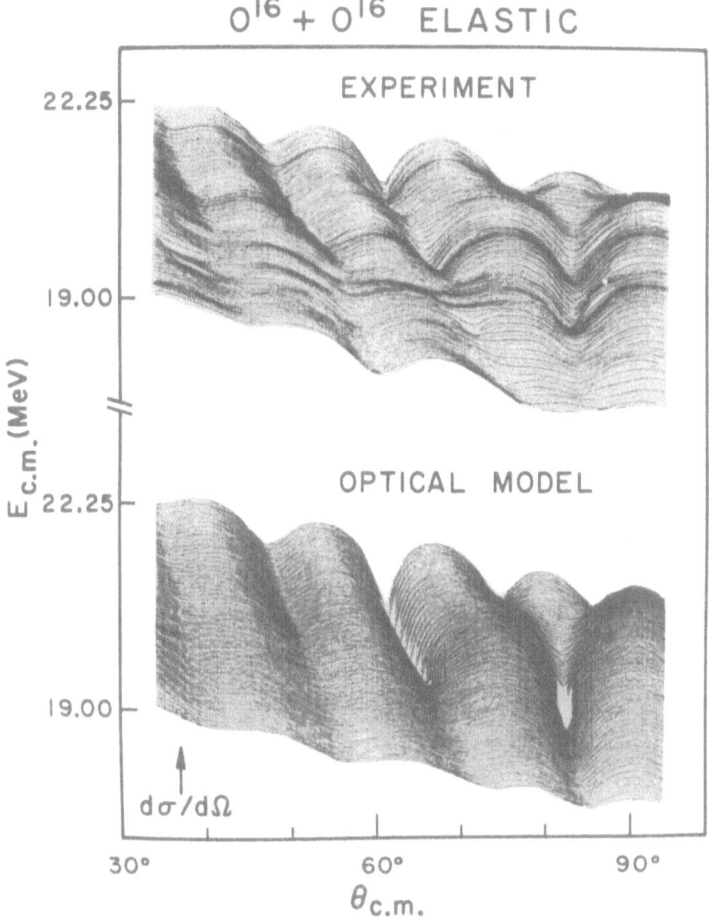

Fig. 17. Comparison of optical model predictions and the
gross structure observed in $^{16}O + ^{16}O$ scattering
(ref. [27]). The data have been averaged and inter-
polated to yield a smoother surface free of resonant sturcture. The overall agreement is quite
striking in this presentation

there are two schools of thought concerning the interpretation of the resonant structure in figs. 15 and 16. One attributes it to compound elestic scattering [35], [59], the other to intermediate mechanisms such as virtual resonances produced by coupling to excited states [34], [51]. The proponents of each of these points of vew are careful to note, however, that both effects may contribute and that the truth may lie somewhere in between these two extremes. Since these two approaches are fundamentally quite different it is appropriate here to briefly outline their relative merits.

It is certainly true that the amount of resonant structure at energies well above the Coulomb barrier corresponds closely with the number of channels open for the angular momenta determined by the grazing partial waves. More quantitatively, the compound elastic contribution to the $^{16}O + ^{16}O$ elastic cross section at 90° ($\sim 10\%$) agrees with Hauser-Feshbach estimates of this quantity [52], [10]. Preliminary calculations for $^{16}O + ^{12}C$ (ref. [10]) and $^{12}C + ^{12}C$ (refs. [26], [10]) reproduce the observed trends in the resonant structure. No cross-correlation for $^{16}O + ^{16}O$ elastic scattering at 90° and the alpha particle yield to the low lying states of ^{28}Si has been observed over the energy range 17.5 - 25 MeV c.m. [52].

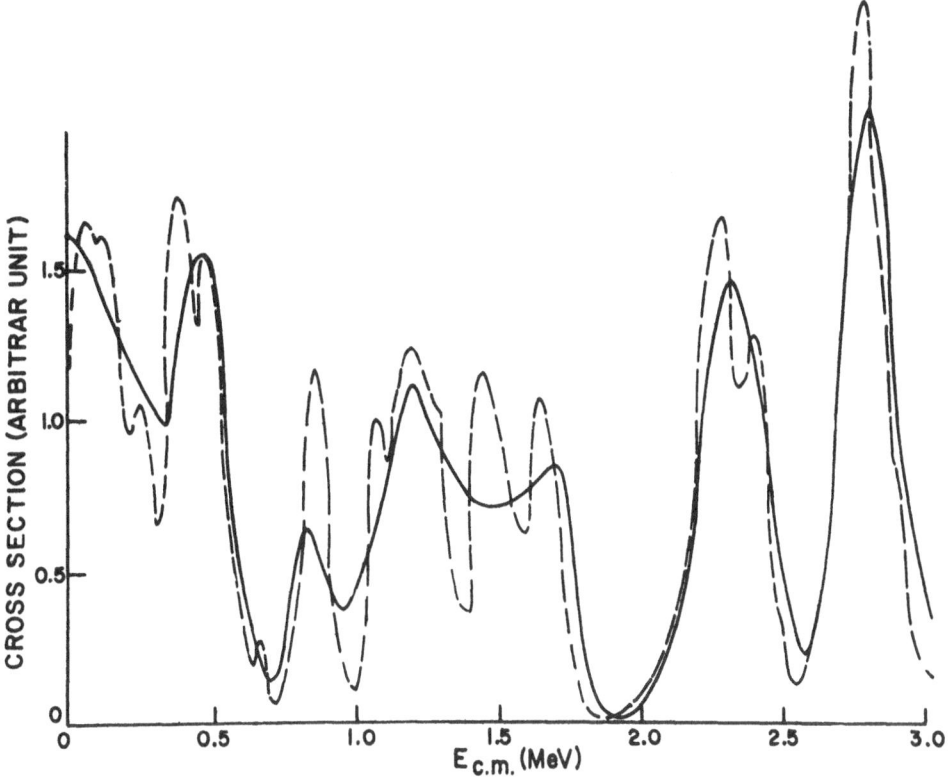

Fig. 18. Fluctuation cross sections calculated in a statistical model using a random number generator (ref.
[35]). No direct amplitude is included. The full
curve is obtained by averaging with a 250 keV
interval

Fig. 18 shows fluctuating cross sections from a statistical calculation by Low and Tamura [35] which evaluates directly the Breit-Wigner
expression for the resonance amplitude. The partial widths for the individual levels are determeined using a random number generator and a
Gaussian distribution with Γ = 100 keV. The total width for all levels
was held constant (100 keV) as was the spacing D between levels (1 keV).
The value of D is not critical as long as Γ/D >> 1. It is well known
[22] that if cross sections arise from strongly overlapping levels
whose partial widths are distributed randomly, then an excitation function will occasionally show structures resembling individual peaks with
widths two or three times the average width of the overlapping levels.
This is borne out in fig. 18 (dashed curve), as is the fact that more
such structures will appear if the data are averaged over an appropriate interval (full curve). Of interest is the probability of actually
observing, for example, three such "peaks" with widths greater than
∿250 keV in a three MeV interval. The evaluation of this probability
would be analogous to performing a large number of calculations such
as those shown in fig. 18, and counting how often three such "peaks"
appeared.

Thus, the evidence supporting an interpretation in terms of compound elastic scattering is still rather fragmentary considering the
few cases [52] in the literature where detailed comparisons have been
made. However, preliminary calculations on $^{12}C + {}^{12}C$ and $^{12}C + {}^{16}O$
elastic scattering show promise of agreement there and, as we shall note later, the statistical model does rather well in predicting the properties of the reaction channels.

Extensive calculations of real and imaginary potentials for heavy
ion scattering have been performed by Scheid, Greiner and collaborators
[51] using both a semi-microscopic approach based on the two-center
shell model and molecular-type potentials derived in the sudden approximation. When coupling to inelastic channels is included, a double-
resonance effect is observed, producing a structure with a width comparable to that observed experimentally. Fig. 19 presents their results
for $^{16}O + {}^{16}O$ scattering at 90° and indicates good agreement with experiment. Calculations have recently been completed for the $^{12}C + {}^{12}C$
system (Fink et al.) [51] including coupling to single and mutual excitations of the 4.4 MeV state. As can be seen in fig. 20, structure
similar to that observed in the data is present up to about 18 MeV.
The inclusion of coupling to higher excited states would produce similar structure at larger energies. An important feature of these coupled
channel calculations [51] is the simultaneous prediction of cross sec-

tions for inelastic scattering (fig. 19). A comparison of the results of Fink et al. for $^{12}C + ^{12}C$ inelastic scattering with measurements by Gobbi et al. [28] is found for the 2^+ state, but not for the higher excited states.

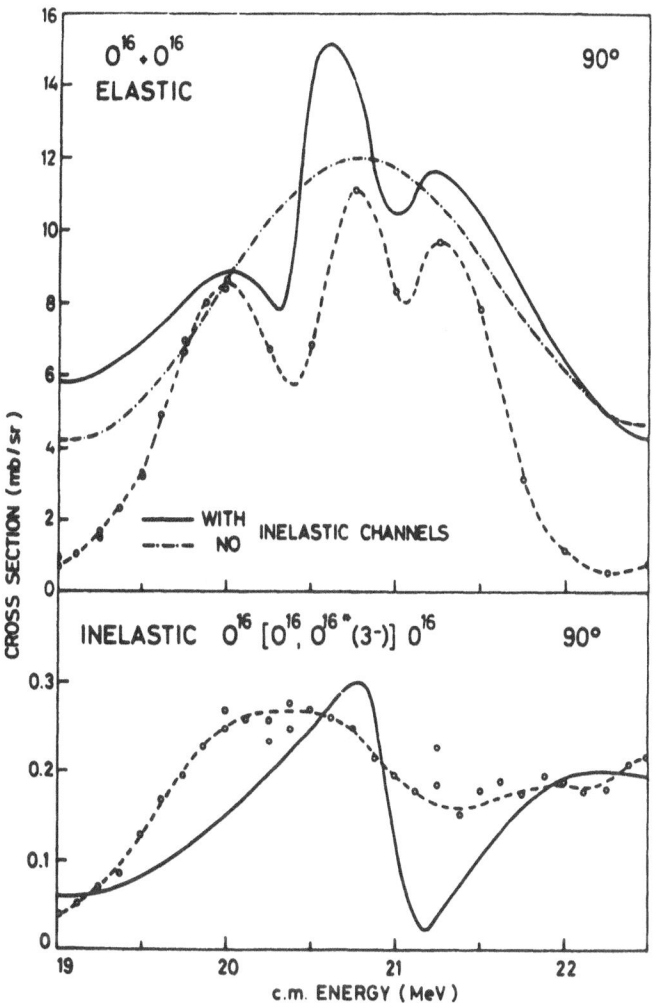

Fig. 19. Coupled channel predictions for $^{16}O + ^{16}O$ elastic
and inelastic scattering. (ref. [51]). The data
(circles) are from ref. [36]. The 6.05 MeV 0^+ and
6.13 MeV 3^- excited states were not resolved

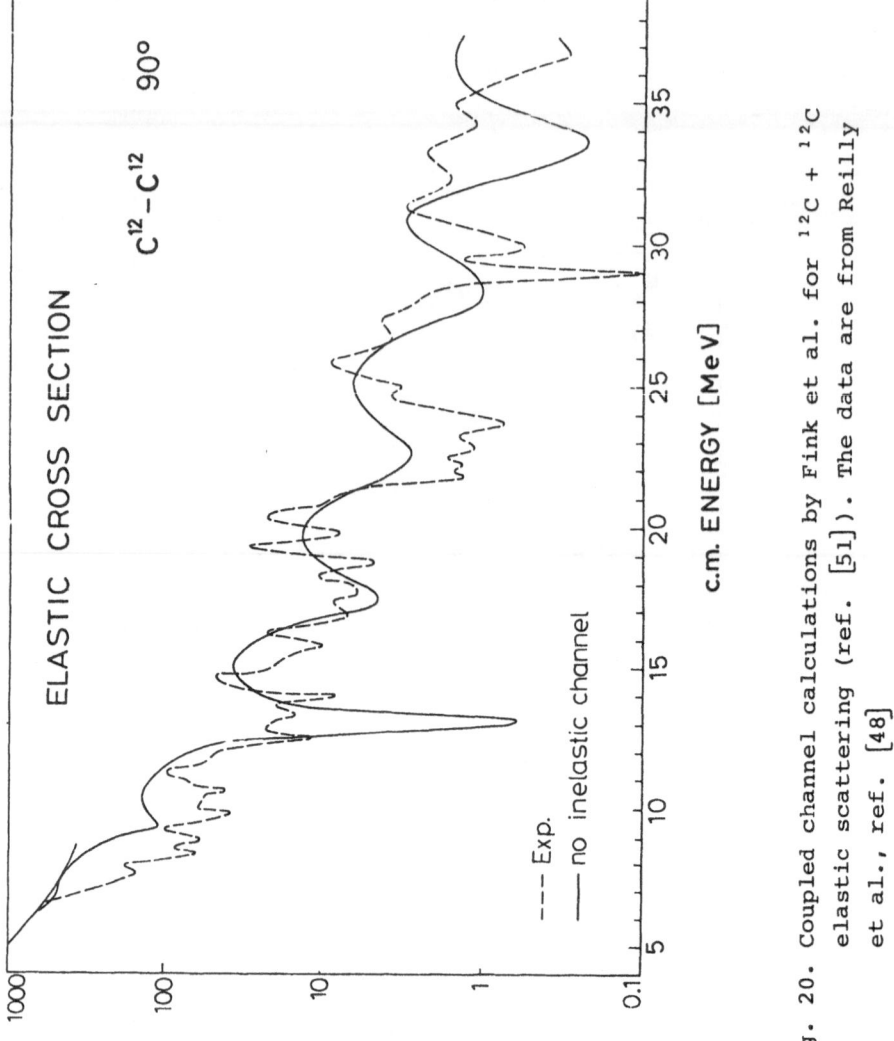

Fig. 20. Coupled channel calculations by Fink et al. for ^{12}C + ^{12}C elastic scattering (ref. [51]). The data are from Reilly et al., ref. [48]

Coupled channel calculations for ^{12}C + ^{12}C by Asciutto (reported by Gobbi [27]) do not show any structure with widths smaller than a few MeV. The different results obtained with these two calculations [27], [51] may be attributed to the strength of the imaginary potential used in the inelastic channel. Asciutto used the same potential as in the elastic channel; Scheid and Greiner suggest that this potential should be very weak and, for ^{12}C + ^{12}C, assume that it vanishes. Thus, the amount of intermediate structure predicted depends critically on a knowledge of the imaginary potential when the system is in the inter-mediate, virtually excited state.

Thus we have a situation in which two quite different mechanisms, compound elastic scattering and coupling to excited states, apparently can account for the experimental observations. The fundamental problem, therefore, is not the lack of an explanation, but whether the observed resonant structure arises in fact from one or both of these effects. This question is very difficult, but one which in principle can be an-swered by experiment. If the resonant structure is produced by an in-termediate mechanism, then the data should exhibit angular cross-cor-relations at the resonance energies. Such analyses, to my knowledge, have not been reported for ^{12}C + ^{12}C or ^{16}O + ^{16}O at energies where the resonant structure is prominent. Although the reduction in infor-mation associated with identical particles in the entrance channel and the pronounced gross structure complicate such an analysis, this never-theless is an area where continued experimental effort might be help-ful in understanding the resonant structure observed in elastic scatte-ring.

Such an approach has been applied by Malmin et al. [54], [37] to their measurements on the elastic scattering of ^{12}C + ^{16}O. Fig. 21 presents excitation functions measured from 12 to 26 MeV c.m. at eight angles. An auto-correlation analysis of these data indicates a fluctu-ation width of ∿110 keV. Except at energies of 13.7 and 19.7 MeV, no significant angular cross correlations are observed. The angular cross-correlation at 19.7 MeV, however, is especially strong as can be seen by visual inspection of fig. 21 and in the analysis given in fig. 22. Here, the quantity $D(E) = \Sigma_i |\sigma_i(E) - \langle\sigma_i\rangle / \langle\sigma_i\rangle$ is plotted as a function of energy. The cross section at energy E and angle θ_i is de-noted by $\sigma_i(E)$, and $\langle\sigma_i\rangle$ is the average value of $\sigma_i(E)$ over a 1.3 MeV wide interval centered at energy E. The width of the resonant structu-re at 19.7 MeV is ∿400 keV and inspection of the angular distribution on resonance indicates a total angular momentum J = 14. Correlations with the cross sections to the excited states of ^{16}O are olso observed. Further experimental results and discussion [57] of this resonance will be included in part 3.3.

In summary, the elastic scattering of heavy ions exhibits resonant phenomena which may be interpreted as intermediate structure. Coupled channel calculations by Scheid, Greiner and collaborators reproduce the $^{16}O + ^{16}O$ data quantitatively and the $^{12}C + ^{12}C$ experiments qualitatively. The intermediate structure models are not a unique explanation, however, and many of the overall features of the experimental results

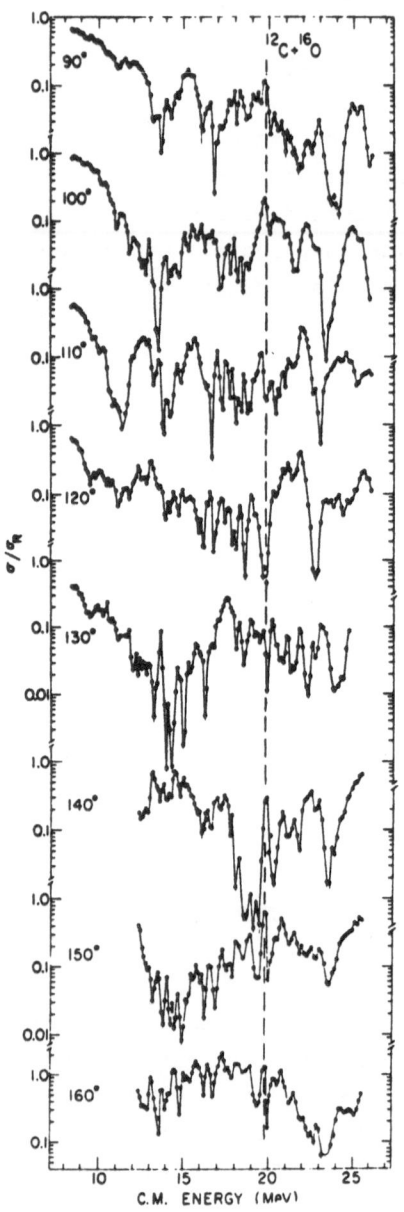

Fig. 21
Cross sections for the elastic scattering of $^{12}C + ^{16}O$ (refs. [54], [37]). The dashed line indicates an anomaly at E c.m.=19.7 MeV

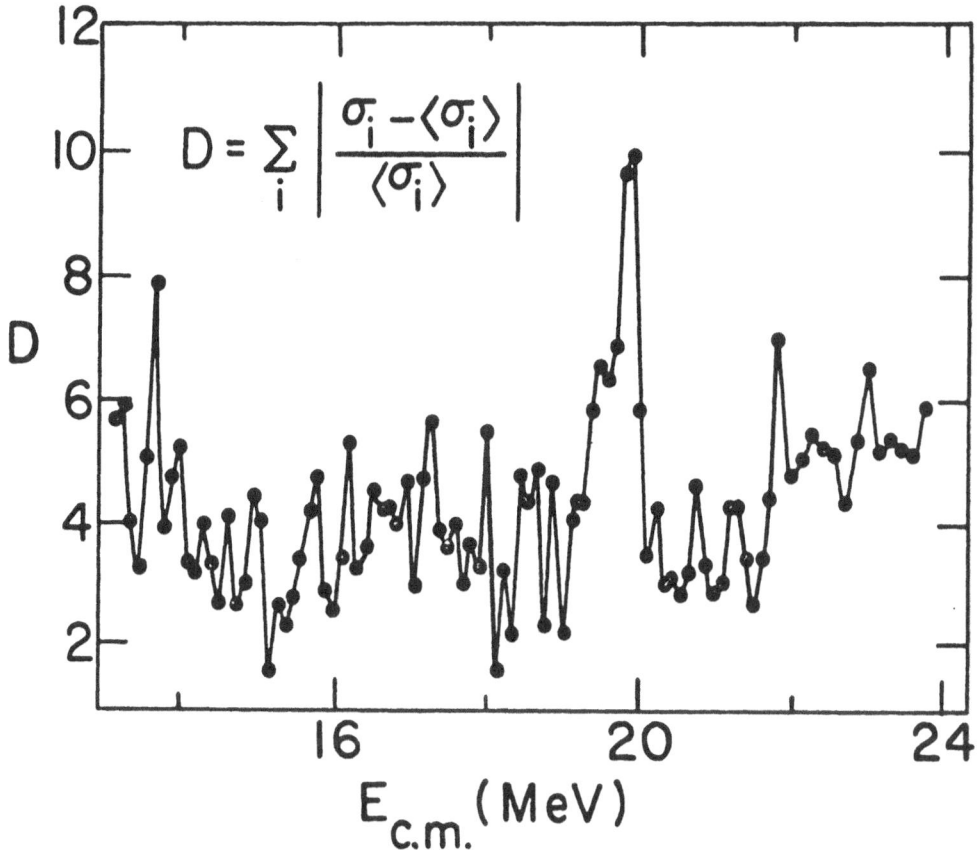

Fig. 22. Angular cross-correlation analysis for $^{12}C + ^{16}O$
elastic scattering (ref. [37])

can be explained in terms of the statistical theory of nuclear reactions. Finally, it is important that further theoretical and experimental work be directed toward a determination of the relative contributions of these different mechanisms.

3.3. The Reaction Channels

Quite detailed experimental studies at energies above the Coulomb
barrier have been made on the reactions $^{12}C(^{12}C,\alpha)^{20}Ne$ (refs. [3], [12]),
$^{12}C(^{16}O,\alpha)^{24}Mg$ (ref. [33]), and $^{16}O(^{16}O,\alpha)^{28}Si$ (ref. [52]). In these
particular investigations, only the alpha-particle yields to the ground
state and lowest few excited states were recorded. The cross sections
and angular distributions show rapid fluctuations and the overwhelming

conclusion of the accompanying statistical analyses [33],[63],[11] was
that the experimental results are consistent with compound nucleus formation (There was one exception; Halbert et al. [33] noted an anomaly
at $E_{c.m.}$=13.6 MeV in the $^{12}C(^{16}O,\alpha)^{24}Mg$ reaction). The deduced coherence widths were in the range 70 - 120 keV, and there was little or no
evidence for a direct component in the reaction.

The recent observation, by Middleton, Garrett and Fortune [40],
of states in ^{24}Mg at high excitation energy, selectively populated in
the $^{16}O(^{12}C,\alpha)^{24}Mg$ reaction, has led to renewed interest in both this
and similar reactions*. What has emerged from the recent work on these
reactions is 1) a further demonstration of the importance of compound
nucleus formation in the reaction mechanism and 2) the discovery, at
isolated bombarding energies or for certain final states, of non-statistical behavior for which intermediate structure is a possible explanation, and 3) an awareness of the possibilities for extracting unique
information on nuclear structure from these isolated phenomena. We will
discuss these points with respect to the reactions initiated by ^{12}C +
^{12}C and ^{12}C + ^{13}C, then ^{12}C + ^{16}O, and finally ^{16}O + ^{16}O.

3.3.1. Reactions induced by ^{12}C + ^{12}C and ^{12}C + ^{13}C

Excitation functions and angular distributions for the $^{12}C(^{12}C,\alpha)$
^{20}Ne reaction leading to states in ^{20}Ne below 8 MeV excitation have
been measured by Middleton et al. [41],[42] at the Univ. of Penn. They
were particularly concerned with two close-lying pairs of states in
^{20}Ne, namely a) the 0^+(6.72 MeV), 2^+(7.42) states and b) the 0^+(7.20),
2^+(7.83) states. Earlier experimental work with one, two and four nucleon transfer reactions, as well as measured alpha particle widths had
suggested that the states of pair (a) were normal shell model states
(i.e. 4 particle, 0 hole) while those of pair (b) were not. The theoretical predictions of Arima et al. [6]for the energies of alpha-cluster or quartet states suggested that states near these energies could
have the configuration 220 , i.e. four nucleons promoted from the 1p
to the 2s - 1d shell to yield an 8 particle-4 hole correlated state.
The results obtained with the multigap-spectrograph at Penn. are shown
in figs. 23,24 and indicate quite different behavior for the two pairs
of states. Although the cross sections fluctuate in all cases, the average values for the differential cross sections at 3.75°(lab) are about
two and five times larger for the 0^+ and 2^+ states, respectively, of

*Reports on most of the recent experimental work in this area can be
found in the Symposium on Heavy Ion Reactions and Many Particle Excitations, Saclay, 1971.

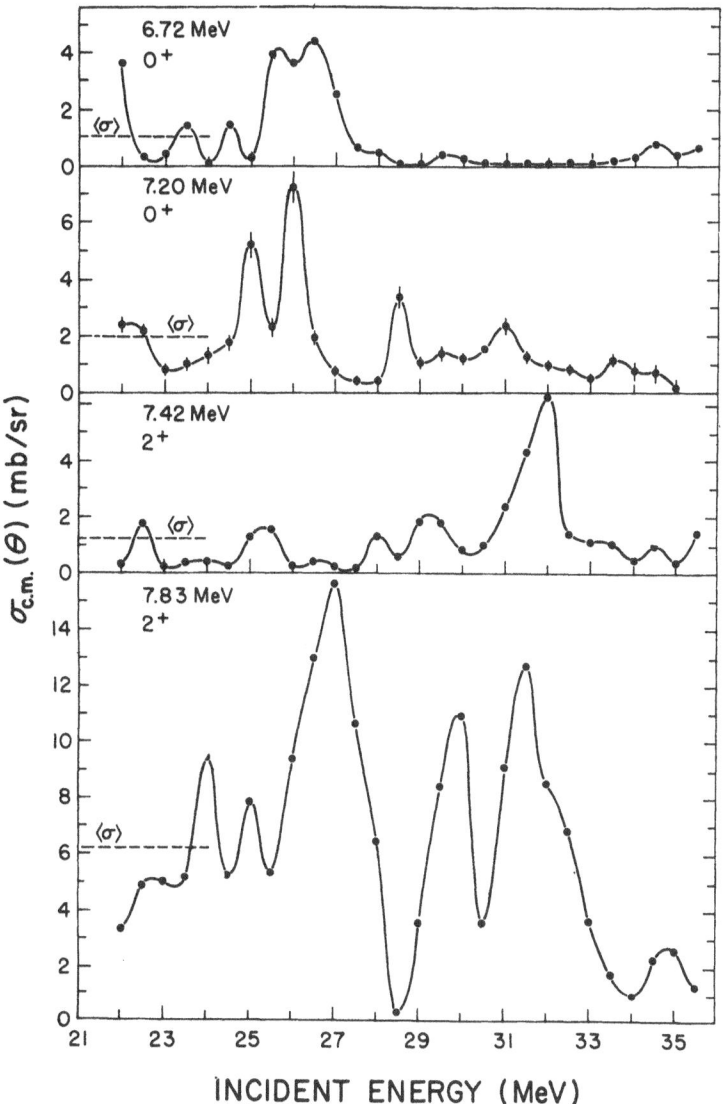

Fig. 23. Differential cross sections measured at 3.7°(lab) for the $^{12}C(^{12}C,\alpha)^{20}Ne$ reaction to states at 6.72, 7.20, 7.42 and 7.83 MeV excitation (ref. [42]). The average cross section is indicated for each excitation function

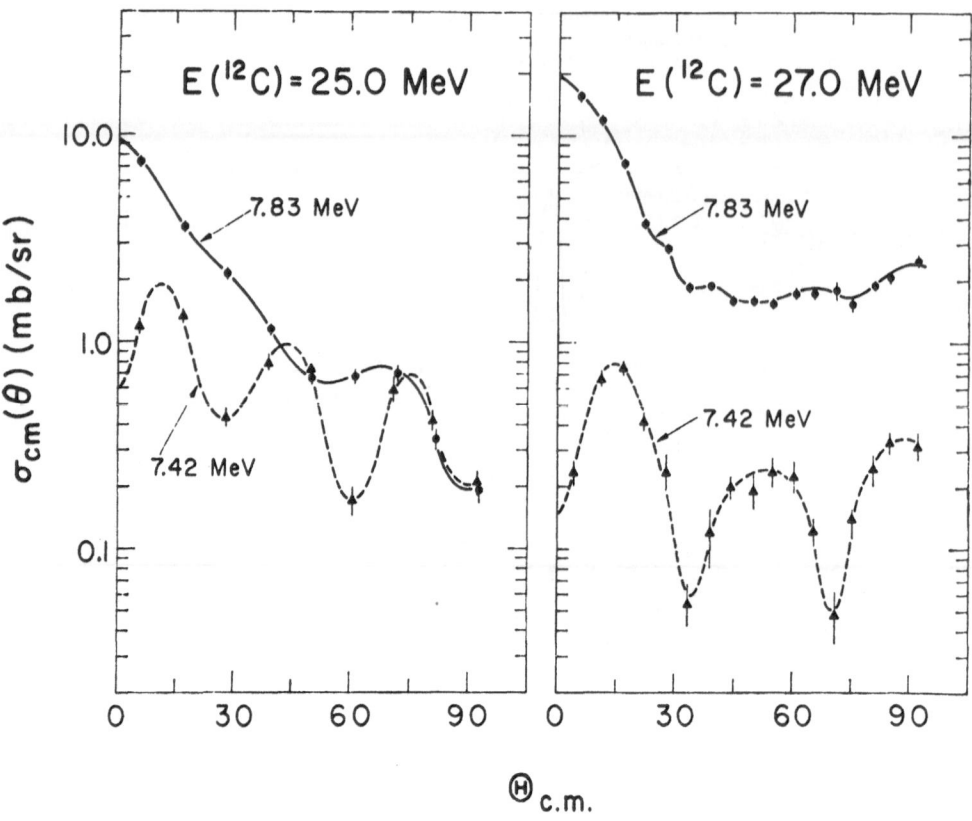

Fig. 24. Angular distributions for the $^{12}C(^{12}C,\alpha)^{20}Ne$ reaction to the 7.42 and 7.83 MeV states in ^{20}Ne
(ref. [42])

pair (b) than for the corresponding states in pair (a). These variations are inconsistent with the statistical model [33] which predicts a
fractional standard deviation of about 1/3 on the average cross section indicated in fig. 23. Angular distributions measured for the 2^+
states (fig. 24) also appear quite different, the 2^+ (7.42) state
(pair a) showing an oscillatory angular distribution similar to that
observed for the 2^+ (1.63) state by Borggreen et al. [12].

Middleton et al. [41],[42] suggest an interpretation of these results in terms of the transfer of two alpha particles from a ^{12}C nucleus to form the 0^+(7.20), 2^+(7.83) states which therefore have a
220 intrinsic configuration. The shell model states of pair (a) are
populated via compound nucleus formation, and the different angular
distributions for the 2^+ states reflect the differences in the respective reaction mechanisms. The fluctuating cross sections to the cluster states could be explained by an intermediate structure in the reaction mechanism involving alpha-particle doorway states in the ^{24}Mg
compound system. Thus, a natural extension to higher energies of the
intermediate structure model, proposed by Michaud and Vogt [39] to explain the resonant phenomena below the Coulomb barrier, may perhaps be
justified on experimental as well as theoretical grounds. Two points
worth noting, however, are 1) that the peaks in the cross sections to
the 0^+ and 2^+ states of pair (b) do not appear correlated as one would
expect if they were both populated by the same doorway state in the
compound system [41], [42], and 2) the amplitude for the reaction may
of course also contain a compound nuclear component.

The interpretation of these results as evidence for intermediate
structure is questioned in a recent letter by Noble [45]. He argues
that a direct reaction involving the transfer of a ^8Be nucleus is sufficiently complicated that the structures in the excitation functions
of fig. 23 are consistent with the interference of several possible direct reaction amplitudes. This complication arises in part because
^8Be can be transferred in several excited configurations. Sample calculations made in the plane-wave Born approximation and for fictitious
states in ^{20}Ne do indeed indicate cross sections which vary almost as
rapidly as those measured.

Additional support for the direct reaction interpretation [45] may
be inferred from the $^{12}C(^{13}C,\alpha)^{21}Ne$ reaction, also studied by Middleton
et al. [41], [43]. Angular distributions, energy averaged over a 250 keV
c.m. interval, for states at E_x = 3.66 and 3.89 MeV in ^{21}Ne show pronounced asymmetries about 90° c.m., one state forward peaked, the other
state backward peaked (see fig. 25).

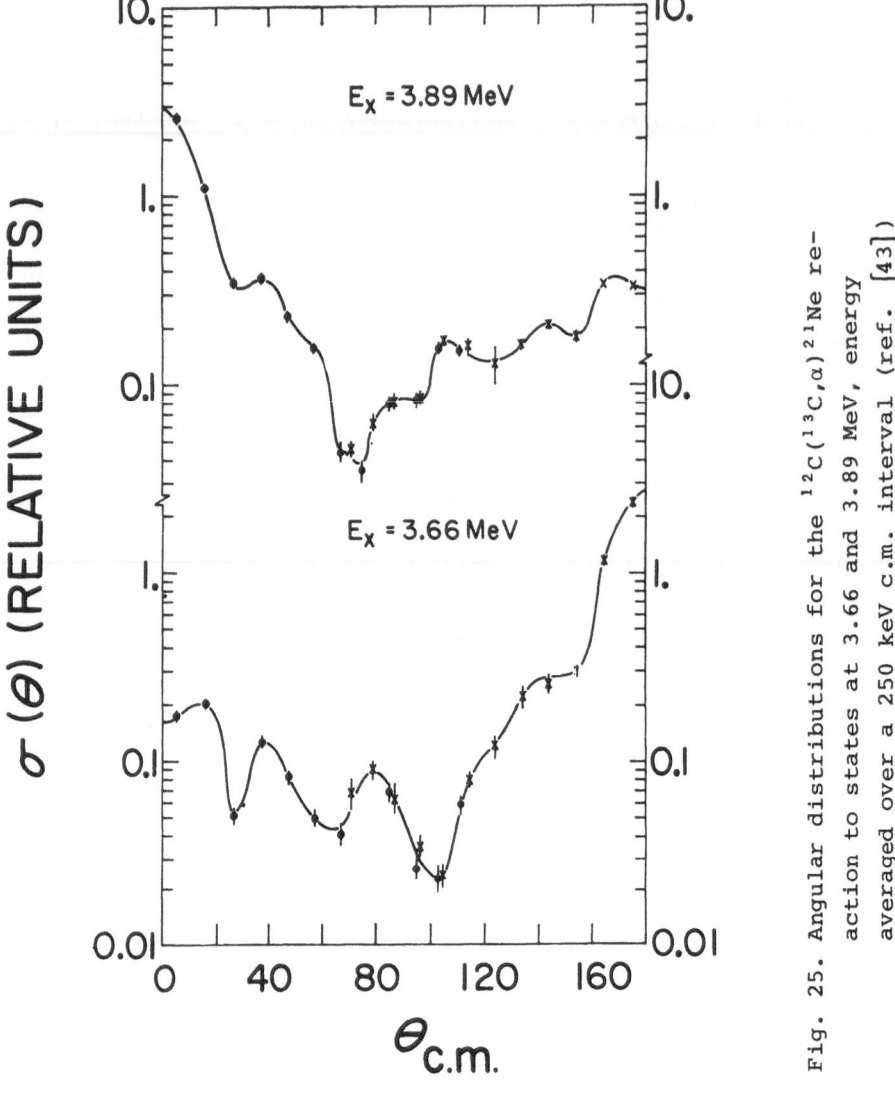

Fig. 25. Angular distributions for the $^{12}C(^{13}C,\alpha)^{21}Ne$ reaction to states at 3.66 and 3.89 MeV, energy averaged over a 250 keV c.m. interval (ref. [43])

(In the $^{12}C + ^{12}C$ reactions, the identical bosons in the entrance channel require that all angular distributions show exact symmetry about 90° c.m.). Since an averaging interval this small cannot sufficiently damp fluctuation effects, excitation functions for the 3.66 MeV states were taken at 5° and 175° c.m. over a 2 MeV c.m. interval. These latter measurements revealed, however, that the forward-backward asymmetry persists, varying from a factor of three to fifteen over this range. It is difficilt to imagine how an intermediate structure mechanism could produce such an effect. Regardless of whether the reaction mechanism is intermediate or direct, however, these measurements are plausibly interpreted as adding two alpha particles to a ^{13}C core and hen-

ce populating states in ^{21}Ne with an 8 particle-3 hole character [41],
[43]. A remaining mystery, however, is why the two angular distributions in fig. 25 are peaked in opposite directions. Further theoretical investigation of the questions raised by the Penn group's measurements will hopefully elucidate these very interesting reactions and
thereby place the conclusions on nuclear structure on a more firm basis.

3.3.2. Reactions induced by ^{12}C + ^{16}O

The reactions initiated by ^{12}C + ^{16}O have recently become the subject of renewed and intensive study. Following the previously noted
measurement at Penn [40], excitation functions for alpha particles have been measured at Argonne [30], Saclay [17],[24],[25] and Yale [15],
and for proton yields at Brookhaven - MIT [15]. The spectra obtained
in these measurements have a common feature, namely the appearance of
sharp states on a continuum background. Fig. 26 is a typical example
[15]. The results of α - α angular correlation experiments by Gobbi et
al. [29] and Balmuth et al. [7] show that the states at high excitation
energies have predominately high spins. Simple angular-momentum matching arguments [15] and more detailed Hauser-Feshbach calculations [18],
[38],[58] also indicate high spins (e.g. 8-10 at ∿17 MeV excitation in
^{24}Mg) and that these states are selectively populated by the grazing
partial waves in the entrance channel*. The cross sections fluctuate
rapidly with energy and the highest resolution measurements [30] yield
a coherence width Γ ∿ 110 keV, in agreement with earlier results [33]
obtained for the states at low excitation energy in ^{24}Mg. Both the
Argonne 30 and Saclay [17],[24],[25] data cover the region ∿19.5 to
∿25 MeV c.m. and exhibit gross structures with widths varying from one
to two MeV; the Saclay group notes correlations in the gross structures
for certain states. but these correlations do not seem to be observed
by Greenwood et al. [30]. Possible explanations offered for the presence of gross structure are coherence effects associated with strongly
absorbed particles [45], the fact that the absorbtion of different partial waves occurs at different energies (see fig. 5 in ref. [52]), and
alpha particle intermediate structure [25].

---

* It is interesting to note a similarity between these reactions and
heavy ion, xn reactions. In each case the light particles in the exit
channel are unable to carry off the large amount of angular momentum
brought in by the grazing partial wave. The residual nucleus is thus
left in a state of quite high angular momentum, very close to or on
the Yrast line, thus offering possibilities for nuclear structure
studies in light nuclei analogous to those currently done in much heavier nuclei.

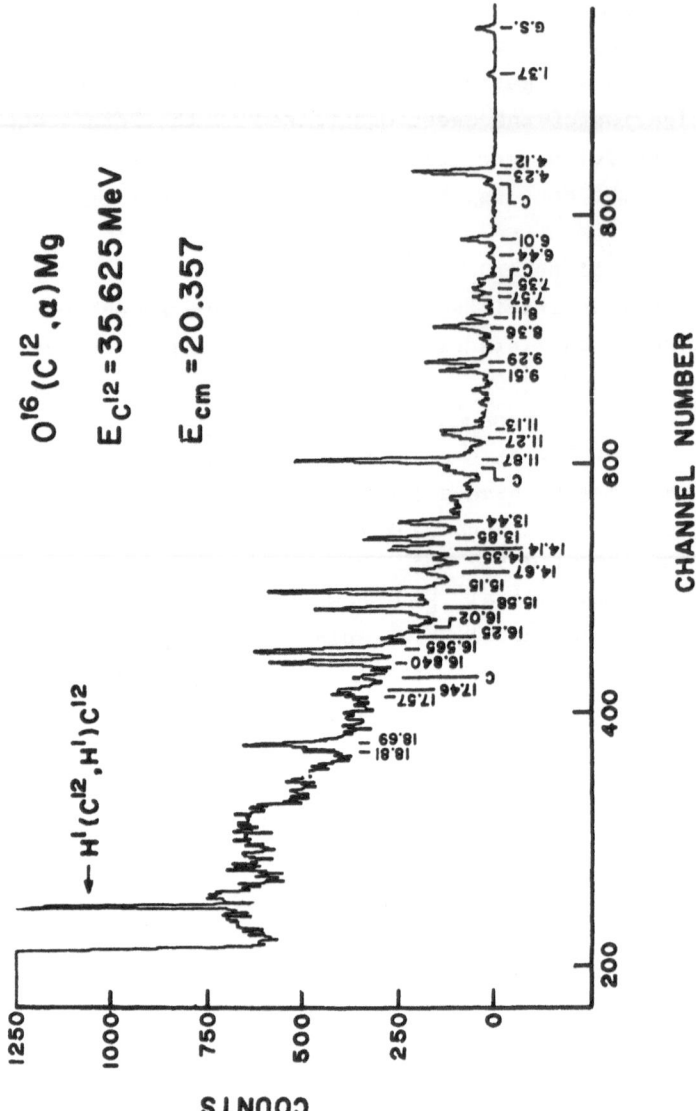

Fig. 26. Spectrum of alpha-particles observed at θ_{lab}
~2° for a ^{12}C beam and a Si O target. Peaks associated with the $^{12}C(^{12}C,\alpha)^{20}Ne$ reaction from
carbon build up on the target are denoted by the
letter C (ref. [15])

The Yale group [15] has measured alpha-particle excitation functions at laboratory angles of $\sim 2^\circ$ for an ^{16}O beam and ^{12}C target ($\theta_{c.m.}$ $\sim 3^\circ$) and ^{16}O target and ^{12}C beam ($\theta_{c.m.}$ $\sim 177^\circ$) over an energy range 18.5 - 21.5 MeV c.m. The results are presented in figs. 27-29 and may be summarized as follows. States in ^{24}Mg below 10 MeV excitation exhibit the fluctuations typical of a statistical compound nuclear process. Many of the higher lying states, however, show a gross structure with a width of 1.5 MeV upon which narrower fluctuations are superimposed; this gross structure is especially pronounced for $\theta_{c.m.}$ = 177° (^{12}C beam) and is correlated, with seven of these states experiencing a common minimum at ~ 19.7 MeV c.m. Excitation functions for elastic scattering at $\theta_{c.m.}$ = 90°, 177° and for inelastic scattering at $\theta_{c.m.}$ = 177° were also measured over this same region [57]. When these latter excitation functions are plotted together with the summed alpha particle yields, as in fig. 30, a striking correlation is evident. There is a sharp dip in the broad structure ($\theta_{c.m.}$ = 177°) at 19.7 MeV which coincides in position and width with the peaks observed in the elastic and certain inelastic channels.

When these results [57] are combined with the detailed elastic scattering measurements of Malmin et al. [37] and the proton yield measurements of Cosman et al. [19] (in which selectively populated states also exhibit a minimum at E c.m. ~ 19.7 MeV), a fairly complete picture emerges which has a possible explanation in terms of intermediate structure. The mechanism suggested [57] involves the formation of a molecular resonance at 36.5 MeV in ^{28}Si consisting of two ^{12}C cores attracted to each other via their mutual attraction for the four correlated valence nucleons*. Thus, the force producing the resonance arises from the exchange of four correlated nucleons which possibly but not necessarily approximate an alpha particle. This force could then give rise to a narrow resonance of the type observed here (~ 350 keV). The total angular momentum of 14 units implied by the angular distributions for elastic scattering [37] is the same as that of the grazing partial wave in the entrance channel. Thus the resonance is associated with the partial wave which would also most favor alpha exchange between the ^{12}C cores.

---

* The importance for heavy ion scattering of single and multinucleon exchange between identical cores is well known. See, e.g., W. Von Oertzen, Nucl. Phys. A148 (1970) 529.

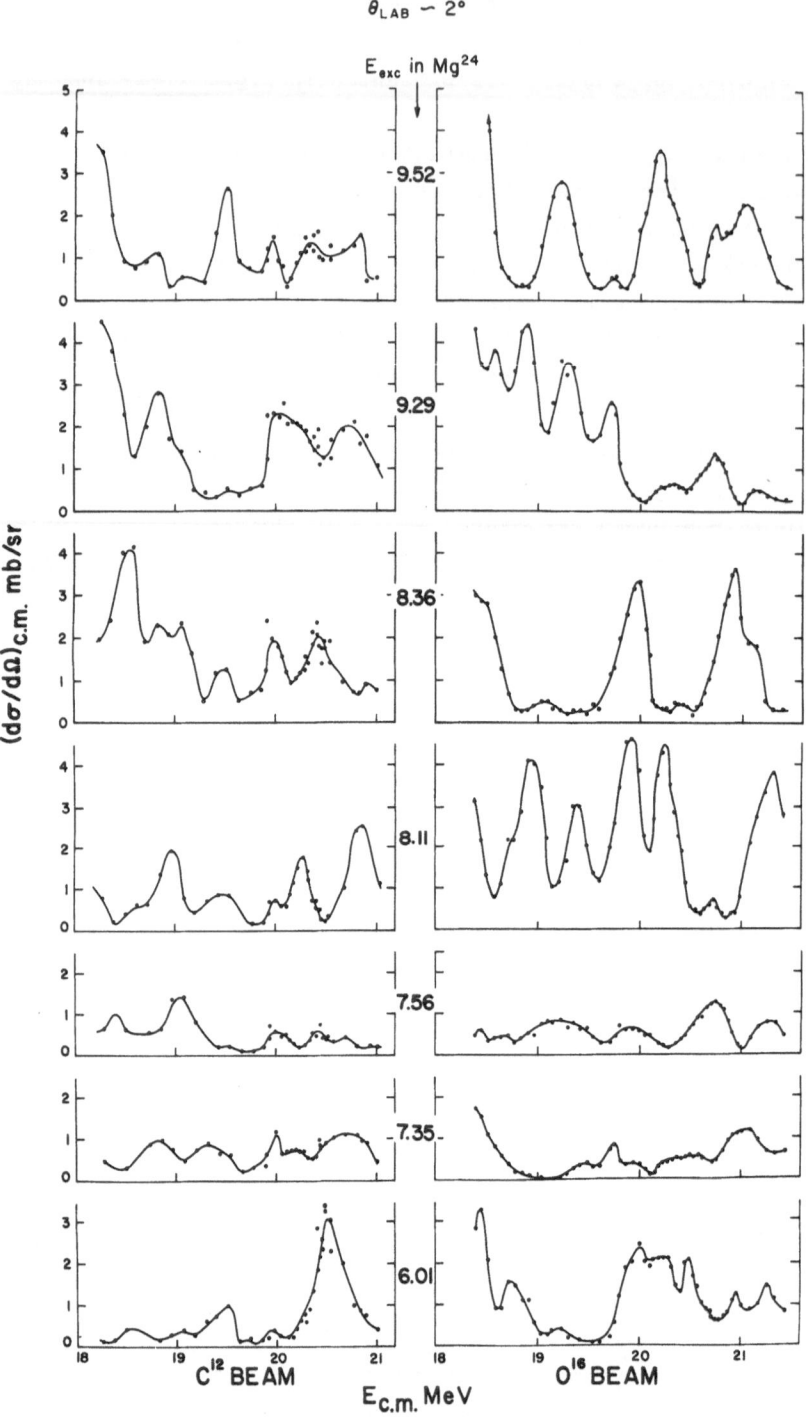

$$O^{16} + C^{12} \longrightarrow Mg^{24*} + \alpha$$
$$\theta_{LAB} \sim 2°$$

E_{exc} in Mg^{24}

Fig. 27. Excitation functions measured at $\theta_{lab} \simeq 2°$ for the $^{12}C(^{16}O,\alpha)$ and $^{16}O(^{12}C,\alpha)$ reactions populating states in ^{24}Mg between 6.0 and 9.5 MeV excitation. The error in the absolute cross section is 30% [15]

227

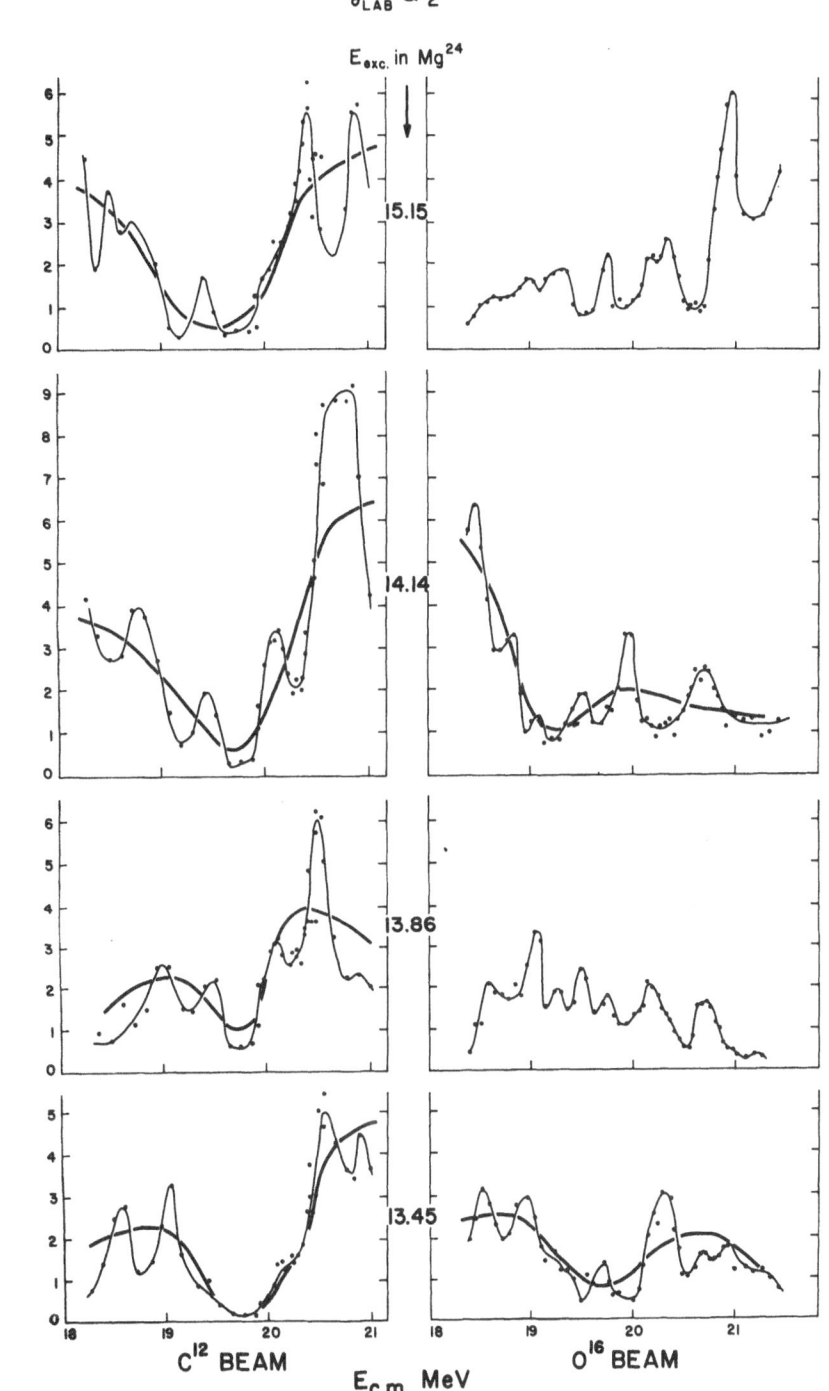

Fig. 28. Same as fig. 27 but for states in ^{24}Mg between 13.4 and 15.5 MeV in excitation. The heavy solid line is only to fuide the eye (ref. [15])

228

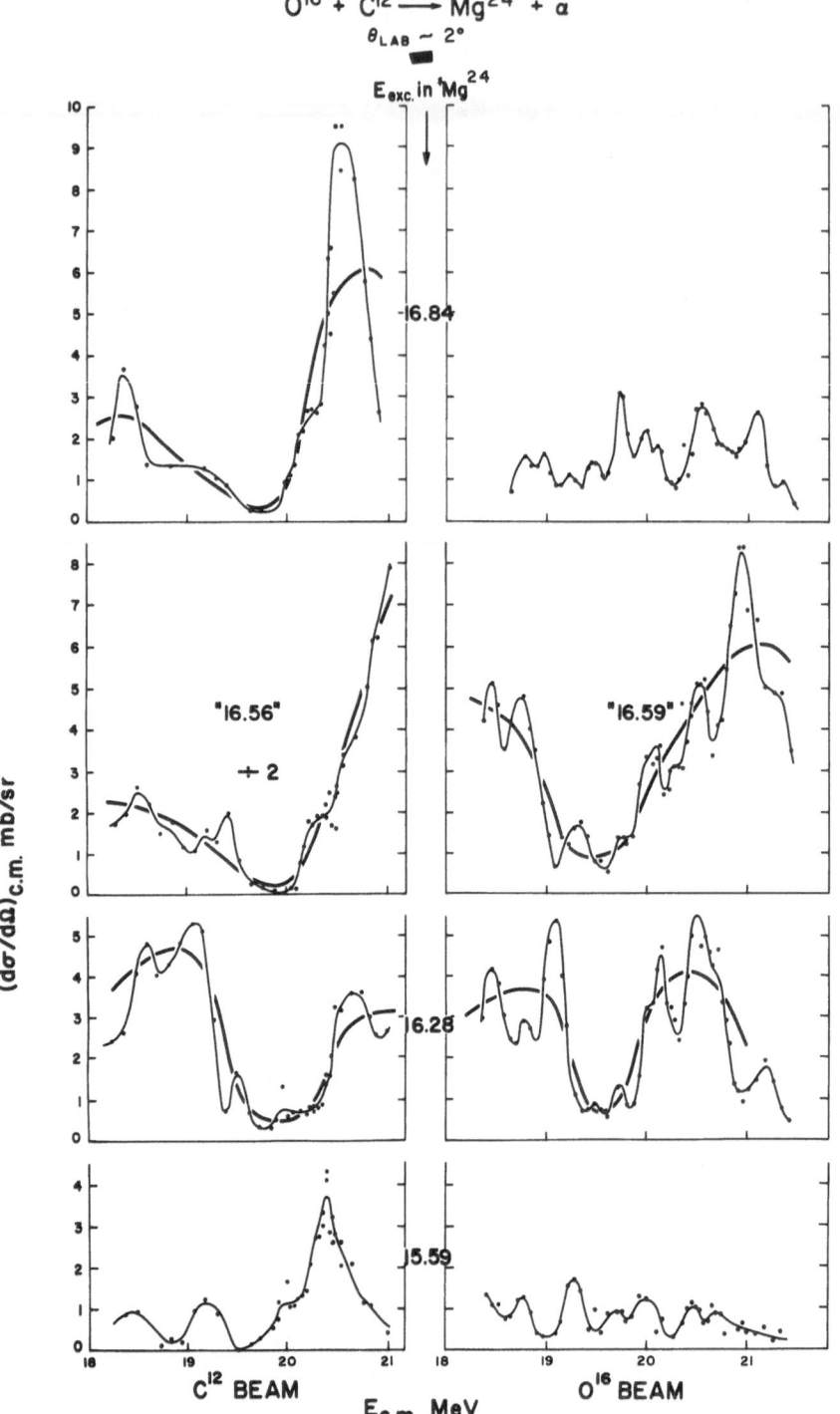

Fig. 29. Same as fig. 27 but for states in ^{24}Mg between 16.3 and 16.84 MeV excitation. The heavy solid line is only to fuide the eye (ref. [15])

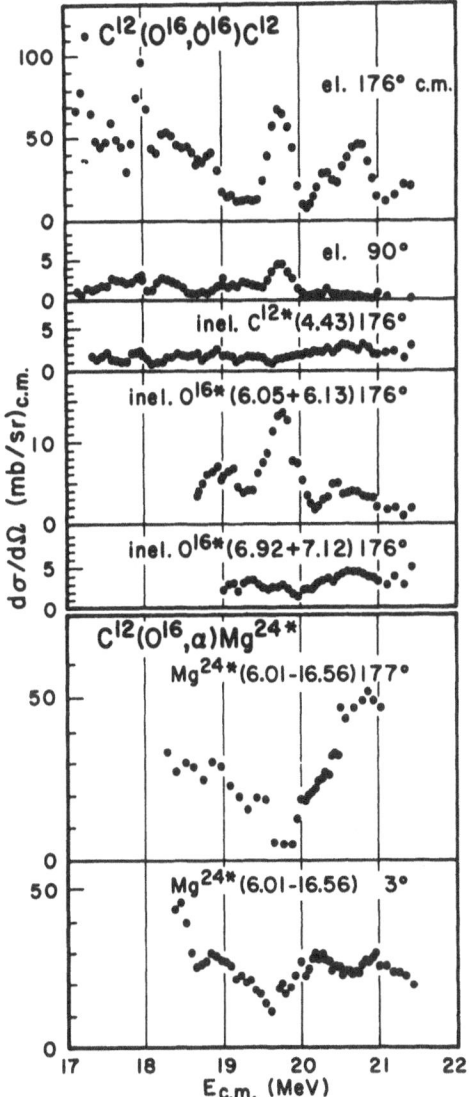

Fig. 30
Comparison of excitation functions for the $^{16}O + ^{12}C$ interaction. (a),
(b) $^{16}O - ^{12}C$ scattering at $\theta_{c.m.} = 176°$ and $90°$; (c),(d),(e) inelastic scattering leading to ^{16}O(g.s. + ^{12}C (4.44 MeV), to ^{12}C(g.s.)+ ^{16}O
(6.05 + 6.13), and to ^{12}C(g.s.) + ^{16}O (6.92 + 7.12); and (f),(g) excitation functions for scattering angles of $\theta_{c.m.} = 177°$ and $3°$ for the reactions $^{12}C(^{16}O,\alpha)^{24}Mg$ summed over twelve separate transitions to states
in ^{24}Mg at E_x=6.01, 7.35, 7.56, 8.11, 9.29, 9.52, 13.45, 13.86, 14.14,
15.15, 16.29 and 16.56 MeV. These latter states have been selected only
on the basis of a clear identification over the full energy range and
specifically not on the basis that they necessarily show the minimum
under discussion (ref. [57])

Furthermore, as shown in fig. 30, the resonance is observed to be especially strong at the extreme backward scattering angles for which, again, alpha-particle exchange is favored. Statistical model calculations [58] indicate that the states selectively populated in the $^{24}Mg + \alpha$ and $^{27}A\ell + p$ exit channels are also fed primarily through the 14th partial wave. Thus, the associated minima in these exit channels may result from the depletion of flux in the 14th partial wave by the $^{16}O + ^{12}C$ exit channels, i.e. the molecular resonance has a small overlap with the former exit channels and a large overlap with the latter. Consistent with this explanation is the observation of resonant yield to an excited state of ^{16}O, but not to the 4.4 MeV 2^+ state in ^{12}C. Recent very high resolution measurements at Yale have now shown that it is the 3^- (6.13 MeV) state in ^{16}O and not the 0^+ (6.05) state which resonates [16] . It is interesting to note that in the $^{12}C(^7Li,t)^{16}O$ reaction, nominally assumed to proceed via alpha transfer, the $3^-(6.13)$ and $0^+(6.05)$ states are populated with equal strength [8] $(E_{6_{Li}} = 15$ MeV).

The mechanism suggested above, which does not exclude the possibility of the 3^- state resonating, apparently offers no simple explanation as to why neither the $0^+(6.05)$ nor $2^+(6.92)$ states in ^{16}O resonates. A resonance associated with the virtual excitation of the 3^- state (double resonance phenomenon [51]) is a possible alterantive explanation, and it would be very interesting, therefore, to have coupled channel calculations for $^{12}C + ^{16}O$ similar to those which Scheid, Greiner, et al. [51] have performed for $^{12}C + ^{12}C$ and $^{16}O + ^{16}O$.

Additional theoretical work on these experimental results would be most welcome as there is now a large amount of information on the properties of this resonance. Its non-statistical behavior in the elastic channel and associated correlations in the various reaction channels are well documented, and the measured width of $\sim$350 keV makes the interpretation in terms of an intermediate mechanism very attractive. Experiments to search for similar phenomena in the $^{12}C + ^{12}C$ reaction are currently in progress at Yale.

3.3.3. Reactions Induced by $^{16}O + ^{16}O$

The $^{16}O(^{16}O, ^{12}C)^{20}Ne$ reaction, recently investigated by Singh et al. [55] at Argonne and by Rossner et al. [49] at Munich, exhibits rather striking phenomena. Figures 31 and 32 are from the Argonne work [55] and present excitation functions and angular distributions for the ground and first excited states of ^{20}Ne, and the unresolved sum of the $4^+(4.25$ MeV) and $2^+(4.43$ MeV) states in ^{20}Ne and ^{12}C, respectively.

There are indications [50] that most of the unresolved yield is to the
4^+ state in ^{20}Ne. Gross structure with widths of ∿1.5 MeV and narrower
resonant structure (∿250 - 300 keV) are evident, the gross structure
being especially prominent for the 4.3 MeV state(s). The situation is
thus qualitatively similar to that observed in the elastic scattering.
However, the remarkable feature of these results lies in the strong
angular and cross-channel correlations for the gross structure which
are immediately apparent in the data. A strong angular correlation is
typical of a process which proceeds mainly through a single partial
wave. Angular correlation for the resonant structure is also indicated,
especially for the 1.63 MeV state.

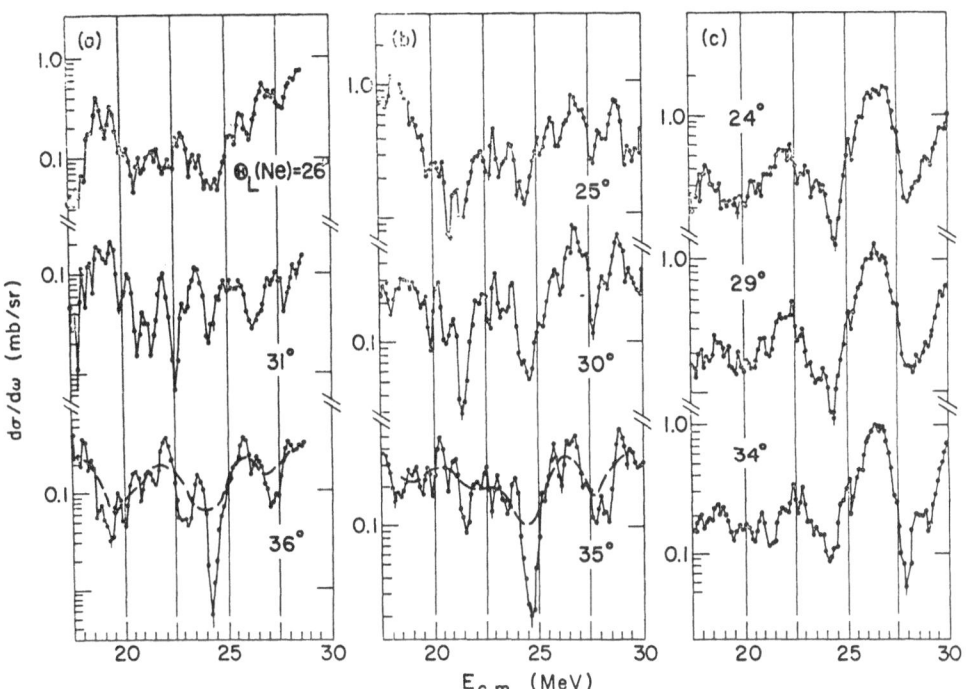

Fig. 31. Excitation functions for the ^{16}O(^{16}O,^{12}C)^{20}Ne reaction populating (a) the g.s., (b) 1.63 MeV, and
(c) 4.25 MeV states in ^{20}Ne (ref. [55]). The 4.43
state in ^{12}C is not resolved from the 4.25 MeV
state. The indicated angles are those of the ^{20}Ne
recoil in the lab system and correspond to angles
of 62°, 76° and 90° c.m. The dotted line in (a)
and (b) represents an average over 1.5 MeV

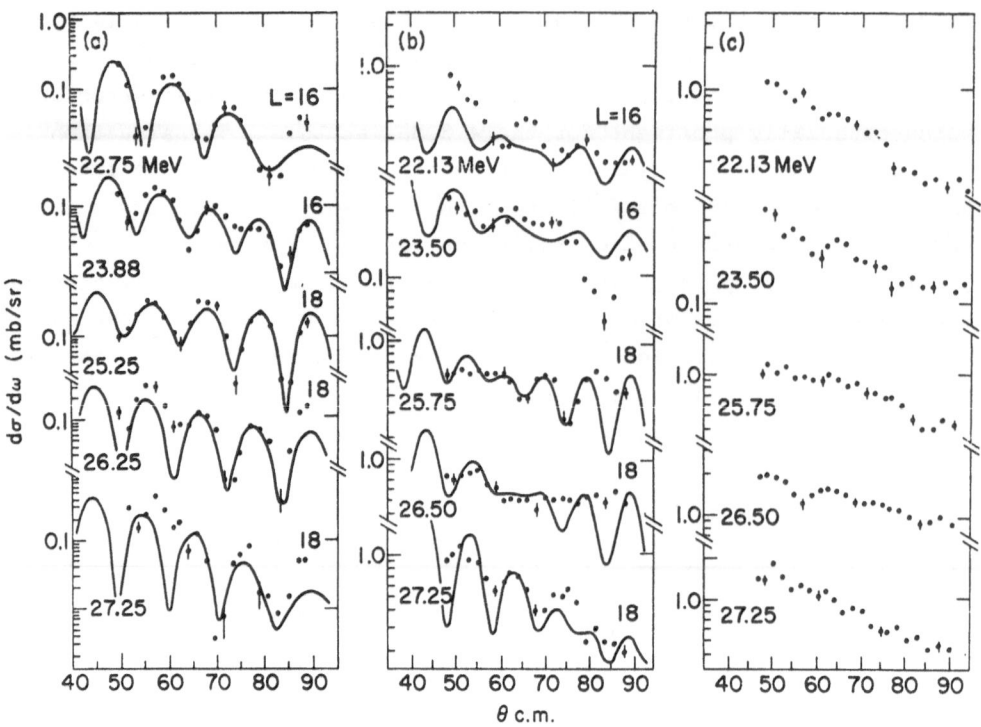

Fig. 32. Angular distributions for the $^{16}O(^{16}O,^{12}C)^{20}Ne$
reaction at various energies, for the states in
fig. 31 (a) and (b) are for the g.s. and 1.63
MeV states, (c) for the sum of the 4.25 and 4.43
MeV states. The solid lines are fits with Legendre polynomials of order L and L-2 (ref. 55)

    The angular distributions (fig. 32) to the lowest states of ^{20}Ne
also indicate that each broad peak in the gross structure is associated
mainly with one partial wave, as is the case in elastic scattering.
There is an important difference between the two cases, however, since
the 90° c.m. maxima at 22.5, 26.5 MeV in the raction channel appear at
nearly the same energies as the minima in the elastic channel. Optical
model calculations [27] indicate that the 90° maxima in the elastic
channel are reached when the respective partial wave is completely absorbed. The maxima found in the reaction channel, however, appear when
the particular partial wave resonates. Furthermore, the ℓ-values from
the measured angular distributions agree with those predicted by the
optical model. Diffractive effects in the reaction channel are also ru-

led out by the lack of structure in the angular distribution for the 4.43 MeV state [55].

Singh et al. [55] interpret the gross structure as evidence for shape resonances occurring in the grazing partial waves of the ^{16}O + ^{16}O system. The evidence for this is clearly strong since the angular distributions for ∿1.5 MeV wide peaks indicate a well defined total angular momentum, and the peaks are cross-correlated in the exit channels, which usually signifies an entrance channel phenomenon. It is furthermore suggested [55] that the resonant structure corresponds to high spin states in the ^{16}O + ^{16}O compound system which are preferentially populated through the resonance in the entrance channel. The shape resonance thus acts as a doorway state, selecting out those states in the compound system having the same angular momentum and which are dynamically related.

The cross correlations in the exit channels need not arise solely from a coherence induced by the entrance channel. The 0^+, 2^+ and 4^+ states of ^{20}Ne all have the same intrinsic structure, i.e. they are themselves dynamically related, and this can produce correlations in the sums of otherwise random transition amplitudes. This possibility has been noted by Shaw et al. [52] who observed strong correlations in the alpha particle yields to the ground and first excited states of ^{28}Si in the $^{16}O(^{16}O,\alpha)^{28}Si$ reaction. Nevertheless, the results presented in figs. 31 and 32 not only confirm much of the knowledge of the ^{16}O + ^{16}O interaction gleaned from studies of elastic scattering, they also stand as interesting examples of evidence for intermediate processes in heavy-ion reactions.

4. Summary and Conclusions*

This review attempts to extract from a wealth of available data those measurements which bear directly on the presence of intermediate structure in heavy ion reactions. It is seen that there is good evidence for intermediate processes in a number of cases. This evidence is usually in the form of a departure from the behavior which would be expected if the compound system were in statistical equilibrium. The smaller number of degrees of freedom associated with the non-equilibrium state are those which determine the nature of the intermediate structure. In the cases of ^{12}C + ^{12}C and ^{12}C + ^{16}O interacting at energies at or below the Coulomb barrier, the experimental results sug-

---

* Detailed references to the experimental results mentioned in the summary are found in section 2. and 3.

gest that these degrees of freedom involve clusters of loosely bound alpha particles. The absense of resonant structure, however, in most of the other reactions studied near the Coulomb barrier is remarkable. Although one can note empirically an approximate correspondence between the observed lack of resonances and an increase in the number of open channels, there is as yet no theoretical approach which has also dealt with these negative results. The single resonance in the gamma-ray yield from the ^{14}N + ^{14}N reaction presents an interesting puzzle as it does not appear to fit with the pattern set by the other reactions.

The observation of a broad gross structure in elastic scattering above the Coulomb barrier with smaller structures of the order of 0.2 - 0.4 MeV width superimposed thereon has stimulated much theoretical interest. The possibility of interpreting this narrow structure by a resonant coupling to excited states is attractive and quantitative comparisons to the data have been rather successful. However, an alternative explanation in terms of compound elastic scattering is possible. Further effort should be devoted to determining the relative contributions of these two processes.

In the last two years there has been a rapid increase of experimental studies on the non-elastic reactions induced by ^{12}C + ^{12}C, ^{12}C + ^{16}O and ^{16}O + ^{16}O, providing interesting and surprising results. The reaction channels are in general dominated by statistical phenomena and it has required detailed measurements and careful searches to uncover the non-statistical effects. In the case of the ^{12}C(^{12}C,α)^{20}Ne and ^{12}C(^{13}C,α)^{21}Ne reactions, non-statistical behavior of cross sections to particular states in ^{20}Ne is observed and it has been possible to derive qualitative spectroscopic information from the results. The ^{12}C + ^{16}O induced reactions exhibit gross structures, and also narrower resonances which are correlated in a number of exit channels. Statistical phenomena are not as prominent in the ^{16}O(^{16}O,^{12}C)^{20}Ne reaction; here, both angular and cross-channel correlations are quite strong.

In both of these latter two reactions the narrow resonances have very high spins with the same value as the orbital angular momentum of the grazing partial wave; J = 14 for ^{12}C + ^{16}O at E c.m. = 19.7 MeV, and J = 16, 18 at E c.m. 22.5, 26.5 MeV for ^{16}O + ^{16}O (It is interesting to contrast these angular momenta with the values J = 2, 4 observed in ^{12}C + ^{12}C near the Coulomb barrier). For ^{12}C + ^{16}O, the resonance raises the elastic scattering to 60 mb/sr at the most backward angle where elastic alpha-transfer is favored. Similarly, the ^{16}O (^{16}O, ^{12}C)^{20}Ne reaciton is one which can proceed via alpha-transfer mechanism. Considering that the remaining system which exhibits intermediate structure, ^{12}C + ^{12}C, is also alpha conjugate, one is tempted to regard this as more than just circumstantial evidence for the importance of alpha-particle clusters in intermediate processes. Finally, there is a whole class of experimental results, not discussed here, which provides additional support for the alpha-cluster mechanism. The backward scattering of alpha particles from ^{16}O, ^{24}Mg and ^{40}Ca is enhanced at excitation energies in the compound system which are spaced roughly as $\ell(\ell + 1)$, where ℓ is the angular momentum of the grazing partial wave [49], [21]. The deduced moment of inertia [49] corresponds approximately to that of an alpha particle orbiting the target nucleus.

Indeed, the concept of alpha-cluster intermediate structure appears to hold great promise for developing a theoretical framework which can unify the diverse phenomena which have been presented here. It is hoped that this review of experimental results will contribute toward this goal.

Acknowledgements

I would like to thank my colleagues at Yale, D.A.Bromley and A. Gobbi, whose knowledge of heavy-ion reactions has been of great benefit to me. A number of people kindly communicated their results before publication or were helpful in other ways - H. Voit, M. Halbert, D. Sink, W. Callendar, K. Dietrich, R. Wieland and D. Shapira come immediately to mind. Mrs. Agnethe Elsving was masterful in her typing of the manuscript and special thanks are due Jacob Bondorf for many stimulating, clarifying and patient discussions.

This review was prepared during a brief stay at the Niels Bohr Institute. The excellent hospitality of the Institute is known to many people, and for me it proved no exception.

Note Added in Proof

The statement (p.232) that the 3^-, 6.13 MeV state in ^{16}O resonates at $E_{c.m.}$=19.7 MeV is correct. However, a high resolution experiment of the reaction $^{12}C(^{16}O,^{12}C)^{16}O(\sim6.1$ MeV) at $E_{c.m.}$=19.7 MeV and θ_{lab}=7.5° shows only that this differential cross section to the 0^+ state at 6.05 MeV in ^{16}O is smaller by at least a factor of 12 than the cross section to the 3^-, 6.13 MeV state. Thus it is presently not known whether the 6.05 state resonates. The decay of the resonance to the 3^- state relative to the 0^+ is favored by a statistical factor and a penetrability factor, the latter having a value $\geqslant10$.

References

References to certain conference proceedings are abbreviated as
follows: "Proc. Heidelberg (1969)", "Proc. Argonne (1971)", and "Proc.
Saclay (1971)" refer respectively to:
- BOCK, R. and HERING, W.R., eds: Proceedings of the International
  Conference on Nuclear Reactions Induced by Heavy Ions, Heidelberg
  1969 (North Holland, Amsterdam, 1970).
- SIEMSSEN, R.H., MORRISON, G.C. and SCHIFFER, J.P., eds.: Proceedings
  of the Symposium on Heavy Ion Scattering held at Argonne National
  Laboratory, March 1971. To be published.
- Proceedings of the Symposium on Heavy Ion Reactions and Many Parti-
  cle Excitations held at Saclay, September 1971 appear in Journal de
  Physique 32, Colloque C-6, Supplement au n° 11-12 (1971).

1. ALMQVIST,E., BROMLEY, D.A. and KUEHNER, J.A., Phys. Rev. Letters
   4, 515 (1960).
2. ALMQVIST, E., BROMLEY, D.A., KUEHNER, J.A. and WHALEN, B., Phys.
   Rev. 130, 1140 (1963).
3. ALMQVIST, E., KUEHNER, J.A., McPHERSON, D. and VOGT, E.W., Phys.
   Rev. 136 B, 84 (1964).
4. ALMQVIST, E., BROMLEY, D.A. and KUEHNER, J.A., Proc. Int. Conf.
   on Nuclear Structure, Kingston, Canada, (North Holland, 1960)
   p. 258.
5. ALMQVIST, E., BROMLEY, D.A., KUEHNER, J.A., Proceedings of Second
   Conference on Reactions between Complex Nuclei, Gatlinburg,
   (Wiley 1960) p. 282.
6. ARIMA, A., GILLET, V. and GINOCCHIO, J., Phys. Rev. Letters 25
   (1970) 1043.
7. BALAMUTH, D.P., HOLDEN, J.E., NOE, J.W. and ZURMUHLE, R.W., Phys.
   Rev. Letters 26, 1271 (1971).
8. BETHGE, K., Proc. Heidelberg 277 (1969).
9. BLANN, M., invited lecture at this conference. Phys. Rev. Letters
   21, 1357 (1968).
10. BONDORF, J., private communication.
11. BONDORF, J.P. and LEACHMAN, R.B., Mat. Fys. Medd. Dan. Vid. Selsk.
    34 no. 10 (1965).
12. BORGGREEN, J., ELBEK, B. and LEACHMAN, R.B., Kgl. Dan. Vidensk.
    Selsk. Mat.-Fys. Medd. 34 No. 9 (1964).
13. BROMLEY, D.A., Proc. Heidelberg 33 (1969). This paper reviews the
    experimental and theoretical work on molecular interactions up to

1969.

14. BROMLEY, D.A., KUEHNER, J.A. and ALMQVIST, E., Phys. Rev. Letters 4, 365 (1960).
    Phys. Rev. 123, 878 (1961).

15. BROMLEY, D.A., CHUA, L., GOBBI, A., MAURENZIG, P.R., PARKER, P.D., SACHS, M.W., SHAPIRA, D., STOKSTAD, R.G. and WIELAND, R., Proc. Saclay 5 (1971).

16. CALLENDAR, W.D. et al., private communication.

17. CHARLES, P., BALLINI, R., DOST, M., FERNANDEZ, B., FOUAN, P. and GASTEBOIS, J., Proc. Saclay 155 (1971).

18. CHARLES, P., DOST, M., FERNANDEZ, B., GASTEBOIS, J. and LEE, S.M., European Conference on Nuclear Physics, Aix-en-Provence , Contribution 11.77 (1972).

19. COSMAN, E.R., SPERDUTO, A., MOORE, W.H., CHIN, T.N. and CORMIER, T.M., Phys. Rev. Letters 27, 1074 (1971).

20. DAVIS, R.H., Phys. Rev. Letters 4, 521 (1960).

21. DESCHLER, H., SCHÖTER, H., FUCHS, H., BAUM, L., GAUL, G., LÜDECKE, H., SANTO, R. and STOCK, R., Phys. Rev. Letters 28, 694 (1972).

22. ERICSON, T., Ann. Phys. 23, 390 (1963).

23. FESHBACH, H., Comments on Nuclear and Particle Physics, vol. I, no. 1, 40 (1967).

24. GASTEBOIS, J., BALLINI, R., CHARLES, P., FERNANDEZ, B. and FOUAN, J., Lettere al Nuovo Comento 2, 90 (1971).

25. GASTEBOIS, J., Proc. Saclay 57 (1971).

26. Such calculations are in progress at Yale; GOBBI, A., SHAPIRA, D., STOKSTAD, R., WIELAND, R., private communication.

27. GOBBI, A., Proc. Argonne 63 (1971).

28. GOBBI, A., SACHS, M.W., REILLY, W., WIELAND, R. adn BROMLEY, D.A., Proc. Heidelberg 99 (1969).

29. GOBBI, A., MAURENZIG, P.R., CHUA, L., HADSELL, R., PARKER, P.D., SACHS. M.W., SHAPIRA, D., STOKSTAD, R., WIELAND, R. and BROMLEY, D.A., Phys. Letters 26, 396 (1971).

30. GREENWOOD, L.R., BRAID, T.H., KATORI, K., STOLTZFUS, J.C. and SIEMSSEN, R.H., Proc. Saclay 199 (1971).

31. GRIFFIN, J.J., Phys. Rev. Letters 17, 478 (1966).

32. HALBERT, M.L. and NAGATANI, K., Bull. Am. Phys. Soc. 17, 530 (1972), and private communication.

33. HALBERT, M.L., DURHAM, F.E. and VAN der WOUDE, A., Phys. Rev. 162, 899 (1967).

34. IMANISHI, B., Phys. Letters 27B, 267 (1968).
    Nucl. Phys. A125, 33 (1968).

35. LOW, K.S. and TAMURA, T., Phys. Letters 40B, 32 (1972).

36. MAHER, J.V., SACHS, M.W., SIEMSSEN, R.H., WEIDINGER, A. and BROMLEY, D.A., Phys. Rev. 188, 1665 (1969).

37. MALMIN, R.E., SIEMSSEN, R.H., SINK, D.A. and SINGH, P.P., Phys. Rev. Letters 28, 1590 (1972).

38. MALMIN, R.E., KATORI, K., GREENWOOD, L.R., BRAID, T.H. and SIEMSSEN, R.H., Proc. Saclay 223 (1971).

39. MICHAUD, G. and VOGT, E.W., Phys. Letters 30B, 85 (1969), Phys. Rev. C5, 350 (1972).

40. MIDDLETON, R., GARRETT, J.D. and FORTUNE, H.T., Phys. Rev. Letters 24, 1436 (1970).

41. MIDDLETON, R., GARRETT, J.D., FORTUNE, H.T. and BETTS, R.R., Proc. Saclay 39 (1971).

42. MIDDLETON, R., GARRETT, J.D. and FORTUNE, H.T., Phys. Rev. Letters 27, 950 (1971).

43. MIDDLETON, R., GARRETT, J.D., FORTUNE, H.T. and BETTS, R.R., Phys. Rev. Letters 27, 1296 (1971).

44. MOLDAUER, P.A., Phys. Rev. Letters 18, 249 (1967).

45. NOBLE, J.V., Phys. Rev. Letters 28, 111 (1972).

46. PATTERSON, J.R., WINKLER, H. and ZAIDINS, C.S., Ap. J. 157, 367 (1969). See also Mazarakis, M.G., DEBOLT, G.O., Jr. and STEPHENS, W.E., Bull. Am. Phys. Soc. 16, 600 (1971).

47. PATTERSON, J.R., NAGORCKA, B.N., SYMONS, G.D. and ZUK, W.M., Nucl. Phys. A165, 545 (1971).

48. REILLY, E., WIELAND, R., GOBBI, A., SACHS, M.W., MAHER, J.V., MINGAY, D., SIEMSSEN, R.H. and BROMLEY, D.A., Proc. Heidelberg 95 (1969).

49. RINAT, A.S., (Reiner), Phys. Letters 38B, 281 (1972).

50. ROSSNER, H.H., HINDERER, G., WEIDINGER, A. and EBERHARD, K.A., European Conference on Nuclear Physics, Aix-en-Provence, Contribution 11.6. (1972).

51. SCHEID, W., invited lecture at this conference.
    FINK, H.J., SCHEID, W. and GREINER, W., Nucl. Phys. A188, 259 (1972).
    GREINER, W. and SCHEID, W., Proc. Saclay 91 (1971). References to earlier works of GREINER, W., SCHEID, W. and collaborators may be found in this comprehensive article.

52. SHAW, R.W., NORMAN, J.C. and VANDENBOSCH, R., Phys. Rev. 184, 1040 (1969).

53. SIEMSSEN, R.H., MAHER, J.V., WEIDINGER, A. and BROMLEY, D.A., Phys. Rev. Letters 19, 369 (1967).

54. SIEMSSEN, R.H., Proc. Argonne 145 (1971).

55. SINGH, P.P., SINK, D.A., SCHWANDT, P., MALMIN, R.E. and SIEMSSEN,
    R.H., Proc. Saclay 279 (1971).
    Phys. Rev. Letters $\underline{28}$, 1714 (1972).

56. SPINKA, H., Ph. D. Thesis, California Institute of Technology,
    1971, unpublished.

57. STOKSTAD, R., SHAPIRA, D., CHUA, L., PARKER, P., SACHS, M.W.,
    WIELAND, R. and BROMLEY, D.A., Phys. Rev. Letters $\underline{28}$, 1523 (1972).

58. STOKSTAD, R.G., SHAPIRA, D., CHUA, L., GOBBI, A., PARKER, P.,
    SACHS, M., WIELAND, R. and BROMLEY, D.A., Bull. Am. Phys. Soc. $\underline{17}$,
    529 (1972), and to be published.

59. VANDENBOSCH, R., Proc. Argonne 103 (1971).

60. VOGT, E. and McMANUS, H., Phys. Rev. Letters $\underline{4}$, 518 (1960).

61. VOGT, E.W., McPHERSON, D., KUEHNER, J. and ALMQVIST, E., Phys. Rev.
    $\underline{136B}$, 99 (1964).

62. VOGT, E. in Advances in Nuclear Physics Vol. 1, Ed. M. Baranger,
    VOGT, E., (Plenum N.Y.) p. 261 , 1968.

63. VOGT, E.W., McPHERSON, D., KUEHNER, J. and ALMQVIST, E., Phys. Rev.
    $\underline{136B}$, 99 (1964).

64. VOIT, H., ISCHENKO, G., SILLER, F. and HELB, H.D., Nucl. Phys.
    $\underline{A179}$, 23 (1972).

65. VOIT, H., HARTMANN, G., HELB, H.D., ISCHENKO, G. and SILLER, F.,
    Preprint and private communication.

66. VOIT, H., private communication.

INTERMEDIATE STRUCTURE IN ISOBARIC
ANALOGUE RESONANCES

M. PETRASCU

Institute for Atomic Physics, Bucharest, Romania

Introductory Remark

In my talk I will refer to topics which I consider most relevant,
at the present state of knowledge, to the subject of isobaric analogue
resonances as intermediate structure. These topics are:"gross structure
of isobaric analogue resonances as intermediate structure" in which
special consideration is given to the fine structure enhancement in
the vicinity of IAR and "substructures in isobaric analogue resonances"
in which the recent experimental results, indicating in some cases the
presence of substructures (different from the fine structure) within
the isobaric analogue resonances gross structures, are discussed.

1. Gross Structure of Isobaric Analogue Resonances as Intermediate
   Structure

   1.1. Introduction

By intermediate structure we understand the structure appearing in
the energy dependence of the cross-section, characterized by a width
much smaller than the giant resonance width, but much larger than the
compound nucleus fine structure width.

It can be reminded that the giant resonance width is of the order
of 1 MeV, and the fine structure width is, for instance in the case of
the resolved neutron resonances, of the order of 1 eV. The intermediate structure width is of the order of 100 keV.

The interpretation of the intermediate structure was given by
Feshbach, Kerman and Lemmer [5], by considering a mode of excitation
of the type 2p - 1h. This mode of excitation is more complex than the
mode (single particle virtual states) responsible for the giant resonances but less complex than the mode (compound nucleus states) explai-
ning the ultimate fine structure.

The mode of excitation responsible for the intermediate structure
has been called doorway states, and they are supposed to be the only
states that couple strongly with the entrance channel. These states,
of course are not eigenstates of the nuclear hamiltonian, for they represent only a small component of the total nuclear wave function. Having the nucleus in a doorway state, the probability of finding the
nucleus in this state would decrease in time. There are two components

contributing to the decay rate. One component is the decay into the
entrance channel. The other component is the decay into the more complicated states, to which the doorway states are coupled. The first
component is usually defined by the width $\Gamma^\uparrow$ called escape width, and
the second component by the width $\Gamma^\downarrow$, called the spreading width or
damping width.

For the isobaric analogue resonances, one of the clearest example
of intermediate structure, various microscopic theories give explicit
formulas for the escape and the spreading width [16], [18], [24]. As a
consequence of the spreading, the fine sturcture levels in the vicinity of an isobaric analogue resonance are enhanced, in a manner that
was quantitatively described by Robson [19] and by Mac Donald and Mekjian [13]. This enhancement is one of the basic aspects of IAR treated
as intermediate structure, and in the following I will focus my attention on this particular problem.

Firstly I will give an outline of theoretical work concerning the
enhancement factor, and subsequently I will describe some recent experimental results.

1.2. Outline of Theoretical Work on Fine Structure Enhancement

The first theoretical interpretation of the fine structure enhancement was given by Robson [19]. Assuming that the mixing of the $T_>$
and $T_<$ states can be considered negligeable inside the nucleus, Robson
developped his theory on the basis of R-matrix theory.

Thus are defined the R matrix states $\chi_{nc\lambda}$ for the $\|nc>$ system
satisfying the usual orthonormality relation

$$\int d\Omega \int^{a_c} dr \chi_{nc\lambda}\chi_{nc\lambda'} = \delta_{\lambda\lambda'}$$

and the boundary conditions at $r=a_c$

$$a_c \left[\partial U_{nc\lambda} | \partial r\right] r = a_c = B_c U_{nc\lambda}$$

B_c is the boundary matching parameter and $U_{nc\lambda}$ is the radial part of
$\chi_{nc\lambda}$. Expanding χ_{nc} in terms of $\chi_{nc\lambda}$ one obtains the R matrix relation
at $r=a_c$

$$U_{nc} = R_{nc} \left[a_c (\frac{\partial U_{nc}}{\partial r})_{r=a_c} - B_c U_{nc}\right]$$

in which

$$R_{nC} = \sum_{\lambda} \frac{\gamma_{nC\lambda}^2}{E_{\lambda} - E_n}$$

and

$$\gamma_{nC\lambda}^2 = (\frac{h^2}{2Ma_C})(U_{nC\lambda})^2$$

For the (p c) system which is not pure in isobaric spin one has the matrix equations at $r = a_C$

$$\begin{pmatrix} U_{>} \\ U_{<} \end{pmatrix} = \begin{pmatrix} R_{>>} & 0 \\ 0 & R_{<<} \end{pmatrix} \begin{pmatrix} a_C[\frac{\partial U_{>}}{\partial r}]_{r=a_C} - B_C U_{>} \\ a_C[\frac{\partial U_{<}}{\partial r}]_{r=a_C} - B_C U_{<} \end{pmatrix}$$

R is diagonal, internal mixing being neglected. Because $U_{>}$ and $U_{<}$ are not physical cahannels one has to consider the R-matrix relation for the physical channels U_{pC} and U_{nA}

$$\begin{pmatrix} U_{pC} \\ U_{nA} \end{pmatrix} = \begin{pmatrix} R_{pp} & R_{pn} \\ R_{np} & R_{nn} \end{pmatrix} \begin{pmatrix} a_C[\frac{\partial U_{pC}}{\partial r}]_{r=a_C} - B_C U_{pC} \\ a_C[\frac{\partial U_{nA}}{\partial r}]_{r=a_C} - B_C U_{nA} \end{pmatrix}$$

The physical R matrix elements R_{pp}, R_{pn}, R_{np} and R_{nn} are related to $R_{>>}$ and $R_{<<}$ by

$$R_{pp} = (2T_O + 1)^{-1} [R_{>>} + 2T_O R_{<<}]$$

$$R_{pn} = (2T_O + 1)^{-1} (2T_O)^{1/2} [R_{>>} - R_{<<}] = R_{np}$$

$$R_{nn} = (2T_O + 1)^{-1} [2T_O R_{>>} + R_{<<}]$$

The collision matrix U can be expressed in terms of R matrix using the definition of Lane and Thomas [10]. In the case of two channels and isolated resonances the following relation can be derived:

$$U_{pp} = \exp[2i(\omega_p - \phi_p)] \{1 + \frac{i\Gamma_A}{E_A - E_p - \frac{1}{2}i\Gamma_A}$$
+ (ctd. next page)

$$+ 2iP_p \left[\frac{E_> -E_p}{E_A -E_p -\frac{iTA}{2}} \right]^2 x \ (1 - \frac{2T_o}{2T_o +1} R_{<<} L_p')^{-1} \frac{2T_o}{2T_o +1} R_{<<} \}$$

in which ω_p and ϕ_p are the Coulomb and hard sphere phase shifts. The analogue state parameters are:

$$T_A = 2P_p \gamma_A^2$$

$$E_A = E_p + \Delta_A = E_\lambda^P \Delta_c + \Delta_A$$

$$\Delta_A = -(S_p^+ - B_C) \gamma_A^2$$

$$\gamma_A^2 = (2T_o + 1)^{-1} \gamma_{nC\lambda}^2$$

The function L_p' is given by

$$L_p' = L_p^o \frac{E_> -E_p}{E_A -E_p -\frac{1}{2} iT_A} = L_p^o f_p$$

with L_p^o defined by $L_p^o = S_p -B_C +iP_p$

The factors f_p in U_{pp} are due to the external mixing. If this mixing would be neglected then the collision matrix is diagonal in isobaric spin and we would obtain U_{pp} but without the multiplying factors f_p.

The enhancement of fine structure levels described by the factor f_p is not symetric as can be easily seen. However the asymetry of the enhancement factor was shown by Mac Donald and Mekjian [13] to be not the general case. They used a K-matrix formulation and a shell model approach to a unified reaction theory. The representation of the K-matrix is obtained by diagonalizing the hamiltonian H_s, which describes the independent particles with residual interactions, on the set of door-way and hall-way states. The eigen vectors are

(a) $|\Psi> = \sum_i |D_i><D_i|\Psi_\lambda> + \sum_i |h_i><h_i\Psi_\lambda>$

satisfying

(b) $<\Psi_\lambda|H_s|\Psi_\mu> = E_\lambda \delta_{\lambda\mu}, <\Psi_\lambda|\Psi_\mu> = \delta_{\lambda\mu}$

The K matrix elements are

$$
K_{cc'} = <c|v|c'> + (2\pi)^{-1} \sum \frac{\Gamma_{\lambda c}^{1/2} \Gamma_{\lambda c'}^{1/2}}{E - E_>}
$$

where

$$
\Gamma_{\lambda c}^{1/2} = (2\pi)^{1/2} \sum_j <D_j|\Psi_\lambda>^* <D_j|v|c>
$$

The distribution of the widths $\Gamma_{\lambda c}$ as a function of E_λ can be found from the matrix equation (b).

In the case when one assumes a single door-way strongly coupled to the continuum and to a set of hall-ways which are also directly coupled to the continuum but with a coupling much weaker than for the door-way, it follows that the width $\Gamma_{\lambda c}$ is the result of the coherent contribution of decay amplitudes.

$$
\Gamma_{\lambda c} = 2\pi |<D|V_c|c,\varepsilon_c> - \sum \frac{<D|V_c|h_i><h_i|V_c|C,\varepsilon_c>}{E_\lambda - \varepsilon_i}|^2 \times |<D|\Psi_\lambda>|^2
$$

The distribution in width, exhibits the assymetry obtained by Robson if the magnitude and the relative phase of matrix elements $<D|V_c|h_i>$ and $<h_i|V_c|C,\varepsilon_c>$ remain the same across the region of enhancement fine structure.

1.3. Recent Experimental Results
A comprehensive review concerning previous experimental results on fine structure enhancement was given by Lane [11]. Data described in that review are based on fine structure resolving measurements. Recently results on ^{59}Co were obtained with the same method [12].

In the following I will present a recent experiment based on a new approach. It is an experiment performed at Rutgers [7] in which it was possible to observe by direct time measurement the fine structure enhancement due to an isobaric analogue resonance. In this experiment the so called blocking effect technique was used. This technique consists in detecting the supression of the elastic or inelastic yield, along a crystallographic axis. The elastic yield can be considered as prompt, being almost entirely Rutherford. The inelastically scattered protons, being delayed by the compound nucleus, would display shallower

dips in the yield along the crystallographic axis than the elastically
scattered protons. The principle of lifetime measurement by the blocking effect is schemetically illustrated in fig. 1.

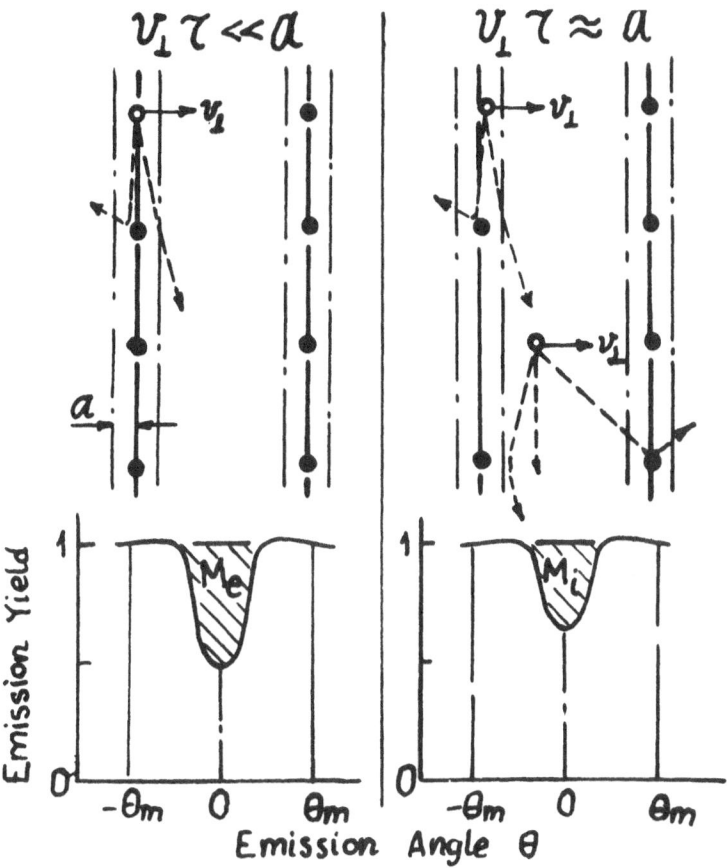

Fig. 1. Diagram illustrating the principle of lifetime
measurement by the blocking effect technique. The
case of a charged particle emitted from an excited nucleus in the atomic row is drawn in the
left part and that of a particle emitted from a
recoiling excited nucleus with a finite value
$v_\perp \tau$ in the right part. Full and open circles
represent the lattice nuclei and the excited nuclei
respectively. The emission angle θ is measured
with respect to the atomic row

In this experiment a germanium crystal 1.5 μ thick, corresponding to 30 keV energy loss of the incident proton beam was used. In order to avoid distortion, the crystal was heated during the bombardment. The scattering angle was chosen as 90°, and the 110 axis of germanium was chosen as a blocking axis. The protons were detected by the aid of a silicon position sensitive detector, located 130 cm from the target. Fig. 2 shows the blocking patterns in the case of two analogue resonances in ^{73}As. The resonance at 5.035 MeV was identified to be an $\ell = 2$ resonance, while the one at 5.110 MeV was identified to be an $\ell = 0$ resonance.

The first IAR gives a bump in the inelastic excitation function to the first 2^+ state in ^{72}Ge, and the second IAR a bump corresponding to the second 0^+ excited state in ^{72}Ge, as can be seen from fig. 3. Thus being on the first IAR, one is at the same time on resonance with the first 2^+ state but off resonance with the second 0^+ state, and being on the second IAR one is at the same time off resonance with the 2^+ state and on resonance with the 0^+ state in ^{72}Ge. Table 1 gives the results of lifetime measurements. As seen from this table, there is a marked difference in transition lifetimes, on and off resonance to the 2^+ state. On the basis of the obtained values the authors obtained crude estimate of the enhancement factor

$$f = 2.5$$

One can also see from table 1 that the lifetime for the proton group feeding the 2^+ state in ^{72}Ge is about 30 percent larger than the group going to the 0^+ excited state of ^{72}Ge. This is expected on the basis of angular momentum considerations.

I have selected this experiment not only because of its relation to the basic aspects of IAR as intermediate structure but also to illustrate the possibilities of a new experimental method.

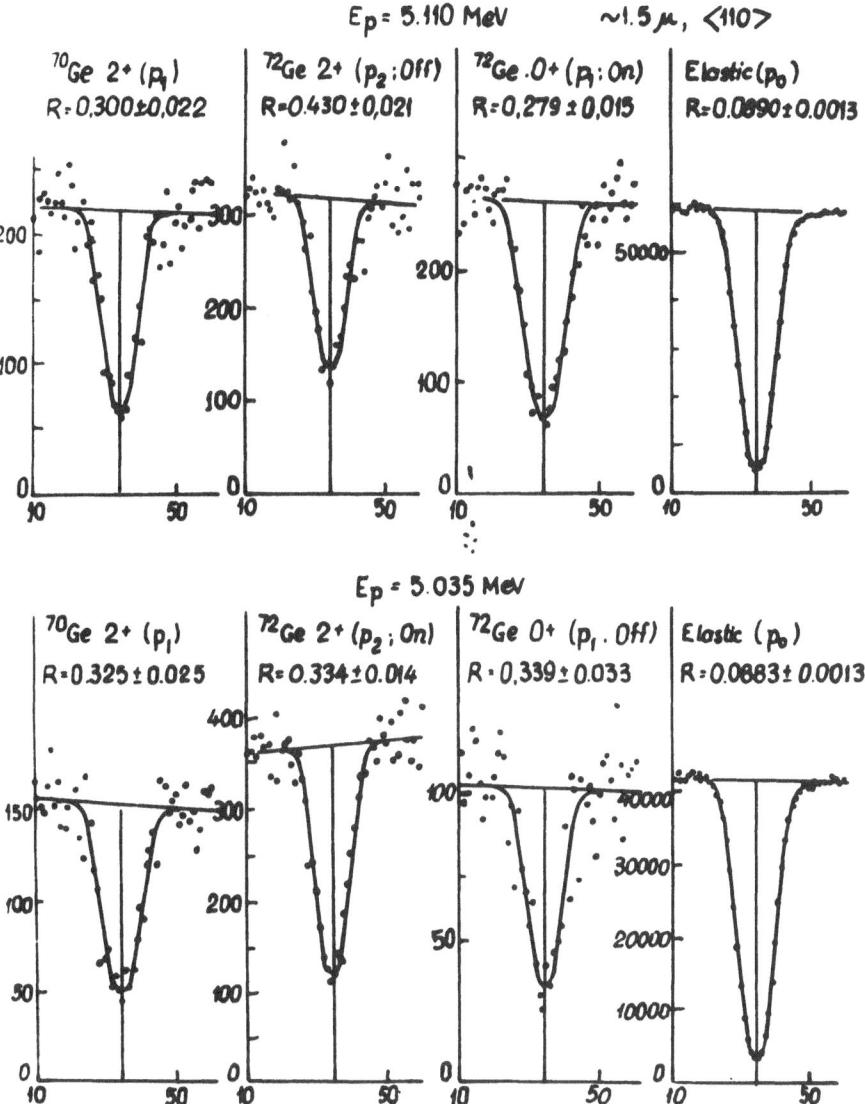

Fig. 2. Blocking "dips" obseved at two proton energies
along the 110 axis, for 1.5 μ thick Ge crystal

TABLE I Blocking lifetime results for Ge + p along the 110 axis

	R or $R_{l'}$	$R - R_{l'}$	v (Å)	(as)[a]
E_p = 4.200 MeV				
Ge(p,p$_o$)	0.0888 ± 0.0025 $R_{l'}$			
E_p = 5.035 MeV				
Ge(p,p$_o$)	0.0883 ± 0.0013			
^{70}Ge(2+)p$_1$	0.325 ± 0.025	0.236 ± 0.025	0.258 ± 0.020	58.8 ± 4.6
^{72}Ge(2+)p$_2$	0.334 ± 0.014 (On)[b]	0.245 ± 0.014	0.265 ± 0.011	62.1 ± 2.6
^{72}Ge(0+)p$_1$	0.339 ± 0.033 (Off)[b]	0.250 ± 0.033	0.269 ± 0.027	63.1 ± 6.3
E_p = 5.110 MeV				
Ge(p,p$_o$)	0.0890 ± 0.0013			
^{70}Ge(2+)p$_1$	0.300 ± 0.022	0.211 ± 0.022	0.238 ± 0.017	53.9 ± 3.8
^{72}Ge(2+)p$_2$	0.430 ± 0.021 (Off)[b]	0.341 ± 0.021	0.349 ± 0.021	81.2 ± 4.9
^{72}Ge(0+)p$_1$	0.275 ± 0.015 (On)[b]	0.186 ± 0.015	0.219 ± 0.011	51.0 ± 2.6
^{74}Ge(2+)p$_1$	0.161 ± 0.017	0.072 ± 0.017	0.132 ± 0.014	31.6 ± 3.3

[a] Obtained with a formula from: W.M. Gibson, K.O. Nielsen, Phys. Rev. Lett. 24 114 (1970) and $r_c = 3a_{TF} = 0.422$ Å. 1 as = 10^{-18} sec.

[b] Refers to on or off the appropriate resonance for this channel. R is the ratio of minimum to shoulder values.

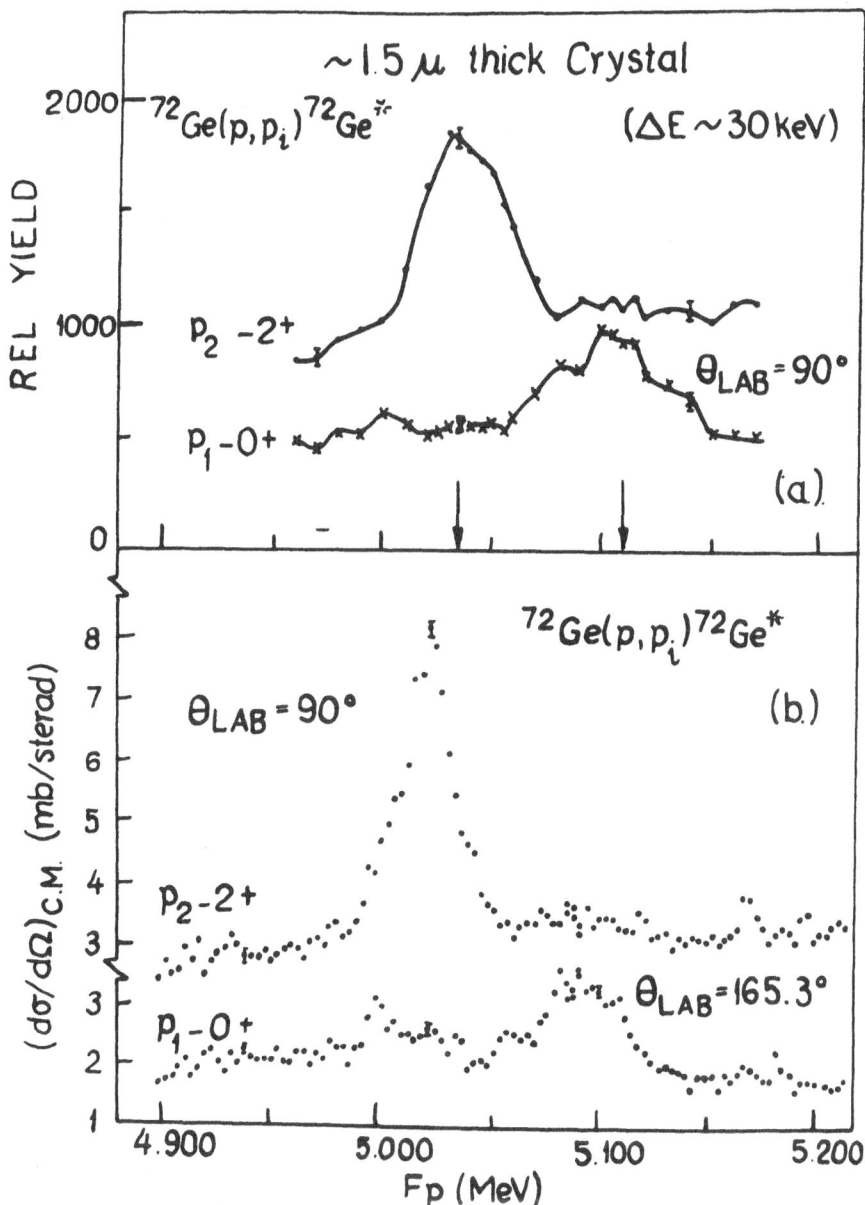

Fig. 3. (a) Excitation curve for 90° inelastic proton scattering on ^{73}Ge to first excited 0^+ and second excited 2^+ states for 30 keV thick crystal used.
(b) Thin target (3 keV) excitation curves for the
same states as in (a). $\ell = 2$ IAR in ^{73}As at 5.022
MeV; $\ell = 0$ IAR at 5.094 MeV

2. Substructures in IAR

2.1. Introduction

Until last year the only known substructure of IAR was the compound nucleus fine structure. Since then, experimental results were
obtained indicating the presence of substructures considerably larger
in width than the width of enhanced fine structure but narrower than
the IAR gross-structure width. By establishing a hierarchy of the particle hole configurations in which at the top are the doorway configurations and at the bottom the compound nucleus configurations, the configuration of the substructures to be discussed here are located somewhere in the middle, that is a rank bellow the doorway but a rank above
the compound nucleus configurations.

In the following I will present the results obtained in ^{70}Ge and
^{66}Zn, which appear to give evidence for such substructures.

2.2. Experimental Data on ^{70}Ge

Experimental results, obtained by studying elastic and inelastic
scattering of protons on ^{70}Ge (enriched 91.4 %, 3 keV thick) were reported last qear [20], [21]. These experiments were started in order to
locate the position of analogue resonances, for lifetime measurements
by the blocking effect. Fig. 4 shows the excitation function for the
inelastic scattering ^{70}Ge (p,p') via a ℓ = 0 IAR at 5.06 MeV in ^{71}As.
In this inelastic scattering the residual ^{70}Ge nucleus is left in its
first 2^+ excited state. The inelastic scattering, leading to the next
2^+, 0^+ and the unresolved 2^+ + 4^+ states shows a completely different
shape, as can be seen from fig. 5.

After observing the striking feature of the inelastic scattering,
we examined more closely the pattern in the case of elastic scattering.
Fig. 6 reveals several bumps within the IAR range.

It is to be noted that one of these small bumps corresponds to
the first peak in the inelastic function for the first 2^+ state. Fig.
7 shows a tentative fit of the structure in elastic scattering at 90°.
A special measurement was done at Rutgers in order that the ℓ = 0 parent state in ^{71}Ge is a single one. The results obtained, together
with data from a (d,p) experiment performed by Goldman [8], prove that
there is no other resonant state responsible for the behaviour of the
excitation function that was shown.

The possibility that these structures are Ericson fluctuations is
ruled out because:

1) they are correlated at different angles;
2) they are correlated in different channels, and
3) when one estimetes the ration, Γ/D, it is at most of the order

of unity. This can be obtained by using the experimental results of
Maruyama [14], concerning Γ and the calculations performed by Huizenga
[9], concerning D.

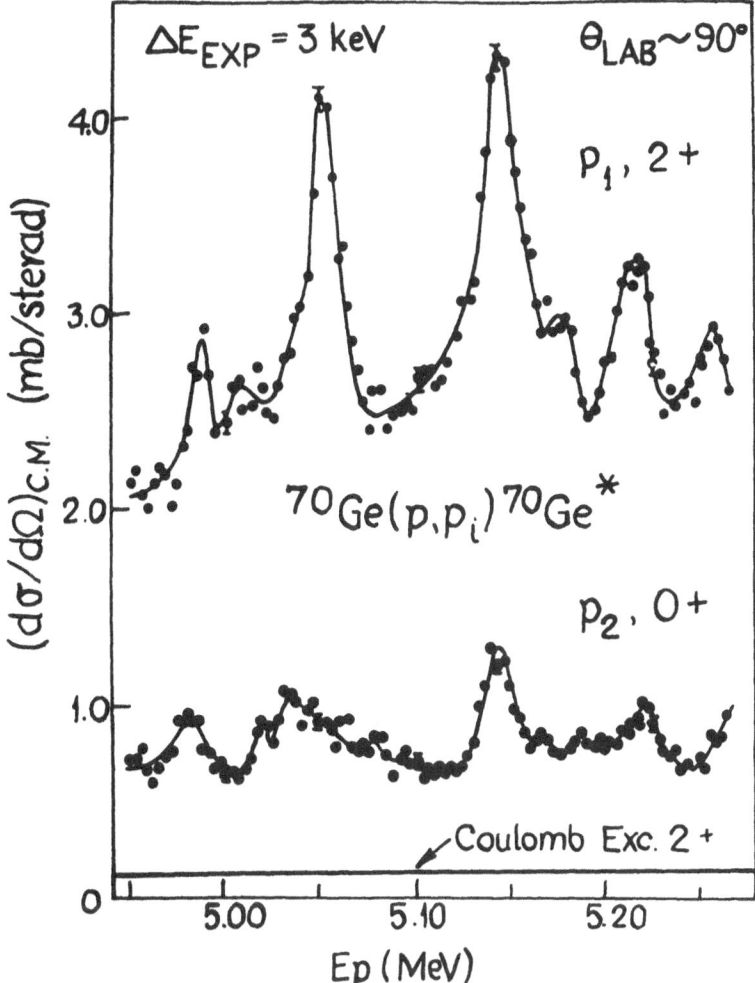

Fig. 4. Differential cross-section vs. proton energy for
inelastic scattering to the first (2^+, 1.040 MeV)
and second (0^+, 1.216 MeV) excited states of ^{70}Ge

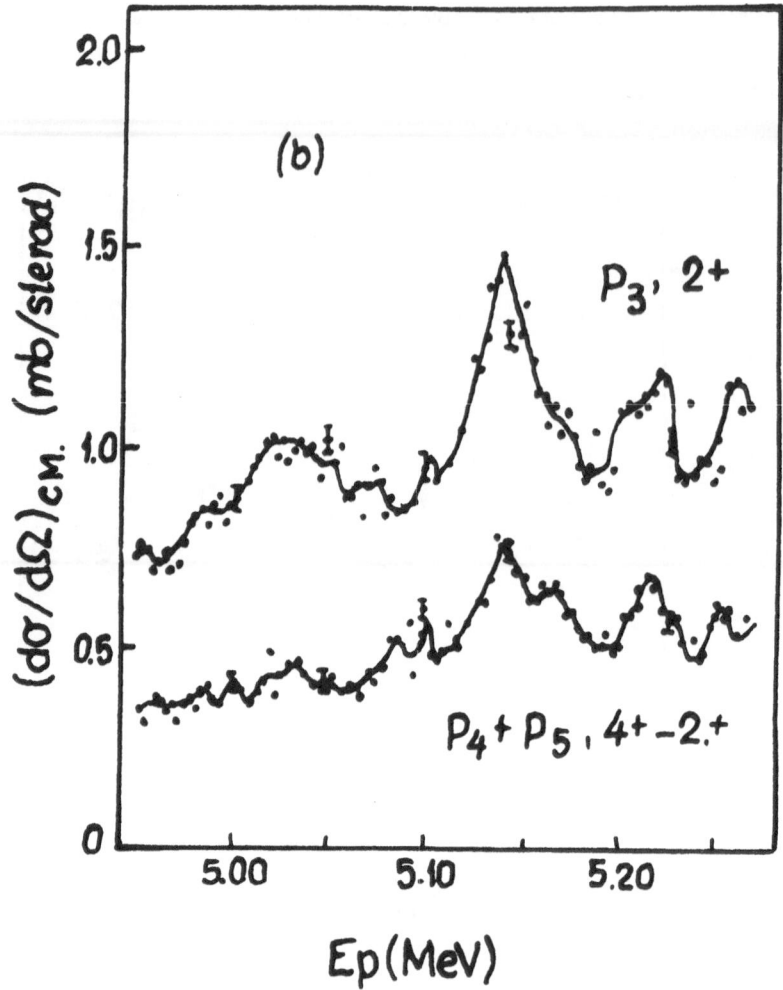

Fig. 5. Inelastic scattering excitation function on ^{70}Ge
leading to the second 2^+ and $2^+ + 4^+$ states

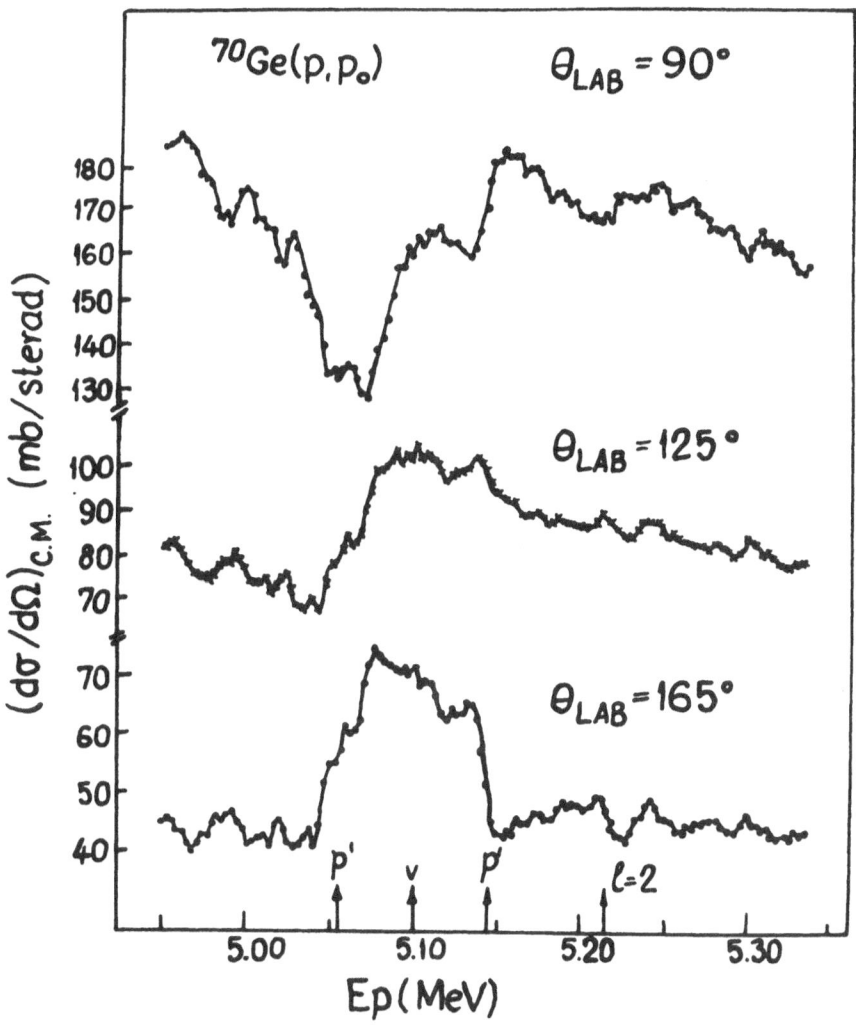

Fig. 6. ^{70}Ge + p elastic differential cross-section vs. proton energy at 90°, 125°and 165°

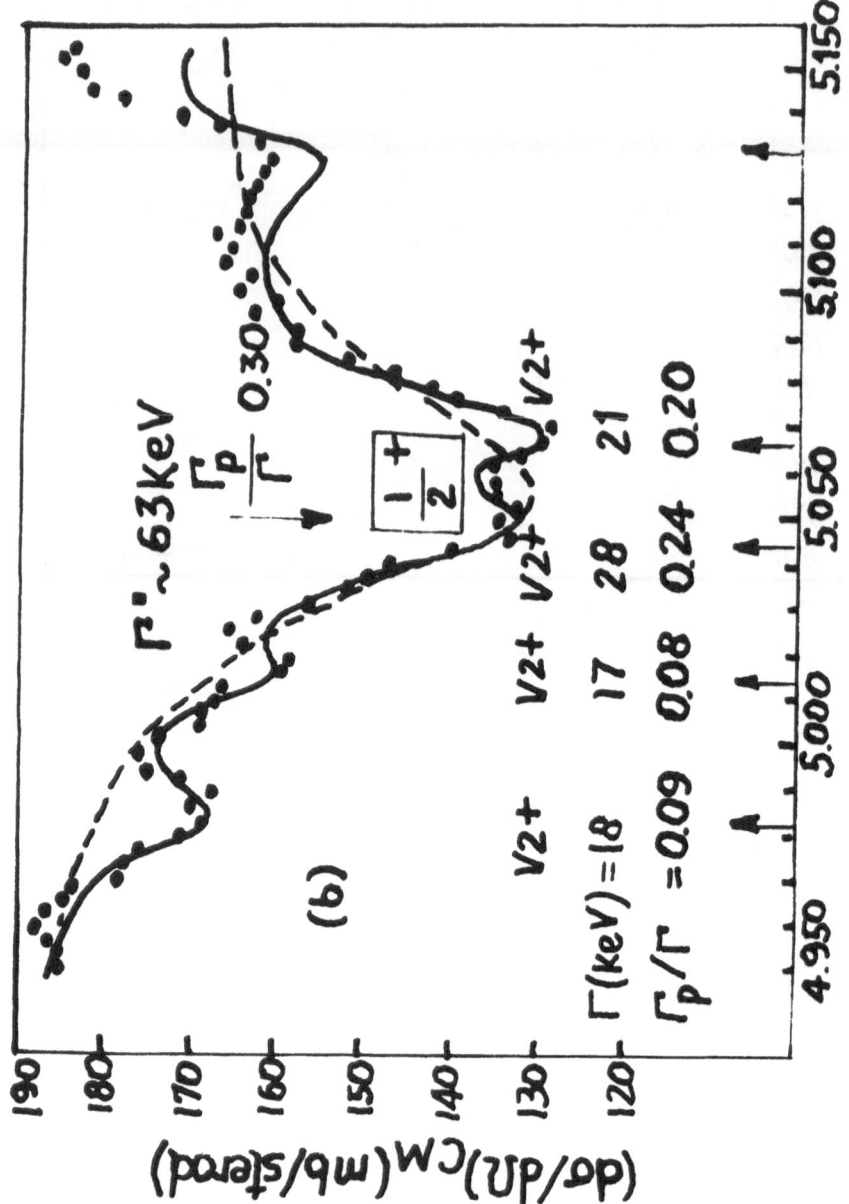

Fig.7. Tentative fit to the 5.06 MeV IAR at 90°

Fluctuations of the Ericson type can be present in the case of an IAR in ^{72}Ge [22] shown in fig. 8. There is an essential difference between the ^{70}Ge and ^{72}Ge cases. The IAR in the case of ^{72}Ge is well above the neutron threshold, while the resonance considered in ^{70}Ge is well below the neutron threshold. A Hauser-Feshbach estimate yields about 2700 eV for the width in the case of ^{73}As, and as concerns level density again the extrapolation of neutron capture data gives about 15 levels per keV. Hence the ratio Γ/D is of the order of 50 and therefore Ericson fluctuations cannot be ruled out.

Here at the conference I have learnt about recent high resolution measurement performed on ^{70}Ge by V. Mayer at Zürich [1]. It seems that these results are in agreement with the data discussed above [23].

2.3. Results Obtained on ^{66}Zn

I will show now some of the preliminary results obtained at Heidelberg [3]. Fig. 9 shows the inelastic excitation function in which protons around 4.400 MeV were scattered by ^{66}Zn through an analogue $\ell = 0$ resonance, leading to the first 2^+ excited state in ^{66}Zn. The IAR corresponds to a parent $\ell = 0$ state at 1.676 MeV in the ^{67}Zn nucleus. This state is excited with a large cross-section in (d,p) reaction as was found out by Ehrenstein and Schiffer [4], whose results are shown in table 2. As follows from these results there is no important state in the vicinity of the 1.676 MeV, $\ell = 0$ state which could explain the behaviour of the excitation function.

A similar pattern is shown by the inelastic scattering through a second $\ell = 0$ IAR, that appears at an energy 750 keV higher than the first one (fig. 10). Again there is no other important state in the vicinity. Apart from this marked structure in the inelastic scattering, no significant structure could be defined in the elastic scattering.

2.4. Possible Interpretation

Experimental data from (d,n) [17] (^{3}He,d) [6] and (^{4}He,t) [15] reactions show that the cross-section corresponding to the first 2^+ state in ^{66}Zn is quite large. Figs. 11, 12 and 13 show these data. One can also see that in the case of (d,n) reactions there is an increase in yield when going form ^{60}Ni to ^{66}Zn. Altogether these data indicate that the first 2^+ excited state is simple in terms of particle-hole configurations.

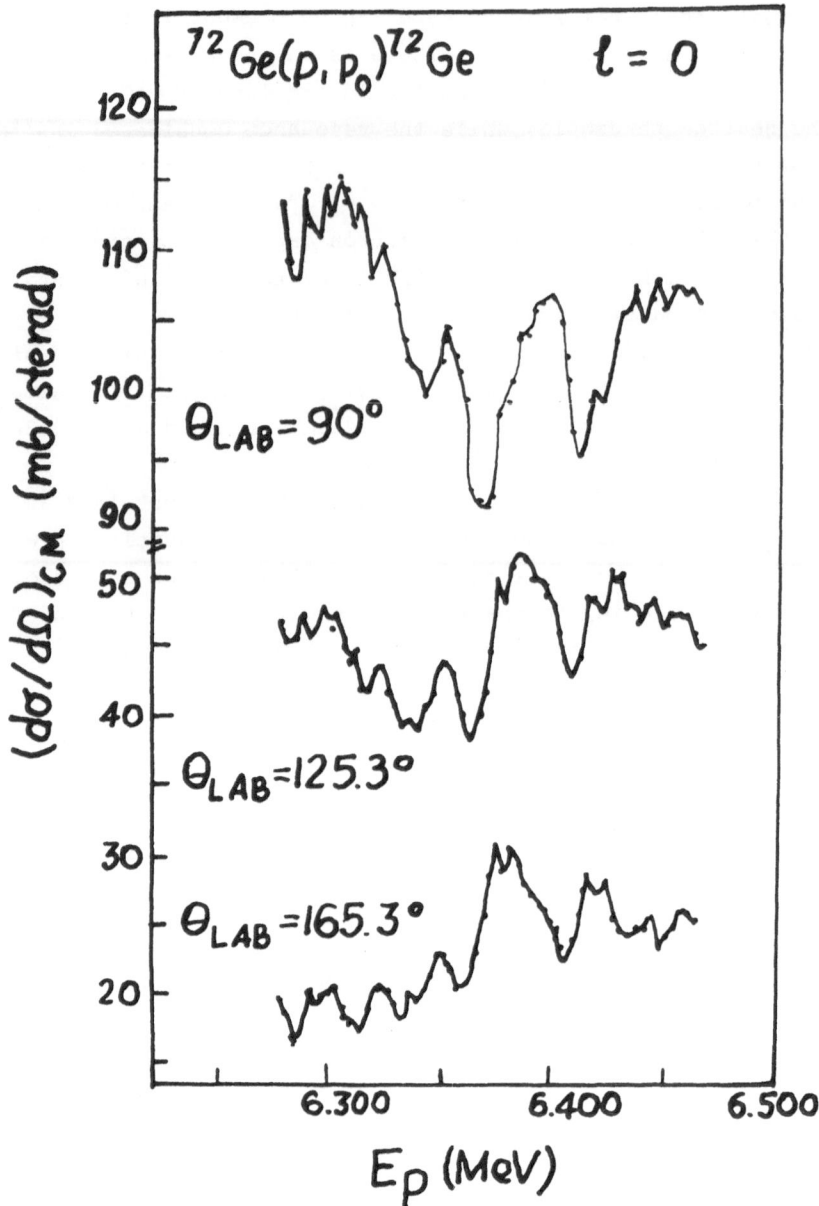

Fig. 8. ^{72}Ge + p elastic differential cross-section
vs. proton energy at 90°, 125.3° and 165.3°

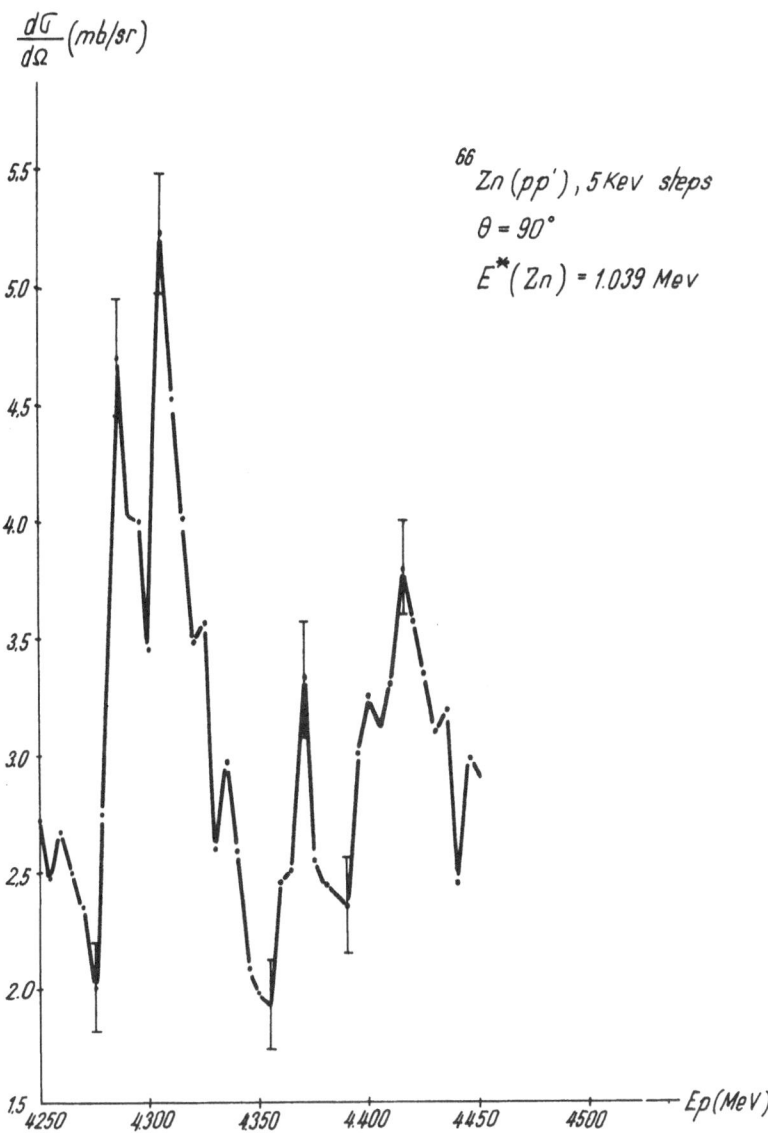

Fig. 9. Proton inelastic scattering excitation function
on ^{66}Zn via the ℓ = 0 IAR at E_p = 4.400 MeV

TABLE II
Summary of results from ^{66}Zn (d p) reactions[a]

E_x (MeV)	l	$J\pi$	σ_{max} (mb/sr)	$(2J+1)S$	$(2J+1)S$
1	2	3	4	5	6
0.0	3	$\frac{5}{2}^-$	0.56	1.7	1.8
0.093	1	$\frac{1}{2}^-$	3.8	0.82	1.10
0.184	1	$\frac{3}{2}^-$	0.29	0.06	0.08
0.390	1	$\frac{3}{2}^-$	3.7	0.75	1.03
0.602	4	$\frac{9}{2}^-$	1.1	5.1	8.4
0.978	2	$(\frac{5}{2})^+$	3.7	1.1	1.55
1.142	1	$\frac{1}{2}^-$	1.5	0.26	0.37
1.444	1	$\frac{3}{2}^-$	0.26	0.04	0.06
1.542	1	$\frac{1}{2}^-, \frac{3}{2}^-$ [b]	0.12	0.02	0.03
1.642...	c		0.09		
1.676	0	$\frac{1}{2}^+$	7.5[d]	0.2	0.23
1.782...	c		0.07		
1.808[e]	(0)	$(\frac{1}{2})^+$	0.06[d]	(0.001)	(0.001)
1.842	(1)	$(\frac{1}{2}^-, \frac{3}{2}^-)$ [b]	0.07	(0.01)	(0.015)
2.172[f]			0.04		
2.246...	c		0.06		

TABLE II ctd.

1	2	3	4	5	6
2.273	2	$(\frac{5}{2})^+$	0.66	0.16	0.22
2.407	2	$(\frac{5}{2})^+$	0.65	0.16	0.21
2.430	0	$\frac{1}{2}^+$	4.3[d]	0.11	0.12
2.609[e]	2	$(\frac{5}{2})^+$	0.19	0.04	0.06
2.648...	c,e		0.1		
2.797	2	$(\frac{5}{2})^+$	0.88	0.19	0.25
2.849	0	$\frac{1}{2}^+$	0.2[d]	0.006	0.006
3.233	2	$(\frac{5}{2})^+$	0.33	0.07	0.09
3.295	0	$\frac{1}{2}^+$	1.4[d]	0.04	0.04
3.480	2	$(\frac{5}{2})^+$	0.3	0.06	0.08
3.538...	c,e,f		(4)		
3.557	0	$\frac{1}{2}^+$	2[g]	0.25	0.27
3.607	0	$\frac{1}{2}^+$	0.85[g]	0.1	0.1
3.651...	f,h		(0.1)		
3.67...	f,h		(0.2)		
3.770...	c,e,f		0.9		
3.882...	f		(0.1)		
3.840...	f		(0.3)		

TABLE II   ctd.

1	2	3	4	5	6
3.863	$(0)^f$	$(\frac{1}{2})^+$	$0.52^{i)}$	(0.09)	(0.1)

a) An additional level at 887.87 ± 0.1 keV with $J^{\pi} = (3/2)^-$ is excited
   very weakly/σ_{max} 0.01 (mb/sr)/ in the present work.
b) No reliable data at backward angles
c) No ℓ assignment possible
d) At $\theta_{lab} = 5^O$
e) Probably several unresolved levels
f) Incomplete data
g) At $\theta_{lab} = 15^O$
h) Unresolved
i) At $\theta_{lab} = 35^O$.

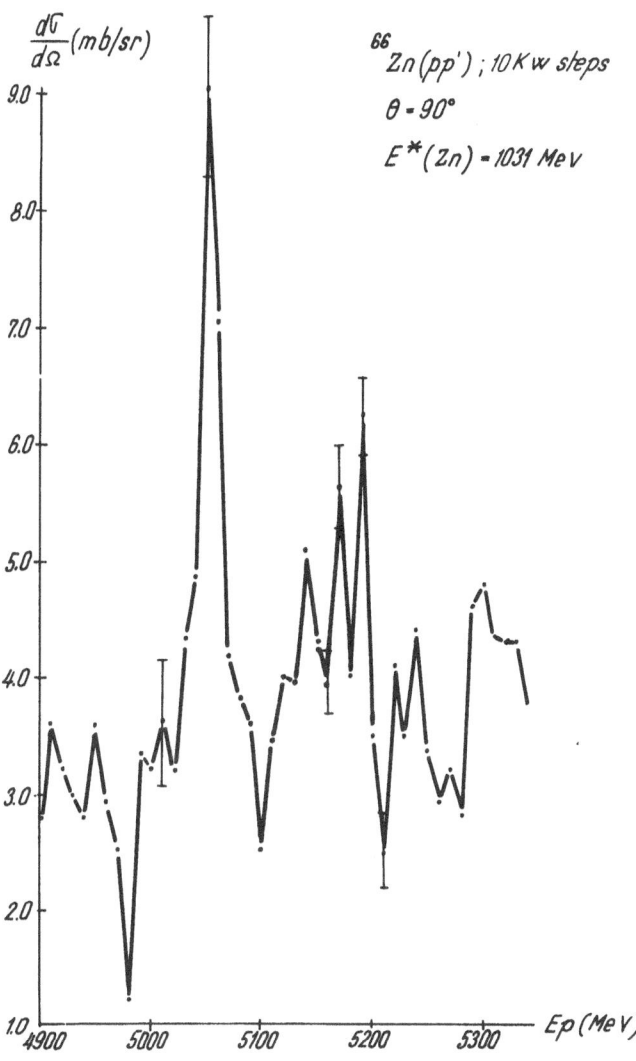

Fig. 10. Proton inelastic scattering excitation function
on ^{66}Zn through the ℓ = 0 IAR at E_p = 5.150 MeV

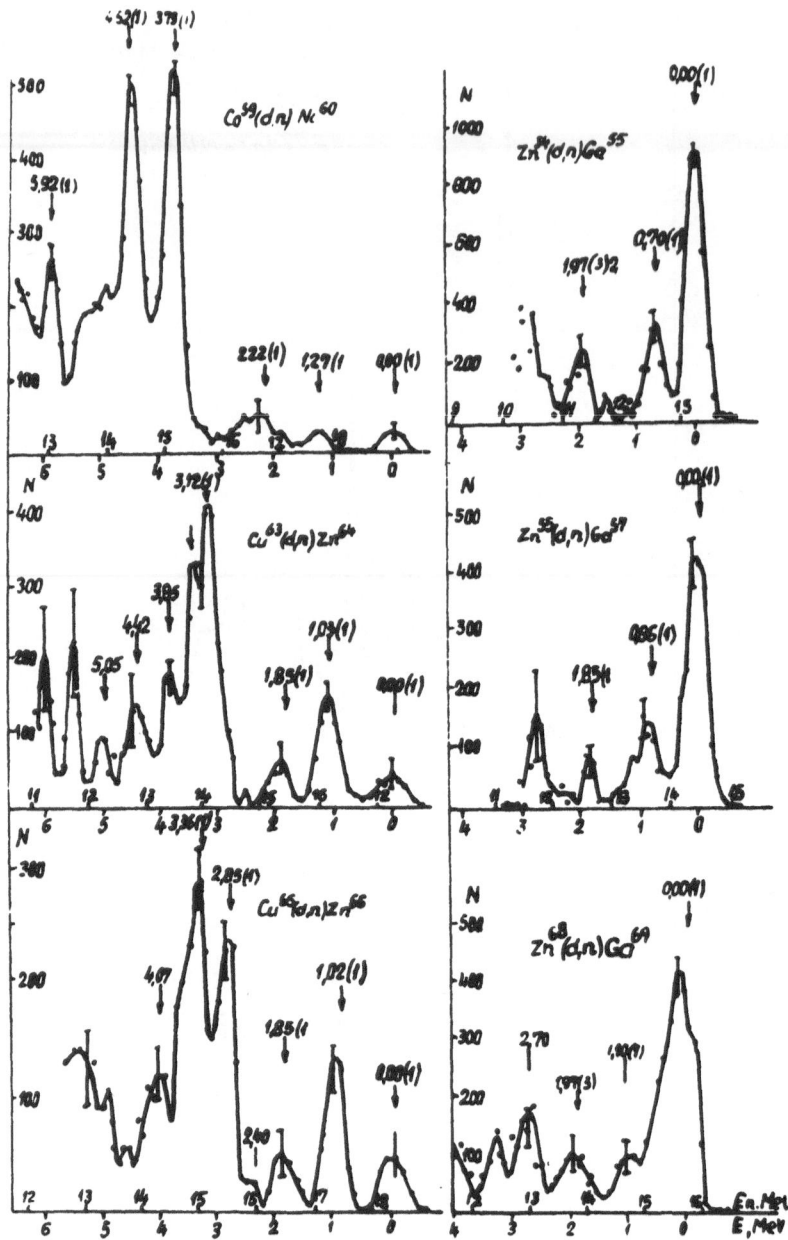

Fig. 11. Spectra from (d,n) experiments (see ref. [3])

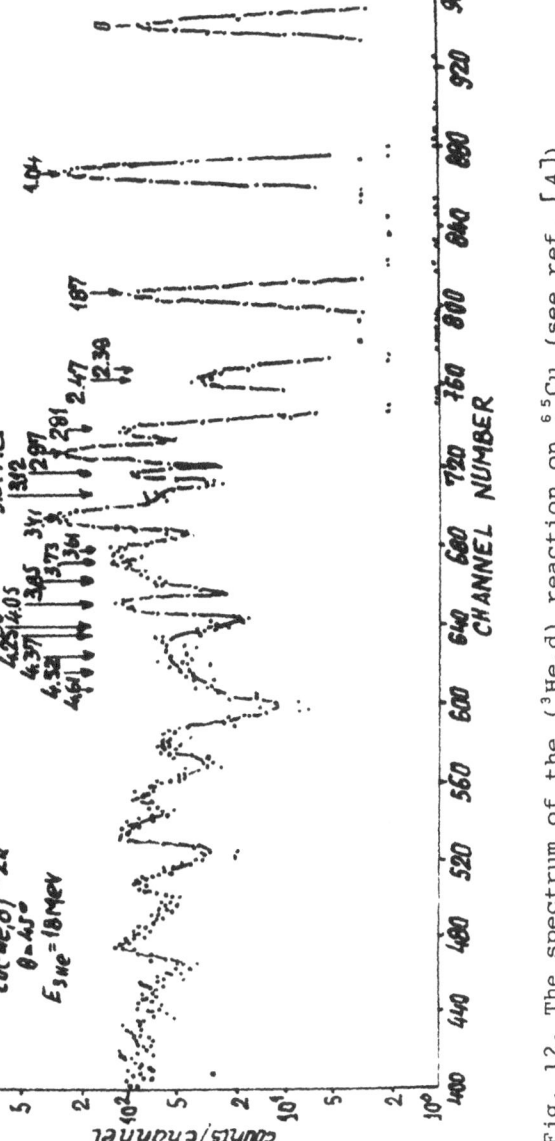

Fig. 12. The spectrum of the (^{3}He,d) reaction on ^{65}Cu (see ref. [4])

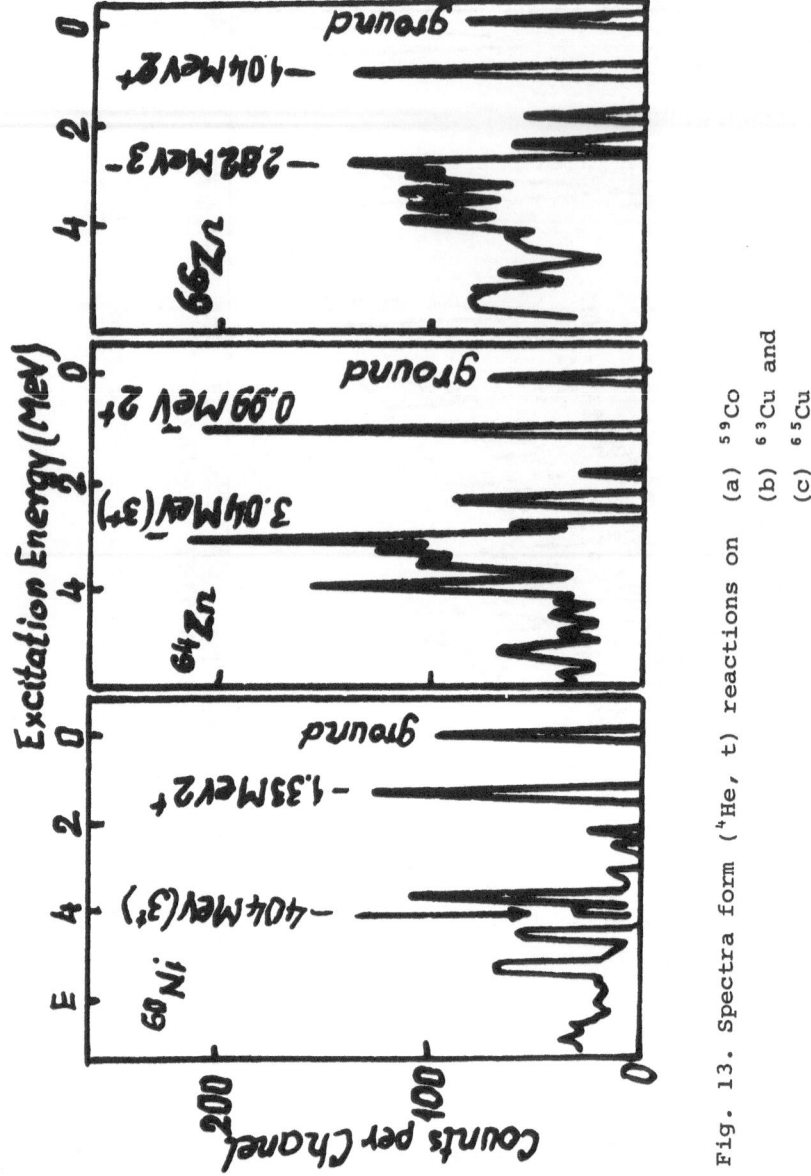

Fig. 13. Spectra form (^{4}He, t) reactions on   (a) ^{59}Co
(b) ^{63}Cu and
(c) ^{65}Cu

In the case of ^{70}Ge I could not find similar transfer reactions
data, but this nucleus being close to ^{66}Zn and being of the same even-
even type one can suppose that the nature of the first 2^+ state is similar. In this case the structure in the inelastic scattering can be
accounted through the picking-up by the residual nucleus of a particular hallway state, which, due to its configuration can couple directly
to the former. For instance, if the residual nucleus excited state is
of the 2p - 2h type, a 3p - 2h type hallway should be picked up by
that state. A theoretical attempt to explain the first 2^+ excited states in medium nuclei was made by Belyaev [2] taking into account correlated particle-hole configurations with an angular momentum J = 2.

In conclusion I would mention that:

1) There exists evidence concerning substructures within the range
of the gross structure of IAR. These substructures are not compound
nucelus fine structure.

2) These substructures connot be explained by the presence of
other parent states.

3) These substructures connot be interpreted as Ericson fluctuations.

4) A possible interpretation of the inelastic data can be given
in terms of coupling of hallway states with configurations in the residual nucleus due to their simplicity, and this, together with the
structure in elastic scattering, can be regarded as a support to the
doorway, hallway picture of nuclear reactions.

Acknowledgement

I would like to thank professon I. Ursu (Institute of Atomic
Physics, Bucharest) professor G.M. Temmer (Rutgers University, New
Brunswick) and professor U. Schmidt-Rohr (Max Planck Institute for
Nuclear Research, Heidelberg) for their support during some of the
work accounted here.

To professor H. Feshbach (MIT) I express my gratitude for kindly
accepting to read the preliminary version of the manuscript of this
talk and for useful comments.

References

1. ALBRECHT, R., MAHAUX, C., private communication.

2. BELYAEV, S.T., JETP, 39 1387 (1960).

3. BERINDE , A., BORZA, A.,GRAMA, N., MIHAI, I., PETRASCU, M., SIMION, V., WURM, J.R., to be published.

4. VonEHRENSTEIN, D. and SCHIFFER, J.P., Phys. Rev., 164 1374 (1967).

5. FESHBACH, H., KERMAN, A. and LEMMER, R., Annals of Phys. 41 230 (1967).

6. FORD, J.L.C. Jr., WARSH, K.L., ROBINSON, R.L., MOAK, C.D., ORNL - Report.

7. GIBSON, W.M., HASHIMOTO, Y., KEDDY, R.J., MARUYAMA, M. and TEMMER, G.M., Phys. Rev. Lett. 28 74 (1972).

8. GOLDMAN, L.H., Phys. Rev., 165 1203 (1968).

9. HUIZENGA, J.R., private communication. This represents the best estimate using a slight extrapolation from the observed neutron capture $1/2^+$ level density at 7.6 MeV excitation.

10. LANE, A.M., THOMAS, R.G., Rev. Mod. Phys. 30 257 (1958).

11. LANE, A.M., Isospin in Nuclear Physics, ed. D.H. Wilkinson (North. Holl. Amst.) p.509.

12. LINDSTROM, D.P., NEWSON, H.W., BILPUCH, E.G., MITCHELL, G.E., Nucl. Phys. A187 481 (1972).

13. MacDONALD, W. and MEKJIAN, A., Phys. Rev., 160 730 (1967).

14. MARUYAMA, M., TSUKADA, K., OZAWA, K.,FUJIMOTO, F., KOMAKI, K., MANNAMI, M. and SAKURAY, T., Nucl. Phys., A145 581 (1970).

15. MASARU MATOBA, J. Phys. Soc. Japan 25 901 (1968).

16. MEKJIAN, A. and MacDONALD, W.M., Nucl. Phys., A121 385 (1968).

17. OKOROKOV, V.V., SEREJIN, V.M., SMOTRIAEV, V.A., TOLCHENKOV, D.L., TROSTIN, I.S., TCHEBLUKOV, Iv. N., Yad. Fiz. (Journal of Nuclear Physics) 4 975 (1966).
See also N. Cindro, invited paper 57° Congresso della Societa Italiana di Fisica, L'Aquila , October 1971.

18. ROBSON, D. and LANE, A.M., Phys. Rev., 162 982 (1967).

19. ROBSON, D., Phys. Rev., 137B 535 (1965).

20. TEMMER, G.M., MARUYAMA, M., MINGAY, D.W., PETRASCU, M., VAN BREE, R., Bull. Amer. Phys. Soc., 16 132 (1971).

21. TEMMER, G.M., MARUYAMA, M., MINGAY, D.W., PETRASCU, M., VAN BREE, R., Phys. Rev. Lett., 26 1341 (1971).

22. TEMMER, G.M., MARUYAMA, M., MINGAY, D.W., PETRASCU, M., VAN BREE, R., unpublished.

23. TEMMER, G.M., private communication.

24. WEIDENMÜLER, H.A., Nucl. Phys. A99 269 (1967).

THE ENERGY-AVERAGED S-MATRIX AND DOORWAY RESONANCES

P. von BRENTANO

Institut für Kernphysik

Universität zu Köln

1. Introduction

The concept of doorway resonances which was originated by Feshbach, Kerman and coworkers at the M.I.T. has been shown to be a very
successful concept. In various experiments beautiful examples of doorway resonances were discovered, and the paper by C. Mahaux gives a
review of theory and experiments in this field. After it was formulated
in the framework of Feshbach theory, the doorway resonances were discussed in many different frameworks of resonance reaction theories.
We should mention in particular the work at the M.I.T. (cf [4],[5]),
the shell model theory of Mahaux and Weidenmüller [9], and the related
theory by Mekjian and McDonald [11]. A unifying theory which comprises
the various approaches has been given by Lane and Robson [8].

In this paper I want to discuss the doorway resonance concept from
a pure S-matrix point of view. The S-matrix point of view will show
why the various theories end up giving the same S-matrix for an isolated doorway resonance. This formulation allows also in a simple way to
discuss the case of many overlapping doorway resonances. The definition
of the doorway resonances which we want to use here is an extension of
the definition of a normal resonance as given by Humblet and Rosenfeld [7]. They define a resonance to be a pole of the S-matrix and we
will define a doorway resonance as a pole in the continued energy-averaged S-matrix. In order to do this, we have to discuss in some detail
the concept of energy-averaged S-matrix which we will refer to as the
average S-matrix below. This concept is the basis of the following paper. As we will note in our discussion below, the concept of the average S-matrix and the concept of doorway resonances,as the poles of
the average S-matrix,can be given a clear definition only if the widths
and spacings of the doorway resonances are very much larger than the
widths and spacings of all poles of the S-matrix. This case is called
the strong coupling case. We want to mention, however, a paper by Shapiro in which doorway resonances are discussed as broad poles in the
normal S-matrix [12].

It does not seem to be possible to define the average S-matrix in
a completely unique way. For the following, however, it is sufficient
to know various general properties of the averaged S-matrix and to pro-

ve them or make them at least plausible. Even this will not be done
for the general case of a scattering problem, but we shall rather consider the average S-matrix for a normal S-matrix consisting of an entire function and a sum of pole terms with constant coefficients. For
this simple case we shall discuss the properties of the energy-averaged S-matrix.

We shall find under suitable assumptions on the pole density function that the average S-matrix $S(Z)$ consists of two branches $S_1(Z)$
and $S_2(Z)$ which are separated by the line of the poles of the S-matrix.
Both functions $S_1(Z)$ and $S_2(Z)$ are analytic functions on the physical
sheet but they can be extended beyond their region of definition and
the branch line into the full complex plane. These extensions of the
functions $S_1(Z)$ and $S_2(Z)$ can then have poles in the unphysical sheets
and we shall associate these poles with the doorway resonances. The
fact that one finds 2 branches S_1 and S_2 rather than one is associated
with the non-hermitian character of the effective hamiltonian H obtained by the energy averaging of the S-matrix.

After we have discussed the concept of the average S-matrix, we
can obtain the properties of the poles of the average S-matrix which
are analogous to the properties of the poles of unitary S-matrices. In
particular one can show that the residues factor also for poles of the
average S-matrix and we will generalize the relation between the sum
of the partial widths and the total width. As was mentioned above,
there are two branches of the average S-matrix rather than one and
thus a doorway resonance corresponds to a pole in both of these branches. Thus we have two pole terms rather than one, and this allows us
to decompose the pole of the doorway resonance into a spreading-pole
$\varepsilon\downarrow$ and a decay-pole $\varepsilon\uparrow$. Thus one can in a natural fashion decompose the
total width of the doorway into a natural decay width $\Gamma\uparrow$, and into a
spreading width $\Gamma^\downarrow$. The preceding arguments are very general and thus
our results are also very general and unspecific. We should mention,
however, that our definition of doorway resonances - general as it
seems to be - covers only doorway resonances which obey the auxiliary
condition $\Gamma^\downarrow > |\Gamma^\uparrow|$. And in this respect the above definition is somewhat
more restrictive than the usual definitions of doorway resonances, even
though it agrees within this restriction with the usual theories.

2. The S-Matrix for Many Resonances, Extension into the Complex Plane

As discussed in the introduction we shall not consider the most
general S-matrices, but rather only S-matrices which can be written as
a sum of pole terms with constant coefficients and an entire analytic

function $S_o(E)$

$$S(E) = S_o(E) - i \sum_n \frac{a_n}{Z-\epsilon_n} \tag{1}$$

As is well known the S-matrix has the following properties (symmetry and unitarity)

$$S = S^t, \quad S_o = S_o^t, \quad a_n = a_n^t \quad \text{or} \tag{2a}$$

$$(S)_{cc'} = (S)_{c'c} \quad (S_o)_{cc'} = (S_o)_{c'c} \quad (a_n)_{cc'} = (a_n)_{c'c} \tag{2b}$$

$$S(E) \cdot S^*(E) = 1 \tag{3}$$

Under the above conditions one can extend the function $S(E)$ as an analytic function into the complex energy plane $Z = E + iI$ and this will be used in the following. At this point we should mention that an S-matrix having the above properties is a good approximation to the true S-matrix for the N-channel scattering case if we consider only energies which are very much above the opening of the last threshold and if we consider the S-matrix only in the vicinity of the real energy axis on the main Riemann sheet. The complex function $S(Z)$ can be shown to obey the complex unitarity, i.e. eq. 4. This unitarity is immediately obtained from the real unitarity eq. 3 by noting that both $S(Z)$ and $S^*(Z^*)$ are analytic functions of the complex variable Z. Thus also their product $F(Z) = S(Z) \cdot S^*(Z^*)$ is an analytic function of Z which is analytic in the whole complex Z-plane. But on the real axes $F(E) = = 1$ according to eq. 3 and thus $F(Z) = 1$ everywhere. Thus we have

$$S(Z) \cdot S^*(Z^*) = S^*(Z^*) S(Z) = 1 \tag{4}$$

2.1. The Definition of the Energy-Averaged S-Matrix:

The extension of the S-matrix into the complex energy plane allows, according to Feshbach, Porter and Weisskopf [3] and Brown [2] to obtain the energy-averaged S-matrix in a particularly elegant way. Namely, one defines, following Brown, the energy average of the S-matrix by eq. (5):

$$<S(E)>_{I_o} \cong \frac{I_o}{\pi} \int_{-\infty}^{+\infty} \frac{dE'}{(E-E')^2+I_o^2} S(E') \tag{5}$$

The integral in eq. 5 can be evaluated by contour integration to give eq. 6:

$$<S(E)>_{I_o} \cong S(E+iI_o) \tag{6}$$

In this way the average S-matrix energy-averaged over an energy interval $\Delta E = I_o$ can be defined for real energies E. The definition of an average S-matrix by eq. 6 has only a reasonable meaning if the average S-matrix does not depend strongly on the averaging energy interval $\Delta E = I_o$. A necessary condition for this is that I_o is much larger than the average width $<\Gamma_k>_k$ and spacing $<D_k>_k$ of the poles $\varepsilon_k = E_k + i/2 \; \Gamma_k$ of S

$$I_o >> <D_k>_k = <E_{K+1} - E_K>_K \tag{7a}$$

$$I_o >> <\Gamma_K>_K \tag{7b}$$

Thus I_o must be larger than a certain minimum value. We shall assume that this minimal value I_o of ΔE is so small that it can be neglected when the properties of the average S-matrix are discussed. E.g. we shall assume that I_o is small compared to the width and spacing of doorway resonances. Apparently the isobaric analog resonances in heavy nuclei are examples of such situations. In such situations we expect that $<S(E)>_I$ does not depend strongly on I for $I \gtrsim I_o$.

In the following it will not be enough to consider the energy-averaged S-matrix $<S(E)>$ as a function of real energies, but we have to extend it as a function into the complex energy plane. Eq. 6 indicates the way how to accomplish this. Namely we define a matrix function $S_1(Z)$, which agrees in the upper plane approximately with S(Z), but which has no poles in the strip $|ImZ|<I_o$. Then the value of $S_1(Z)$ will be given by analytic continuation in the lower halfplane and it may have poles there. Similarly we introduce a second matrix function $S_2(Z)$, which agrees approximately with the S-matrix S(Z) in the lower halfplane $ImZ< -I_o$ and which has no poles in the strip $|ImZ|<I_o$. We assume in addition that both functions are meromorphic functions of Z, i.e. that they are analytic and single valued functions of Z except for poles. These requirements are written explicitly:

$$S_1(E+iI) \approx S(E+iI) \quad \text{for} \quad I > I_o \tag{8}$$

$$S_2(E-iI) \approx S(E-iI) \text{ for } I > I_o \tag{9}$$

$$S_1(Z) \text{ and } S_2(Z) \text{ have no poles in strip } |Im(Z)| < I_o \tag{10}$$

$S_1(Z)$ and $S_2(Z)$ are meromorphic functions and can be expanded in a Mittag-Leffler-series with constant čoefficients.

$$S_1(Z) = S_{10}(Z) - i \sum_k \frac{A_{1K}}{Z-\epsilon_{1K}} \tag{11}$$

$$S_2(Z) = S_{20}(Z) - i \sum_k \frac{A_{2K}}{Z-\epsilon_{2K}} \tag{12}$$

where $S_{10}(Z)$ and $S_{20}(Z)$ are entire functions of Z. We further assume that also S_1 and S_2 are symmetric matrices

$$S_1^t = S_1 \tag{13}$$

$$S_2^t = S_2 \tag{14}$$

and that they are related by the generalized unitarity relation which corresponds to eq. 4.

$$S_1(Z) \; S_2^*(Z^*) = 1 \tag{15}$$

In general it is not clear and probably also not true that such functions $S_1(Z)$ and $S_2(Z)$ with these properties exist. But we will try to make in this paper a simple model of doorway resonances and this model just is that the S-matrix poles and residues are so distributed that the functions $S_1(Z)$ and $S_2(Z)$ with the above properties exist. Clearly it is only useful to talk of doorway resonances if the functions $S_1(Z)$ and $S_2(Z)$ are reasonably independent of the energy averaging interval I_o. This requires in particular that

$$|Im\epsilon_{1K}| \gg I_o \; , \quad |Im\epsilon_{2K}| \gg I_o \tag{16}$$

$$|Re(\epsilon_{1K+1}-\epsilon_{1K})| \gg I_o \tag{17}$$

Namely that the width and spacing of the doorway resonances must be
large compared to the averaging interval I_o. In general, the functions
$S_1(Z)$ and $S_2(Z)$ are not uniquely determined by our requirements, but
under the assumptions (7a), (7b), (16) and (17) it seems reasonable to
assume that the differences of the various approximating functions S_1,
S_1' etc. are small. Summing up, we have found that the averaging of the
S-matrix leads to two average S-matrices $S_1(Z)$ and $S_2(Z)$ rather than
one. But these two matrices are related by the generalized unitarity
relation so all physical information is contained already in S_1 which
is usually called the average S-matrix. Still it will be useful to discuss these two functions in a symmetrical way.

2.2. Properties of the Poles of S_1 and S_2

The definition of the average S-matrix allows us to introduce
doorway resonances as the poles of S_1 and S_2 in analogy with the definition of resonances as poles of the unitary S-matrix. In particular
one can thus define doorway resonances also in a region where they are
strongly overlapping. In the following we shall consider the location
of the poles of S_1 and S_2 and we shall show that the residues of the
poles factor also in this case.

From the above definition it is obvious that the poles ε_{1K} of S_1
can be only in the lower halfplane and the poles ε_{2K} of S_2 can be only
in the upper halfplane. That is we have the relations:

$$\text{Im}\varepsilon_{1K} < - I_o < 0 \tag{18a}$$

$$\text{Im}\varepsilon_{2K} > I_o > 0 \tag{18b}$$

The fact that S_2 poles in the upper half of the complex energy plane
seems to be strange, but we must remember that S_1 and S_2 do not belong to a hermitian hamiltonian operator, and thus the usual proofs
that the poles lie in the lower half plane are not pertinent here. In
order to discuss the factorisation of the residues of these poles, we
follow the discussion in the unitary case ([5], [10], [11]). We shall also
make another assumption for S_1 and S_2, namely we shall assume that the
poles of det $(S_1(Z))$ and of det $(S_2(Z))$ are simple. Similarly we shall
assume that all eigenvalues of S_1 and S_2 are different. In the case
that these conditions are not fulfilled, one can fulfill them by adding an arbitrarily small matrix to S_1 and S_2 and thus we shall not
worry about this assumption. This assumption, however, greatly facili-
tates the following consideration, because it means that the matrices
are of simple structure and that they can be diagonalized by a complex

orthogonal matrix.

In order to show that the residues factorize, we shall expand S_1 in the vicinity of the pole ε_{1K} as given by eq. 19:

$$S_1(Z) \cong A(Z) - i \frac{A_{1K}}{Z-\varepsilon_{1K}} \quad \text{for } z \approx \varepsilon_{1K} \tag{19}$$

From the extended unitarity eq. 9, we obtain

$$A(Z) \; S_2^*(Z^*) + \frac{A_{1K} \cdot S_2^*(Z^*)}{z-\varepsilon_{1K}} \approx 1 \quad \text{for } z \approx \varepsilon_{1K} \tag{20}$$

The second term of this equation will clearly go to infinity at the pole unless we have

$$A_{1K} \cdot S_2^*(\varepsilon_{1K}^*) = 0 \tag{21}$$

From the extended unitarity a pole in S_1 is connected with a zero in S_2. Our assumption that the zeros of det $S_2^*(Z^*)$ are simple implies, however, that there is up to a factor one and only one solution of the homogeneous equation

$$r_{1K} \cdot S_2^*(\varepsilon_{1K}^*) = 0 \tag{22}$$

or written explicitly

$$\sum_c (r_{1K})_c \cdot (S_2^*(\varepsilon_{1K}^*))_{cc'} = 0$$

The vector r_{1K} which solves the homogeneous equation is essentially the wavefunciton of the resonance. Comparing eq. 21 and eq. 22, we find $(A_{1K})_{cc'} = \alpha_c \cdot (r_{1K})_{c'}$ and from the symmetry of the S-matrix we find then $(A_{1K})_{cc'} = \beta \, (r_{1K})_c (r_{1K})_{c'}$. If we absorb the constant β into the definition of the vector r_{1K}, we finally obtain eq. 23:

$$A_{1K} = r_{1K} \cdot r_{1K}^t \tag{23}$$

Thus the residues of the S-matrices S_1 and S_2 factor just as the residues of normal unitary S-matrices do.

3. The S-Matrix for an Isolated Doorway Resonance

In the following we want to discuss in some detail the S-matrix for an isolated doorway resonance. In this case one obtains some additional information on the S-matrix if we consider the extended unitarity. These considerations are again in close analogy to the unitary case. The most simple results are obtained if we assume that S_1 consists of a constant background term S_{10} and of a single resonance (24):

$$S_1(z) = S_{10} - i \frac{r_1 r_1^t}{z - \varepsilon_1} \tag{24}$$

In the following we shall show that under these assumptions S_2 has also a pole ε_2 and we shall give a relation between the partial widths and the two pole terms. In the case of normal resonances, such relations are usually obtained for the case of a diagonal background matrix S_{10}:

$$(S_{10})_{cc'} = e^{i2\delta c} \cdot \delta_{cc'}$$

A generalization to non-diagonal background matrices has been given by McVoy [10] and Baranger [1]. We shall essentially follow this procedure with a small change. Namely we shall introduce explicitly the concept of a square root matrix. This square root matrix is defined as follows: according to our assumptions the background matrix S_{10} has simple structure, that is: all its eigenvalues are different. In this case one can diagonalize S_{10} by a complex orthogonal matrix R [6]. Thus we have

$$S_{10} = R D R^t \quad \text{where} \quad R R^t = 1 \quad \text{and}$$

$$D_{cc'} = \exp(2i\delta_c) \cdot \delta_{cc'} \tag{25}$$

where the eigenphases δ_c are in general complex numbers. The square root matrix is then uniquely defined by eq. 20:

$$S_{10}^{1/2} = R \cdot D^{1/2} R^t \quad \text{where} \quad (D^{1/2})_{cc'} = \exp(i\delta_c) \cdot \delta_{cc'} \tag{26}$$

By the help of this concept we can rewrite eq. 19 in the form (21):

$$S_1(z) = S_{10}^{1/2} (1 - i \frac{\gamma_1 \cdot \gamma_1^t}{z - \varepsilon_1}) S_{10}^{1/2} \tag{27}$$

where $\gamma_1 = S_{10}^{-1/2} \cdot r_1$

It is now easy to show that S_2 has also the form of the last equation. We introduce the matrix $A(z)$:

$$A(z) = S_{10}^{-1/2} (1+i \frac{\gamma_1 \gamma_1^t}{z-\varepsilon_2^*}) S_{10}^{-1/2} \qquad (28)$$

with $\varepsilon_2^* = \varepsilon_1 + i \sum_c \gamma_c^2$

Then one can show by explicit computation that $S_1(z)A(z) = 1$ and therefore $A(z) = S_1^{-1}(z)$. Remembering that $S_2(z) = S_1^{-1*}(z*)$, we find for S_1 and S_2 the following formulas:

$$S_1(z) = S_{10}^{1/2} (1-i \frac{\gamma\gamma^t}{z-\varepsilon_1}) S_{10}^{1/2} \qquad (29)$$

$$S_2(z) = S_{20}^{1/2} (1-i \frac{\gamma^*\gamma^{*t}}{z-\varepsilon_2}) S_{20}^{1/2}$$

with $S_{10} \cdot S_{20}^* = 1$

The partial decay amplitudes can be decomposed (30) into a phase and an absolute value:

$$\gamma_c = \exp(i\phi_c) \cdot |\gamma_c| \; ; \; |\gamma_c^2| = \Gamma_c \qquad (30)$$

The real numbers ϕ_c and Γ_c are called mixing phase and partial width (cf [4],[5],[9]). In the case of a unitary S-matrix, γ_c is real and thus there is no mixing phase for a unitary S-matrix. The connection between the two pole terms given in eq. (28) can be written in terms of the mixing phases as follows:

$$\varepsilon_1 - \varepsilon_2^* = -i \sum_c \gamma_c^2 = -i \sum_c \exp(2i\phi_c)\Gamma_c \qquad (31)$$

3.1. The Decomposition of the Total Width into a Spreading Width and a Decay Width

The fact that we have here two poles associated with a resonance rather than one allows in a natural fashion to decompose the total width into a spreading width and a natural decay width. These concepts have been introduced by Feshbach and they have been proven to be very useful.

Essentially we just rewrite the two complex numbers ε_1 and ε_2 in terms of two other complex numbers $\varepsilon^{\uparrow}$ and $\varepsilon^{\downarrow}$:

$$\varepsilon_1 = E_1 - i/2 \ \Gamma_1, \ \varepsilon_2 = E_2 - i/2 \ \Gamma_2 \qquad (32)$$

$$\varepsilon^{\uparrow} = E^{\uparrow} - i/2 \ \Gamma^{\uparrow}, \ \varepsilon^{\downarrow} = E^{\downarrow} - i/2 \ \Gamma^{\downarrow}$$

In order to find the proper relation between ε_1, ε_2 and $\varepsilon^{\uparrow}$, $\varepsilon^{\downarrow}$, which agrees with the usual definitions, we use the relations $\Gamma_1 = \Gamma^{\uparrow} + \Gamma^{\downarrow}$ and we infer from eq. 31 the relation $\Gamma_1 + \Gamma_2 = 2\Gamma^{\uparrow}$. From these relations it is evident that one must write (33):

$$\varepsilon_1 = \varepsilon^{\uparrow} + \varepsilon^{\downarrow} \qquad \varepsilon_2 = \varepsilon^{\uparrow} - \varepsilon^{\downarrow} \qquad (33a)$$

$$\varepsilon^{\uparrow} = \frac{1}{2} (\varepsilon_1 + \varepsilon_2) \qquad \varepsilon^{\uparrow} = \frac{1}{2} (\varepsilon_1 - \varepsilon_2) \qquad (33b)$$

The preceding definitions are rather formal. They are given a much more intuitive meaning by the investigation of concrete models of the mixing between doorway states and the fine structure states. These models show that in these cases one obtains approximately a doorway resonance with an average S-matrix which agrees essentially with eq. 23. The spreading width is then the width into which the doorway is spread out, i.e. this concept has its literal meaning. Thus it must be a positive number. It is very nice, therefore, that one can obtain this result also in the approach outlined in this paper. Namely, if we combine eq. 12 and 33, one obtains eq. 34:

$$\Gamma^{\downarrow} > |\Gamma^{\uparrow}| > 0 \qquad (34)$$

This formula shows also that the concept of doorway resonances as poles of the average S-matrix is slightly more restrictive than the usual concept of doorway resonances which allows also that $\Gamma^{\downarrow} < |\Gamma^{\uparrow}|$ [9]. It is interesting to investigate where exactly this difference comes from. The reason is that one usually requires eq. 7a but not eq. 7b. The latter equation excludes the case that one of the fine structure widths is very much larger than the rest and it requires that all the widths of the fine structure resonances are small compared with the width of the IAR. It is obvious, however, that one can define an average S-matrix in the sense of eqs. 10 and 11 only if both equations 7a and 7b are fulfilled.

References

1. BARANGER, M. and DAVIES, K.T.R., Ann. Phys. (N.Y.) $\underline{19}$, 383 (1962).

2. BROWN, G.E., Revs. Mod. Phys. $\underline{31}$, 893 (1959).

3. FESHBACH, H., PORTER, C.E. and WEISSKOPF, V.F., Phys. Rev. $\underline{96}$, 448 (1954).

4. FESHBACH, H., Ann. Phys. (N.Y.) $\underline{5}$ p 357 (1958), $\underline{19}$ p 287 (1962), $\underline{43}$, 410 (1967).

5. FESHBACH, H., KERMAN, A.K. and LEMMER, R.H., Ann. Phys. (N.Y.) $\underline{41}$, 230 (1967).

6. GANTMACHER, F.R., Matrizenrechnung I, II, (VEB Ceutscher Verlag der Wissenschaften, Berlin, 1966).

7. HUMBLET, J. and ROSENFELD, L., Nucl. Phys. $\underline{26}$, 259 (1961)
   HUMBLET, J., Fundamentals in Nuclear Theory (International Atomic Energy Agency, Vienna, 1967).

8. LANE, A.M. and ROBSON, D., Phys. Rev. $\underline{151}$, 774 (1966)
   LANE, A.M. and ROBSON, D., Phys. Rev. $\underline{161}$, 982 (1967).

9. MAHAUX, C. and WEIDENMÜLLER, H.A., Shell-Model Approach to Nuclear Reactions, North Holland Publ. Company, Amsterdam-London, 1969.

10. McVOY, K.W., Fundamentals in Nuclear Theory (International Atomic Energy Agency, Vienna, 1967).

11. MEDJIAN, A. and McDONALD, W.M., Nucl. Phys. $\underline{A121}$, 385 (1968).

12. SHAPIRO, I.S., Nucl. Phys. $\underline{A122}$, 645 (1968).

13. SMIRNOV, W.I., Vol. III, Lehrgang der Höheren Mathematik, VEB Deutscher Verlag der Wessenschafter Berlin 1963.

THE MECHANISM OF FAST NEUTRON RADIATIVE CAPTURE

F. CVELBAR

J. Stefan Institute and Faculty for Natural Sciences and
Technology, University of Ljubljana, Yugoslavia

1. Introduction

This contribution will be mainly devoted to the description of the radiative capture of neutrons if the intermediate system is excited to the region of the giant dipole resonance (GDR) i.e. about 20 and 15 MeV for light and heavy nuclei, respectively.

It is well known that when bombarding the nucleus in its ground state with gamma-rays of such energy, the probability of their absorption is increased. According to the liquid drop model of the nucleus, the collective oscillation of neutrons against protons is excited and the state decays by particle emission.

If, on the other hand, the same state is excited by neutron absorbtion, it follows from the principle of detailed balance that an enhanced gamma-ray transition to the ground state of the final nucleus is expected. This means that also in this process the GDR state is excited.

In the liquid drop model, the dipole excitation, being a collective one, cannot be excited by the average (optical) potential but by some sort of residual interaction between the incoming particle (e.g. neuron) and the target nucleus. Due to this interaction in the capture process the initial (core) nucleus A is excited into dipole motion and the neutron is captured into the ground state of the nucleus A + 1. According to this model capture gamma-rays are the result of the deexcitation of the core dipole state (fig. 1a). The excitation function for this process should then resemble the GDR curve. Maximum cross section values should be reached when the initial neutron energy E_n plus the neutron binding energy B_n equals E_R, the energy of the GDR.

The capture into the different single particle excited states could be expalined in a similar way. In this process only a part of the excitation energy of the intermediate system is available for the core excitation and the rest remains with the neutron to enter into available bound single particle states. Such a capture process is schematically illustrated in fig. 1b. The excitation curves should be similar to that discussed for the (n,γ_o) process except that they should be displaced to higher neutron energies to account for the final state (single particle) excitation energy E_f.

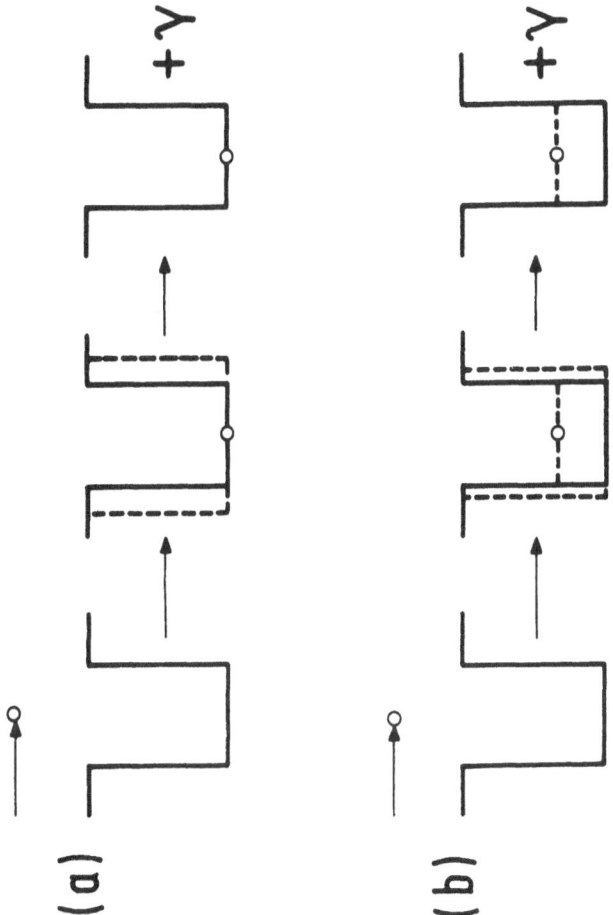

Fig. 1. Semidirect radiative capture process. Capture to
the ground state of final nucleus a). Capture to
the excited single particle state of final nucleus
b)

2. The Semidirect Capture Model

The basis for the qualitative picture described below is the Clement, Lane and Rook semidirect (collective) model [8] of the high energy nucleon radiative capture.

In this model, which is in fact an' intermediate structure model, the true Hamiltonian H is split into the model part H_o which represents an average nucleon-nucleus interaction and the residual part, describing the interaction between the incident nucleon and all nucleons in the nucleus

$$H = H_o + V$$

In the semidirect radiative capture model only the part of the residual interaction which describes the interaction between the particle and dipole vibration mode of nuclear excitation H' is taken into account. Its effect on the system is treated as a perturbation.

The perturbed wave function $|\Psi_i\rangle$ can be written as

$$|\Psi_i\rangle = |\chi_i\rangle + \sum_\lambda \frac{|\Phi_\lambda\rangle\langle\Phi_\lambda|H'|\chi_i\rangle}{E - E_\lambda + \frac{1}{2} i\Gamma_\lambda}$$

where χ_i is the solution of the model hamiltonian H_o

$$H_o|\chi_i\rangle = E_o|\chi_i\rangle$$

$|\Phi_\lambda\rangle$, and Γ_λ being respectively the wave functions, resonant energies and widths of the dipole states built on the ground state or different excited states.

The matrix element for the dipole transition to the final state $|\Psi_f\rangle$ is also a sum

$$\langle\Psi_f|E|\Psi_i\rangle = \langle\Psi_f|E^n|\chi_i\rangle + \sum_\lambda \frac{\langle\Psi_f|E^t|\Phi_\lambda\rangle\langle\Phi_\lambda|H'|\chi_i\rangle}{E - E_\lambda + \frac{1}{2} i\Gamma_\lambda}$$

where the separation of the dipole operator $E = E^n + E^t$ into a part E^n acting only on single particle coordinates and a part E^t acting only on collective coordinates, is performed. The complete (direct-semidirect) matrix element M^D_{if} and the semidirect matrix element M^{SD}_{if}

$$M^{DSD}_{if} = M^D_{if} + M^{SD}_{if}$$

To calculate the semidirect matrix element, the weak coupling approximation, qualitatively treated in the introduction (i.e. that in the excitation of the intermediate system the core dipole state of energy E_R and width Γ is always excited and the nucleon is captured into the unoccupied single particle states of excitation energy E_f), is introduced in the following form:

$$|\Phi_\lambda> = |1>|\phi_f>$$
$$|\chi_i> = |0>|\phi_i>$$
$$|\Psi_f> = |0>|\phi_f>$$

where

$$E_\lambda = E_R + E_f$$
$$\Gamma_\lambda = \Gamma$$

The symbols $|1>$ and $|0>$ mean respectively the dipole state and ground state wave functions of the core nucleus, while $|\phi_i>$ and $|\phi_f>$ represent respectively the optical model initial state wave function and the single particle wave function of the state populated in the capture process.

Introducing these relations in the expression for the semiderect matrix element, its value reduces to only one term

$$M^{SD}_{if} = \frac{|\phi_f|<0|E^t|1>|\phi_f><\phi_f|<1|H'|0>|\phi_i>}{E - (E_R + E_f) + \frac{1}{2} i\Gamma}$$

as all other terms are equal to zero.

The particle-vibration coupling (residual interaction) H', is treated microscopically in the semidirect capture model. Its main contribution comes from the isospin part of the two-body interaction. Assuming that this interaction is of the short range type, we obtain:

$$H' \sim \sum_i P \vec{\tau} \vec{\tau}_i \delta (\vec{r}-\vec{r}_i)$$

Here P means a constant closely connected with the strenght of the Heisenberg exchange potential. Symbols $\vec{\tau}$ and $\vec{\tau}_j$ stand for the isospin of the projectile and i-th nucleon of the nucleus, respectively. After the approximate summation, the introduction of a collective coordinate $\vec{\eta} = \vec{R}_n - \vec{R}_p$ and the separation of centroids of the neutron and proton system, Clement, Lane and Rook obtain

$$H' = (-\frac{V_1}{4}) \cdot 2 \cdot (\frac{N \cdot Z}{A^2}) \cdot \frac{d\rho(r)}{dr} \frac{\vec{r}\vec{\eta}}{|\vec{r}|} \tau_3$$

where the meaning of the symbols is as follows:

V_1 - the strength of the isotopic spin term of the optical potential

r - the position of the incident particle

$\rho(r)$ - the nuclear density function; usually a Saxon-Woods shape is taken.

As the operator E^t can be written in the form

$$E^t = \text{const} + e(\frac{3}{4\pi})^{1/2} \frac{N \cdot Z}{A} \eta_z$$

one obtains for the semidirect matrix element

$$M_{if}^{SD} = |<1|\eta_z|0>|^2 <\phi_f|F(r)|\phi_i> \cdot \frac{1}{E - (E_R + E_f) + \frac{1}{2} i\Gamma}$$

where F(r) contains the constants and the radial and angular dependence of H'.

As the whole dipole strength is concentrated on the GDR, it follows from the dipole sum rule that

$$|<1|\eta_z|0>|^2 = \frac{1}{E_R} \frac{A}{N \cdot Z} \frac{h^2}{2M} (1 + 0,8\ x)$$

where x stands for the charge exchange factor. In such a way the calculation of the semidirect contribution reduces to calculating the single particle matrix elements between the initial and the final states: $<\phi_f|F(r)|\phi_i>$. After performing the angular integration one obtains

$$|M_{if}^{DSS}|^2 \alpha |<u_f|E^n|u_i> + \frac{<u_f|h(r)|u_i>}{E - (E_R + E_f) + \frac{1}{2} i\Gamma}|^2$$

where u_i and u_f are the radial parts of the single particle wave functions.

As the cross section for the capture process is proportional to $|M_{if}^{DSD}|^2$, thus , besides the direct and semidirect terms, the interference term will contribute too in the final expression. Although the semidirect process is dominant, all three contributions have to be taken into account in the calculation (direct-semidirect (DSD) calculation).

Our introductory qualitative description of the capture process is essentially given by the semidirect term. This is even more evident when realizing that E_γ is the difference between the excitation of the intermediate system $E = E_n + Q$ (Q is the ground state neutron binding energy) and the energy of the excitation of the final state E_f

$$E_\gamma = (E_n + Q) - E_f$$

Introducing this relation into the simidirect term, one obtains the resonance dependence of the cross section as a function of the neutron energy for a fixed final state (excitation function). Furthermore, a similar dependence appears if E_n is fixed and the cross section is observed as a function of the excitation energy of the final states (spectrum of primary gamma-ray transitions). This last result is very important because until very recently the radiative capture in the region of GDR was studied mainly by measuring the spectra of primary gamma-ray transitions to the bound states of final nuclei after the capture of 14 MeV neutrons.

3. Comparison with Experimental Results

The experimental data on the radiative capture of energetic neutrons are of two different kinds: the activation analysis neutron capture cross sections σ_{act} which cover all possible gamma-transitions following the capture of neutrons by the nucleus, and the spectra of prompt gamma-rays following the primary deexcitations to the bound states of final nuclei. The energy-angle integral of these spectra is σ_{int}. As σ_{int} does not cover all capture processes, it is expected that $\sigma_{act} > \sigma_{int}$. However the difference should not exceed 10-20%, since the dipole transition probability is proportional to E_γ^3 and furthermore, $\Gamma_\gamma << \Gamma$ particle. However, a large difference between σ_{act} and σ_{int} has been reported to exist for the capture of 14 MeV neutrons in some nuclei [12] (fig. 2). Most recently Kantele and Valkonen [17] reported that by an apparently improved activation technique it was found that

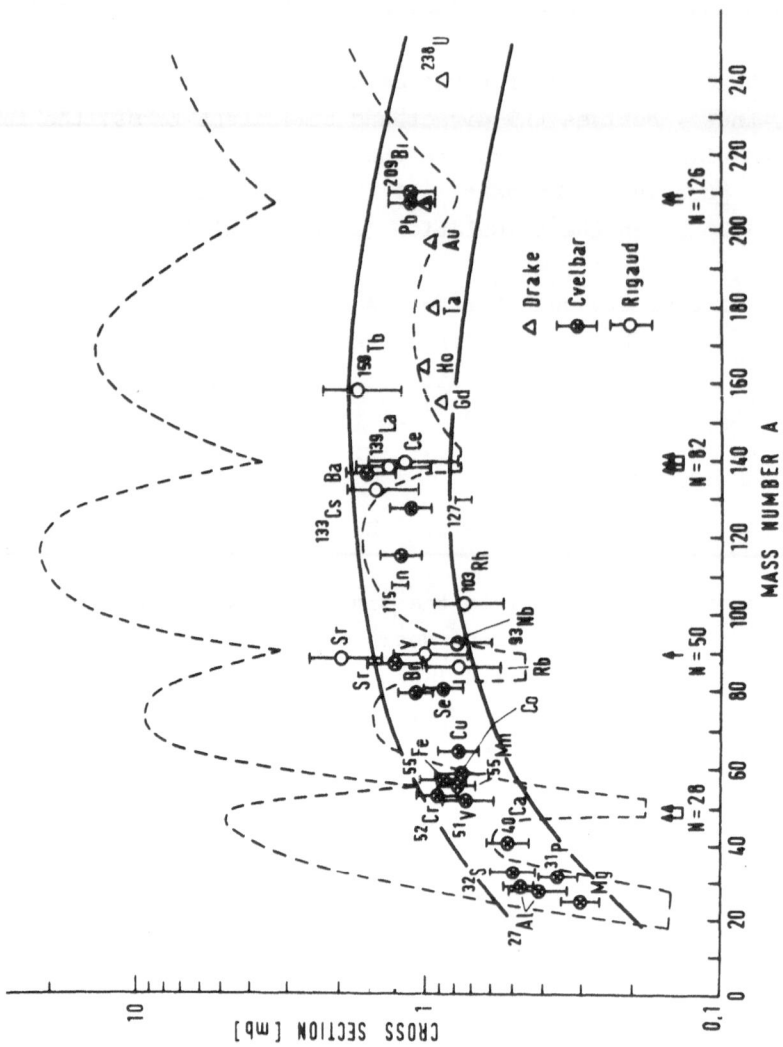

Fig. 2. Mass dependence of the integrated cross section
for radiative capture of 14 MeV neutrons (area
limited by solid lines). In the region limited
by broken lines the reported activation analysis
data appear (see refs [6], [12])

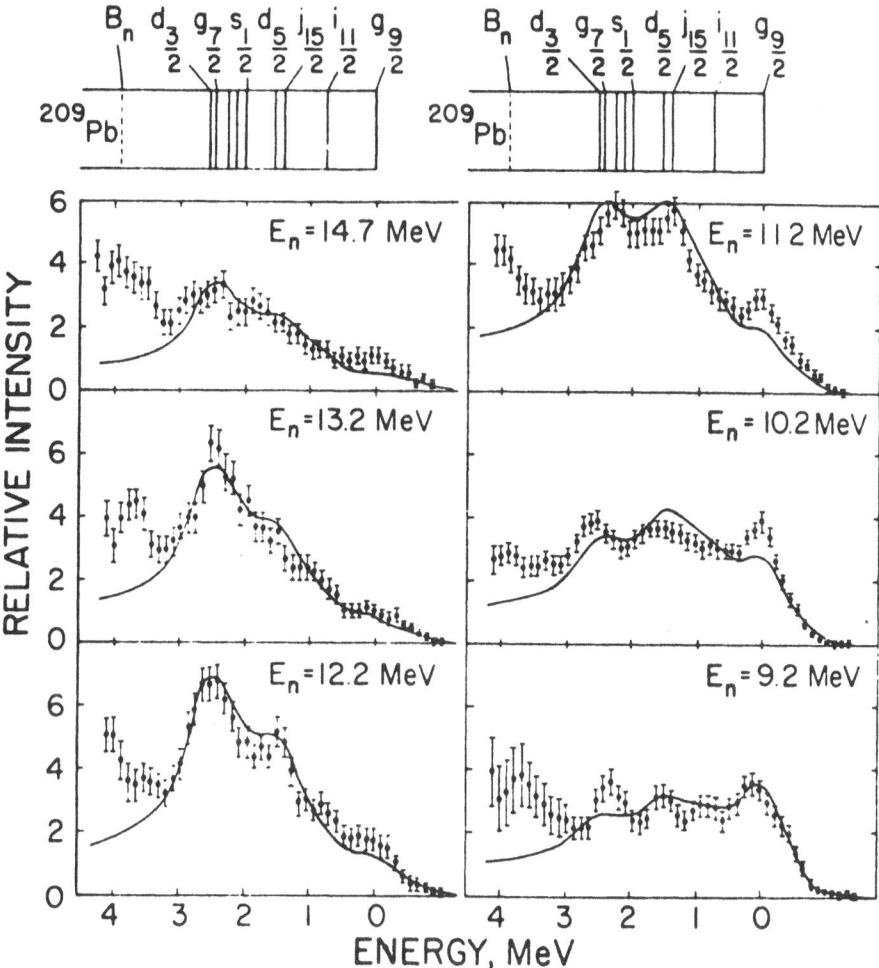

Fig. 3. Experimental and semidirect model spectra of
γ-rays from the radiative capture of neutrons
in ^{208}Pb (taken from ref. [6])

for several nuclei for which previous σ_{act} values appreciably exceeded the σ_{int} data, new σ_{act} results agreeed with σ_{int} values within the experimental error.

At the time when the semidirect theory was developed (1964-65), only few σ_{act} values were experimentally known. Such values* were not appropriate to test the new theory in detail. To do this, the prompt capture gamma-ray spectra and the excitation functions for the capture into different final states were needed. These spectra, measured at 14 MeV and below 8,5 MeV neutron energy became available in the following years (cf [2],[3],[9-11], [14], [15], [26]). Spectra systematically measured over the whole giant dipole resonance region were published in 1972 by Bergqvist et al. [6]. The results for $^{208}Pb(n,\gamma)^{209}Pb$ are shown in fig. 3, where the shape of the experimental spectra is compared to calculations using the semidirect model of Clement et al. (without the direct and the interference terms). The agreement appears to be very good. Both the calculated and the experimental spectra show some primary structure, which is, due to the differences in the matrix elements, modulated by the resonance effect of the denominator of the semidirect contribution.

The corresponding excitation functions for the transition to different final states are shown in fig. 4. The scale on the figures is an absolute one; we see thus that the semidirect capture model reproduces well the shape of the excitation functions, but yields cross sections which in the case of $^{208}Pb(n,\gamma)^{209}Pb$ are about 2-3 times lower than the experimental ones. The agreement in shape became worse if in the calculation direct and interference terms were also taken into account [25] (see Discussion).

The absolute agreement in absolute value between the experimental and calculated spectral intensities is, at least for high energy transitions, better for light nuclei. For these, however, only measurements with 14 MeV neutrons exist. For some of them the calculated and experimantal spectra are compared in fig. 5. The experimental spectra were measured by a telescopic scintillation pair spectrometer [10] and were (experimentally) integrated over a solid angle of 4π. These spectra are, therefore, directly comparable with the results of the semidirect capture model without speculations about the angular distribution of gamma-rays. For more reliable calculations the spectroscopic factors of excited states of final nuclei have to be known. As these are known usually only for the low lying states, only the intensity of high ener-

* Most cf these references are collected in CVELBAR et al. [12].

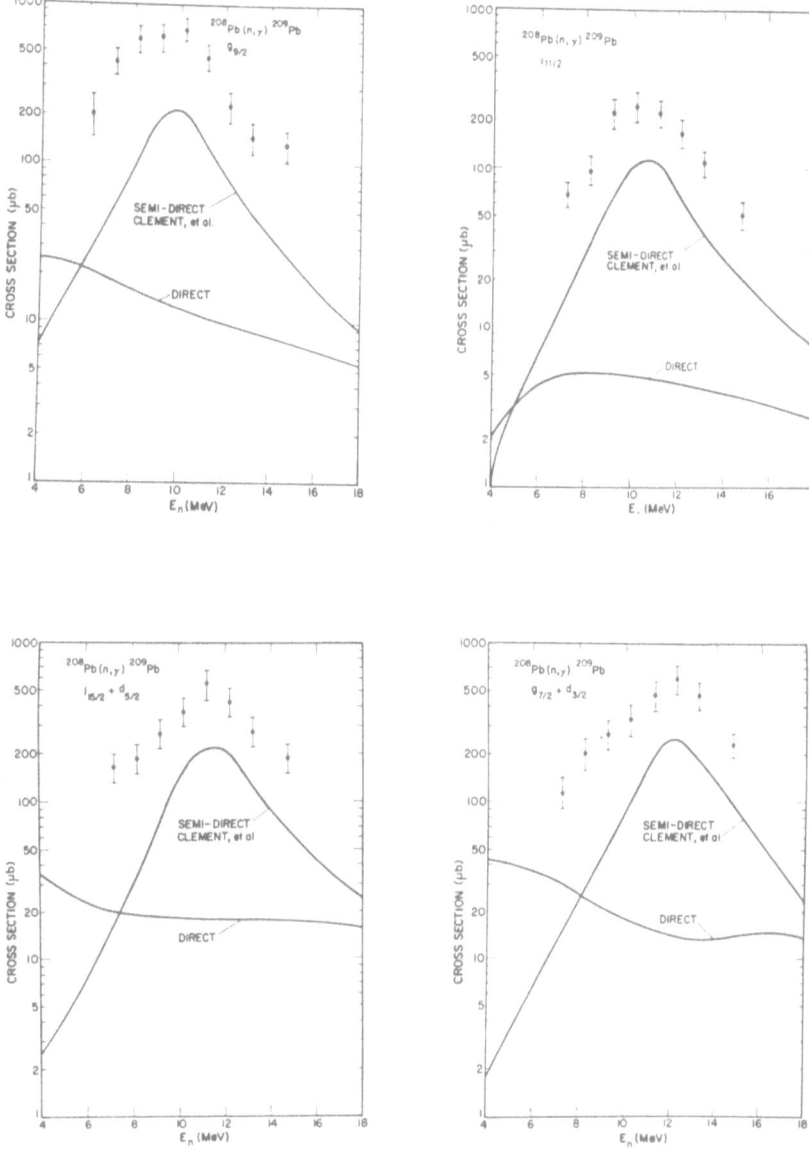

Fig. 4. Experimental and semidirect model excitation functions for the radiative capture of neutrons in
^{208}Pb (taken from ref. [6]). Calculations performed using E_R = 13,5 MeV, Γ = 3,5 Mev, V_1 = 160
MeV

288

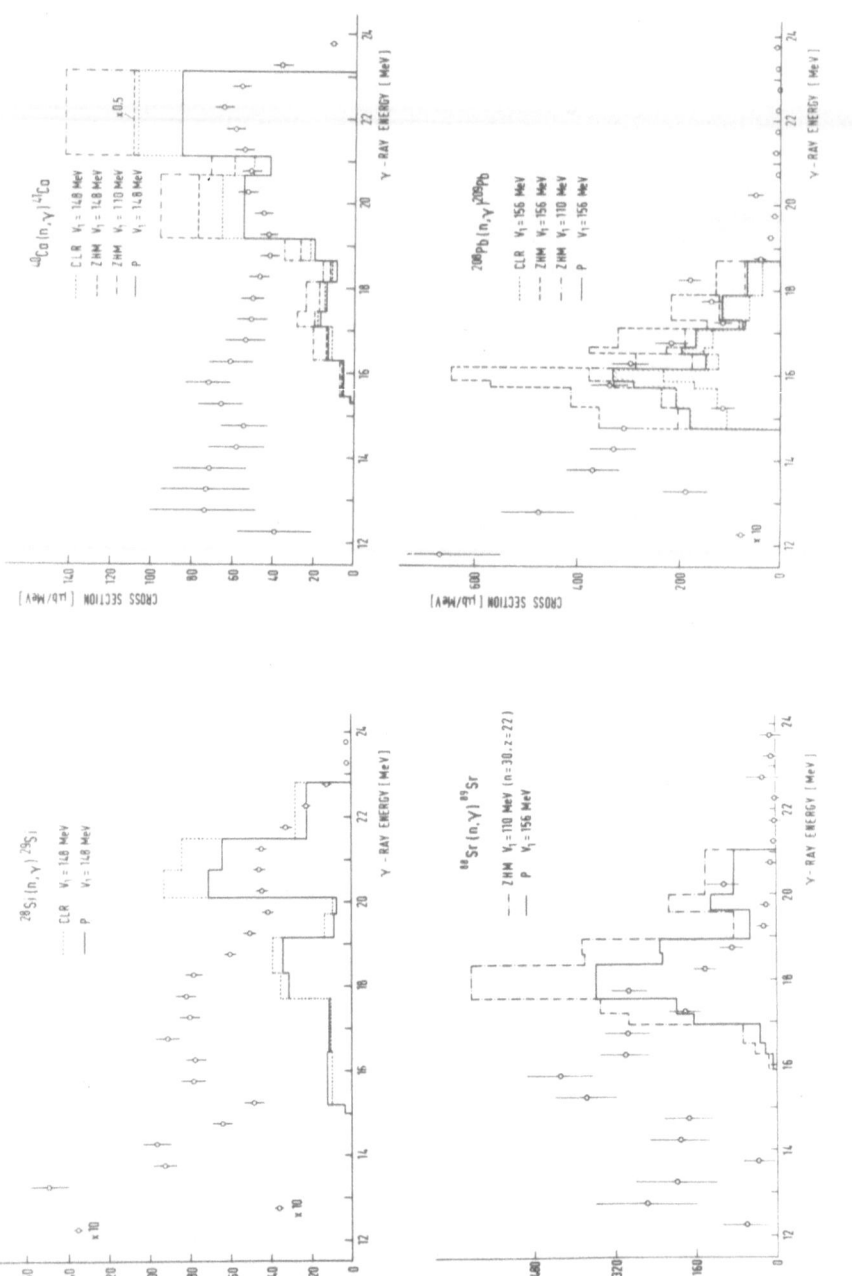

Fig. 5. Experimental (from natural targets) and calculated spectra of γ-rays from the radiative capture of 14,1 MeV neutrons in ^{28}Si, ^{88}Sr and ^{208}Pb: CLR - Clement, Lane and Rook 8, ZMH - Zimanyi, Halpern and Madsen 28, P - Potokar 24, (taken from ref. [25])

gy gamma-ray transitions could be calculated. In addition to the Clement, Lane and Rook (CLR) approach to the semidirect model, calculations were performed also following the approach of Zimany et al. [28] (ZHM) and the approach of Potokar [24] (P). The former (ZHM) includes besides the surface peaked radial dependence of H' (i.e. $d\rho(r)/dr$), also the volume interaction of the form $r\rho(r)$. The latter (P) instead, takes into account only the volume interaction. A similar approach was also used recently by Longo et al. [21] . In all these calculations, not only the semidirect, but also the direct and interference contributions were taken into account. This was done with the aim of seeing which of the approaches reproduces best the high energy spectral intensity for both light and heavy nuclei. The CLR approach reproduces the spectra for light nuclei (^{28}Si, ^{40}Ca) but fails, as already shown, for ^{208}Pb. The approach of Potokar reproduces within about ±30% the spectra of both light and heavy nuclei without any adjustment of parameters.

In the results shown in fig. 5. careful attention was paid to the strength of the isotopic spin term of the optical potential V_1. Its value was taken to be 148 MeV for ^{28}Si and ^{40}Ca and 156 MeV for heavier nuclei [18], [19].

In recent analyses of the radiative capture process in the region of GDR two other approaches, i.e. those of Brown [7] and Lushnikov et al. [22] were used. In both cases the effective charge factor was state independent. The approach of Brown yields too high capture cross section for light nuclei by a factor of 2. Calculations according to Lushnikov and Zaretsky reproduce well the gamma-ray spectra from 14 MeV neutron radiative capture in light nuclei [10] , but underestimate (cf [25]) its strength in heavy nuclei by a factor of 4.

It follows from the present state of the investigation of the radiative neutron capture in the region of the dipole giant resonance that the semidirect capture model calculations describe fairly well the high energy part of the spectra for the radiative capture of 14 MeV neutrons. These gamma-ray transitions belong to the high energy slope of the corresponding excitation functions. In this energy region the direct-semidirect interference contribution is constructive and improves the agreement between the theoretical and experimental spectral intensities.

On the other hand the inclusion of the direct and interference effects into the calculated excitaiton functions for the neutron capture in ^{208}Pb worsens the agreement in shape with the experimental data [25] . To show this the result of the complete direct-semidirect calculation (DSD) in the approach of Clement, Lane and Rook for the

^{208}Pb(n,γ)^{209}Pb excitation function is presented in fig. 6. The destructive interference makes the excitation function asymmetric: steeper at low energies than in the high energy part. Experimental excitation functions generally do not show such an asymmetry. The agreement between the experimental and calculated data is improved only in the region of constructive interference.

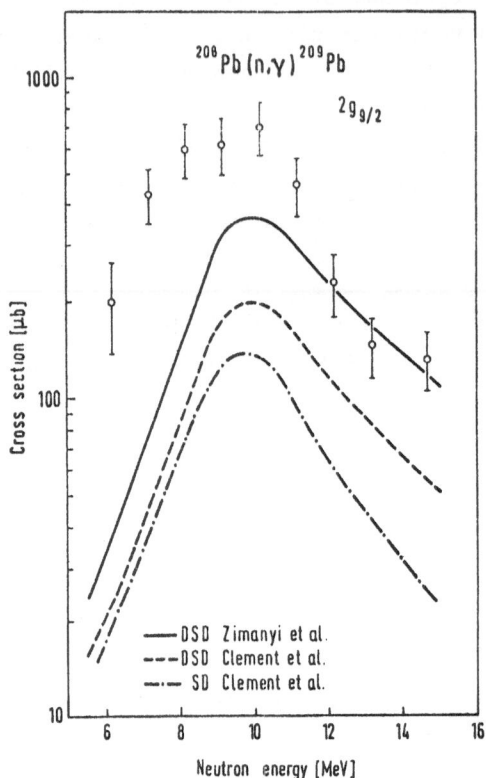

Fig. 6. The ^{208}Pb(n,γ)^{209}Pb ground state ($2g_{9/2}$) excitation function. Calculations were performed 25 using E_R=13,4 MeV and Γ=4 MeV. Values of the symmetry potential V_1=160 MeV and 110 MeV were used in CLR and ZHM approaches respectively. The semidirect (SD) curve in this figure is lower from the corresponding one in fig. 4 due to a larger Γ value

The need to examine this effect in detail will probably induce new complete measurements (spectra at different neutron energies and possibly also different angles). From the results of such measurements on heavy nuclei the adequacy of the semidirect model will be established. Measurements on heavy nuclei are particularly important since the statistical model calculations [20],[21] of the capture cross section were a few orders of magnitude lower than the experimental ones for the capture of 14 MeV neutrons. An exception were the nuclei having nearly closed neutron shells. For these nuclei the compound nucleus process might contribute up to several tens of persent to the experimental cross section [19].

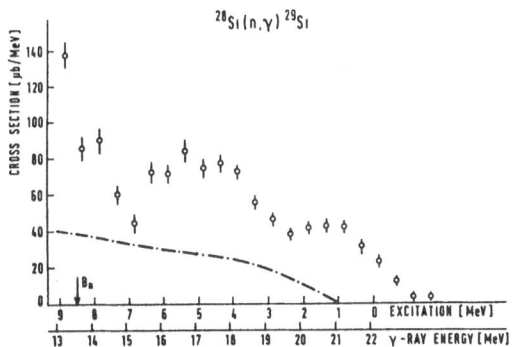

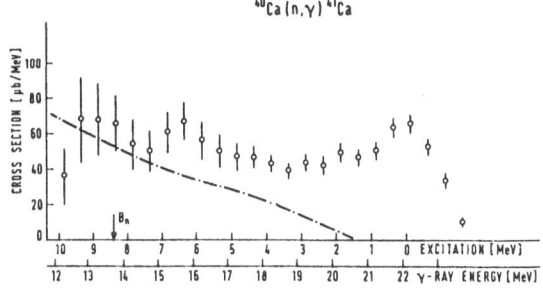

Fig. 7. Experimental and statistical model spectra of gamma-rays from the radiative capture of 14 MeV neutrons in ^{28}Si and ^{40}Ca (form ref. [13])

On the contrary the contribution of statistical process connot be ignored in light nuclei (cf [1], [19], [20])(see fig. 7). This is particularly true for transitions to higher excited states and for the capture of neutrons of energy lower than 14 MeV. Complete measurements, as mentioned before, will help to understand the capture mechanism for such cases.

Attention should be also paid to calculations by Sperber [27] which give the compound nucleus contribution to the capture process. His calculations yield for some heavy nuclei (integrated) cross section values, which do not differ appreciably from the experimental ones. On the other hand, they do not reproduce the shape of the gamma-ray spectra [5].

At this point one could comment on the role of the study of (p,γ) reactions in the region of the GDR in discussing the semidirect capture model. Though the semidirect theory of CLR was first applied [8] to calculate the excitation function measured by the activation technique for proton capture in heavy nuclei, this theory was not used later for the calculation of the prompt proton capture gamma-ray spectra or excitation functions. The main reason probably is that such measurements are limited to light nuclei.

Complete measurements of proton capture spectra in nuclei with A∿60 were carried out very recently [16] . In this case, as in the case of $^{208}Pb(n,\gamma)^{209}Pb$, the excitation functions were rather symmetric, i. e. contrary to what was expected on the basis of complete DSD calculations.

The angular distribution of gamma-rays from the radiative capture of energetic neutrons was not measured so far. Presently, an indication of its anisotropy could be obtained by the comparison of the spectra measured at 90° relative to the neutron direction and those experimentally integrated over a solid angle of 4π (or 2π for heavy nuclei) (measurements by the Ljubljana group). Due to very high experimental errors the anisotropy could be established only in those cases where really pronounced differences appear to exist. Until now only in the case of $^{88}Sr(n,\gamma)^{89}Sr$ was such a difference found and should be worthwhile studying in detail [14]. Values of other integrated cross sections measured either at 90° or integrated over a solid angle of 4π mutually agree within the experimental error of about ±20%.

293

References

1. BENZI, V. and BORTOLANI, M.V., Nuovo Cimento $\underline{38}$ 216 (1965).

2. BERGQVIST, I., LUNDBERG, B., NILSSON and STARFELT, N., Phys. Lett. $\underline{19}$ 670 (1966).

3. BERGQVIST, I., LUNDBERG, B., NILSSON, L. and STARFELT, N., Nucl. Phys. $\underline{A120}$ 161 (1968).

4. BERGQVIST, I., LUNDBERG, B. and NILSSON, L., Nucl. Phys. $\underline{A153}$ 553 (1970).

5. BERGQVIST, I., private communication (1972).

6. BERGQVIST, I., DRAKE, D. and McDANIELS, D.K., Nucl. Phys. $\underline{A191}$ 641 (1972).

7. BROWN, G.E., Nucl. Phys. $\underline{57}$ 339 (1964).

8. CLEMENT, C.F., LANE, A.M. and ROOK, J.R., Nucl. Phys. $\underline{66}$ 273 and 293 (1965). See also LONGO, G. and SAPORETTI, F., Nuovo Cimento $\underline{56B}$ 264 (1968) and Nucl. Phys. $\underline{A127}$ 503 (1969).

9. CVELBAR, F., HUDOKLIN, A., MIHAILOVIĆ, M.V., NAJŽER, M. and RAMŠAK, V., International Conference on Nucl. Struct. Study with Neutrons, Antwerpen (North Holland), p. 563 (1965).

10. CVELBAR, F., HUDOKLIN, A., MIHAILOVIĆ, M.V., NAJŽER, M. and RAMŠAK, V., Nucl. Phys. $\underline{A130}$ 401 and 413 (1969).

11. CVELBAR, F., HUDOKLIN, A. and POTOKAR, M., Nucl, Phys. $\underline{A138}$ 412 (1969).

12. CVELBAR, F., HUDOKLIN, A. and POTOKAR, M., Nucl. Phys. $\underline{A158}$ 251 (1970).

13. CVELBAR, F. and HUDOKLIN, A., Nucl. Phys. $\underline{A159}$ 555 (1970).

14. CVELBAR, F., BUNDAR, M., HODGSON, E. and POTOKAR, M., Contribution D-21 to the Conference on Nuclear Structure Study with Neutrons, Budapest 1972.

15. DINTER, H., Nucl. Phys. $\underline{A111}$ 360 (1968).

16. DRAKE, D., WHETSTONE, S.L. and HALPERN, I., (to be published).

17. KANTELE, J. and VALKONEN, M., Phys. Lett. $\underline{39B}$ 625 (1972).

18. KRUTOV, V.A. and SAVUSHKIN, L.N., J. Phys. (London) $\underline{A2}$ 463 (1969).

19. LANE, A.M. and LYNN, J.E., Proc. of the Second UN International Conference on the Peaceful Uses of Atomic Energy, Geneva (1958), Vol. 15, page 47.

20. LANE, A.M. and LYNN, J.E., Nucl. Phys. $\underline{11}$ 646 (1959).

21. LONGO, G. and SAPORETTI, F. - private communication - see also contribution D-5 to the Conf. on Nuclear Structure Study with Neutrons, Budapest (1972).

22. LUSHNIKOV, A.A. and ZARETSKY, D.F., Nucl. Phys. $\underline{66}$ 35 (1965).

23. MILLENER, D.J. and HODGSON, P.E., Phys. Lett. $\underline{35B}$ 495 (1971)

24. POTOKAR, M., private communication - see also (POTOKAR et al, 72).

25. POTOKAR, M., CVELBAR, F., to be published; see also Contribution D-20 to the Conf. on Nucl. Struct. Study with Neutrons, Budapest 1972.

26. RIGAUD, F., ROTURIER, J., IRIGARAY, J.L., PETIT, G.Y., LONGO, G. and SAPORETTI, F., Nucl. Phys. $\underline{A154}$ 243 (1970), $\underline{A173}$ 551 (1971) and $\underline{A176}$ 545 (1971).

27. SPERBER, D., Phys. Rev. $\underline{184}$ 1201 (1969).

28. ZIMANYI, J., HALPERN, I. and MADSEN, V.A., Phys. Lett. $\underline{33B}$ 205 (1970).

SIMPLE STRUCTURES IN THE EXIT CHANNEL

L. PAPINEAU

C.E.N. Saclay

91190 Gif-sur-Yvette, France

1. Introductory Remarks

The purpose of this talk is to discuss the role of simple configurations in nuclear reactions and the important role they play in the
occurence of intermediate structures.

Let me first give some (trivial) remarks that an experimentalist
could make on intermediate structure in 1972.

1.1. The shell model is now believed to be a fact and it is often
underlined that the success of the shell model is one of the most surprising features of nuclear physics in the last decade.

At this point it is also worthwhile to recall that single particle
states can be bound or unbound and that unbound single particle states
can give rise to narrow or broad resonances; this means that we must
not forget that a narrow resonance can be due to the simplest configuration of the shell model: a single particle state. The spectroscopic
factor then is not obtained from the model comparison with the cross-
section but with the partial (reduced) width in a given channel.

1.2. Simple States and Hierarchy of Shell Model Configurations

Speaking of intermediate structure I prefer Bloch's formulation
[4] of simple states and shell model configurations, keeping in mind
the important notions:

i) that a state is simple with respect to some thing;

ii) that different classifications can be used depending on the
problem under investigation.

This formulation seems to me more convenient in 1972. Ten years
ago, when the possibility of existence of intermediate structure was
pointed out, the shell model was essentially identified with single
particle states; the next step was normally to consider the role of 2p
-1h configurations. Today we know a great variety of states deriving
from shell model + residual interactions, as collective states, pairing
vibrations, quartet states, etc.

Both experiment and theory are more and more dealing with important perturbations of the ground state, so that the next step of complexity (with respect to a certain core) is not only the 2p-1h one.

1.3. Quasi-direct Reactions

I have a few comments on the opposition between direct interaction and compound nucleus formation.

I heard that intermediate structure is a deviation from direct reactions and I heard also that intermediate structure is a deviation from the statistical model. I think that these interpretations depend essentially on whether one has the opportunity to use a DWBA or an Hauser-Feshbach code, and that it is not serious. We know at least since 1961 when Weisskopf [23] gave the description of successive steps in a nuclear reaction that there is continuity from the single particle scattering state to the many-particle many-hole complicated compound nucleus. What is new is that we know today many situations where only a few steps contribute to the formation of the compound nucleus.

Two conditions are necessary for a reaction to be described by the direct interaction formalism: i) the time (short!) ii) the non-perturbation of the core.

Speaking in general about direct reactions, one immedeately thinks spectroscopy and one forgets the time condition.

If the core remains unperturbed (weak coupling in a channel) but for any reason the reaction time leads to an observable increase and decrease of the cross-section i.e. a resonance, the reaction will be a Quasi-direct one. If one prefers Bohr and Mottelson's [6] formulation one can call it a Direct resonance. It is then no longer the cross-section that provides the structure information but the partial width in a given channel. We see immediately that weak coupling in a given channel will give rise to large spectroscopic factors and that the knowledge of special overlapping conditions in a given channel will be particularly important for heavy ion reactions. The analysis of a partial width with respect to a model gives the direct condition on the overlap of an initial and final wave functions. The only difference with the classical direct reaction (and, for that, a major difficulty) is that one of the two related states is an intermediate stage of the whole reaction process and lies in the continuum.

If we think once more in terms of time, with respect to 10^{-22} sec. for Direct Interactions for light projectiles and medium energies it is obvious that

i) for heavier projectiles at same energies, velocities will be smaller, collision times may increase leading to narrower resonances than the single particle ones.

ii) for light projectiles but higher energies even complicated configurations may be reached more rapidly and have a very short life time due to a large number of open channels.

Concluding these preliminary remarks, I would like to emphasize,

following Lane, that intermediate structure is one of the line broadenings observed in nuclear physics.

Underlying any observed broadened line is a discrete state which
is "special" in that it has a very large overlap on one or more observation channels.

2. Simple Structures in Residual States

It was at the winter meeting in Villars, (1963) that I heard for
the first time about intermediate structure. I always remember Feshbach's drawing (fig. 1): the example he gave was not an excitation
function but an evaporation spectrum. At low excitation energies were
the peaks observed in direct reactions corresponding to single particle levels; the evaporation spectrum, instead of being smooth, could
present structures due to the expected 2p-1h states. Nevertheless people began to search these structures not in residual spectra but in
excitation functions. At this time research has been done essentially
for elastic scattering or total reaction cross sections that means
without any selectivity for a particular channel (and for overlap conditions between a ground state and very high excited states of a different nucleus). The result of a survey on intermediate structure in
Villars 1967 gave a picture of hopeless bumpology and the question
arises: was the hypothesis of intermediate structure justified?

Let's turn a moment to the past. How structure information has
been obtained from nuclear reactions and what kind of structure information has been obtained first?

Pure shell model (single particle) states have been identified by
direct one-nucleon transfer reactions, that means essentially by the
large overlap between a ground state (target) and a low-lying final
state. Introduction of a treatment of residual interactions taking into account configuration mixing produces a dissolution of these simple
configurations into more complicated ones. The sharing of single particle strength over several low lying states has been, in fact found.
Intermediate structure may occur by taking into account not only the
first configuration produced in the interaction of the incoming nucleon with the target but also configurations in which (at least) one of
the nucleons of the target is excited. These states will lie at higher
excitaiton energies and have no reason to have a specifically good
overlap with the ground state.

It was the merit of Gillet [14] and of Bloch, Cindro, Harar [5]
to overcome the complication in the compound system in looking for
structures at higher excitation energies in the spectra of outgoing

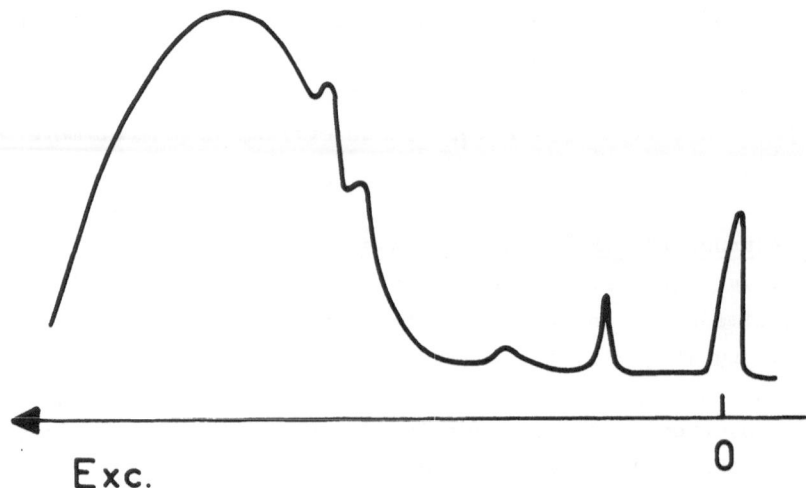

Exc.   0

FESHBACH
VILLARS January 1963

Fig. 1. Schematical spectrum of the outgoing particle as a
function of the excitation energy of the residual nucleus

particles, where the identification of individual bumps was much easier,
and with the important point to look for particular nuclei and reactions in which a relative simple situation is likely to occur. With their
review article [5] on "excitation of simple particle-hole configurations in the residual nucleus" the occurence of intermediate structure
was finally seriously recognized as a very general phenomenon for bound
states.

I will recall only a few typical examples.

Existence of particle-hole structures was clearly demonstrated by
the spectra of the (^{3}He,d) reaction on $In^{113,115}$ [8] (fig. 2), that
means by a transfer of a proton on a proton-hole target, and, I under-
line it once more, by the favorable overlap of the initial and final
states.

Now if we look for example on the level scheme of ^{58}Ni [3] obtained from (pp'γ) measurements (fig. 3) we see that between 3 and 8 MeV
of excitation about 50 levels are known. Six of them concentrated between 6 and 8 MeV decay essentially to the ground state and were interpreted as due to proton core excitations. The inelastic scattering of

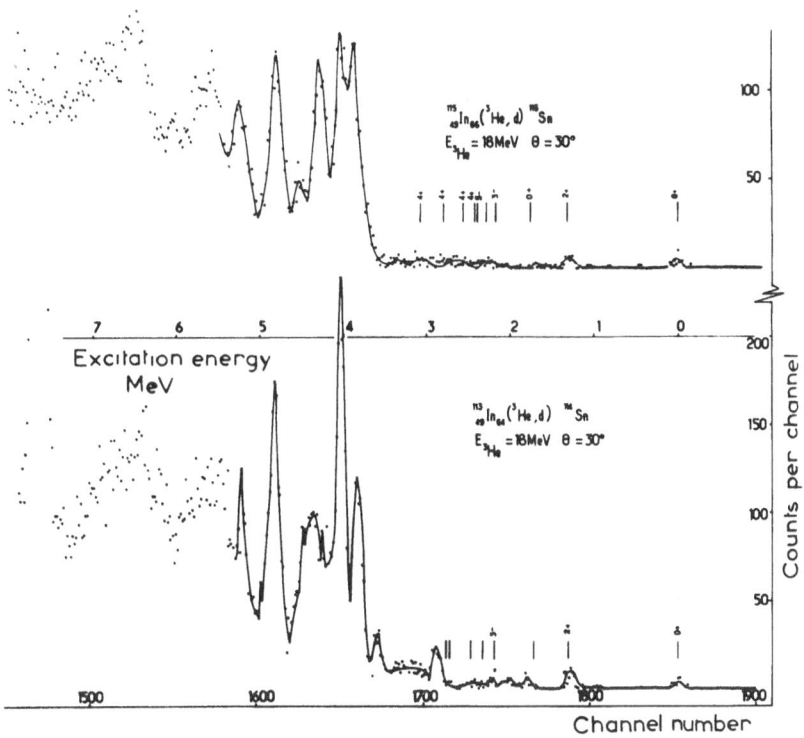

Fig. 2. Deuteron spectra from 18 MeV ^{3}He bombardement of ^{113}In and ^{115}In. Note the similarity of the two spectra

alpha particles on ^{58}Ni [7] (excites in the 3 to 8 MeV energy range ∿ 25 levels and among them 10 very strong, while four-particles trans-

300

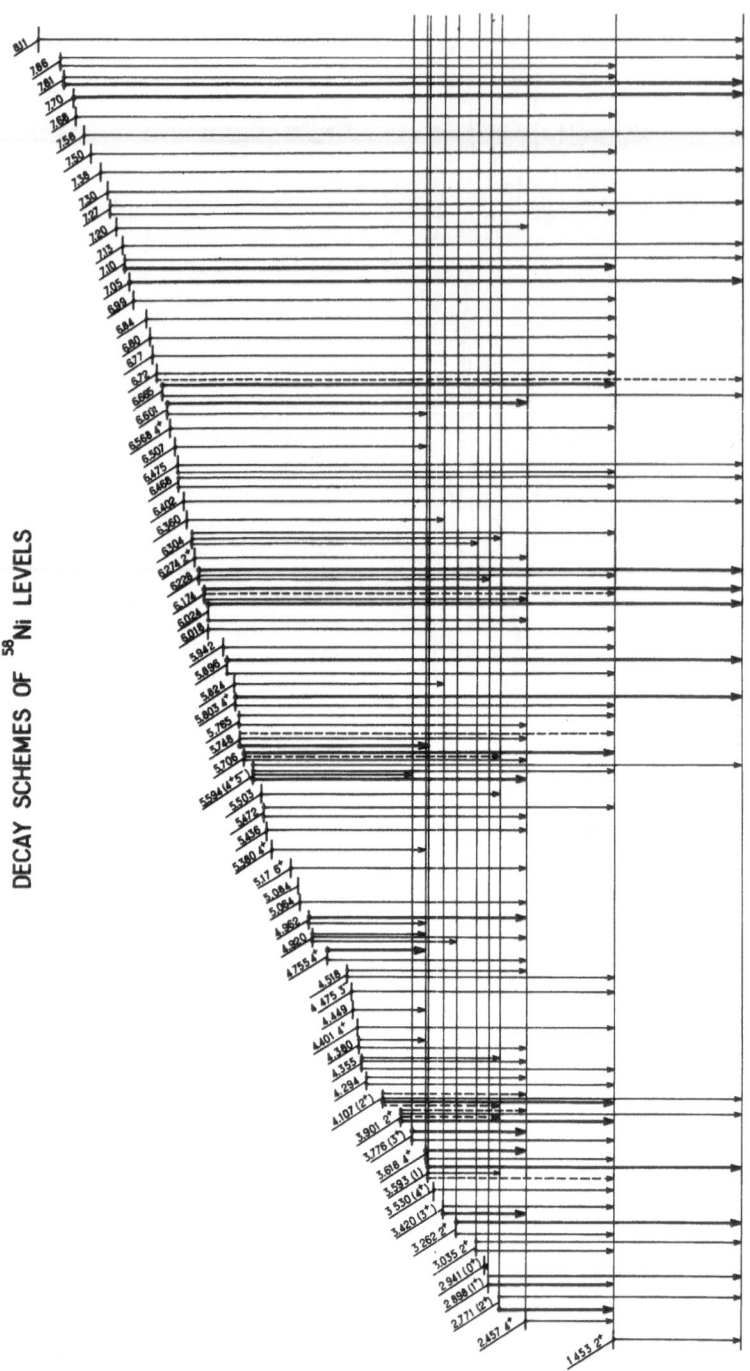

Fig. 3. Gamma decays of the ^{58}Ni levels. Relative intensities
are indicated in decreasing order by thick solid, thin and
dashed lines

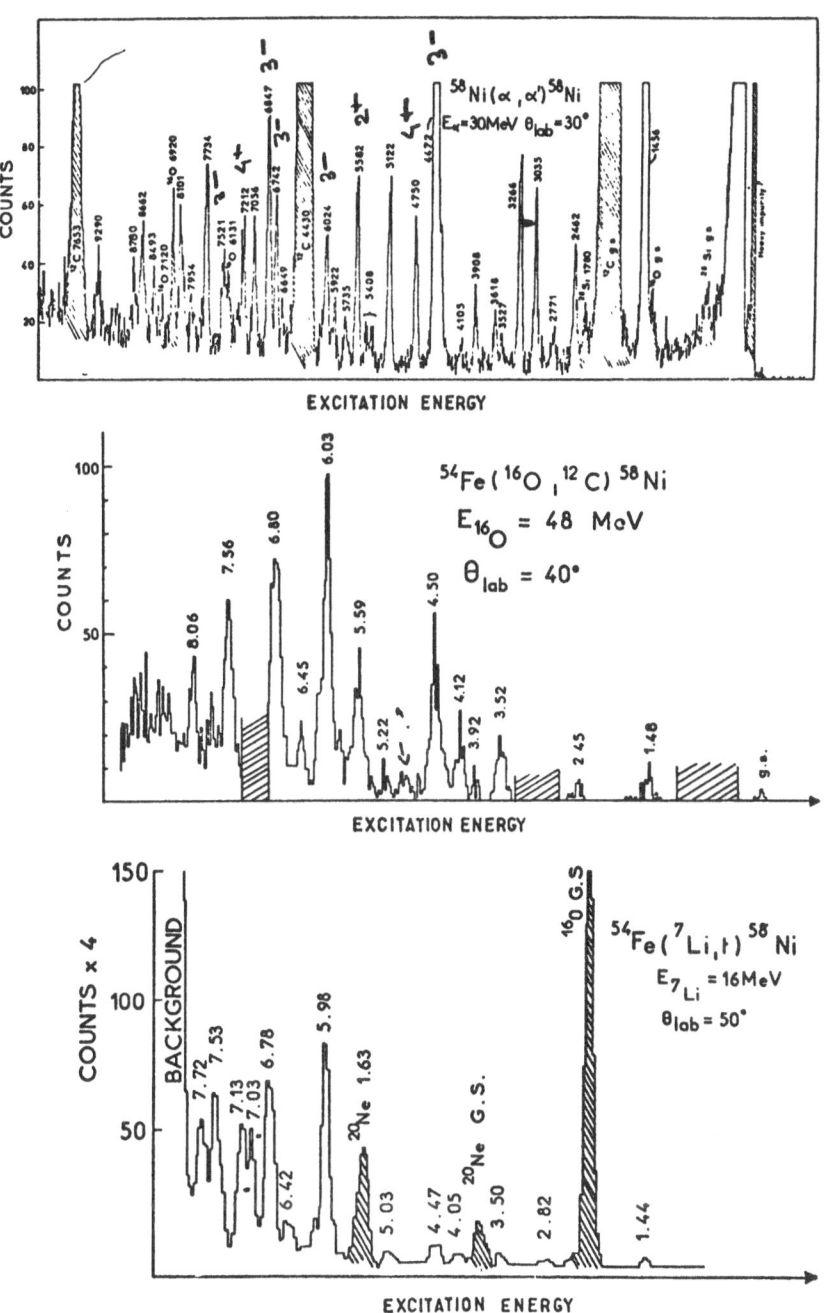

Fig. 4. For comparison, $^{58}Ni(\alpha\alpha')$ ^{58}Ni, $^{54}Fe(^{16}O,^{12}C)$ ^{58}Ni and $^{54}Fe(^7Li,t)$ ^{58}Ni spectra are shown. The shadowed areas in the (^{16}O, ^{12}C) spectrum obtained with a magnet correspond to the position of impurity peaks. The spin assignments on the ($\alpha\alpha'$) spectrum were obtained from angular distributions

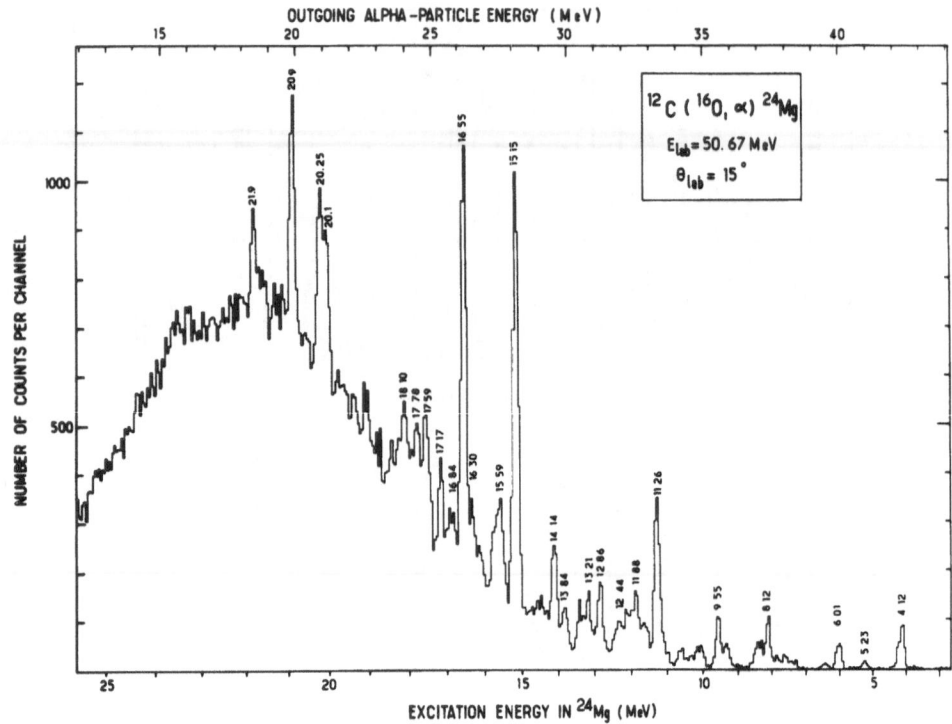

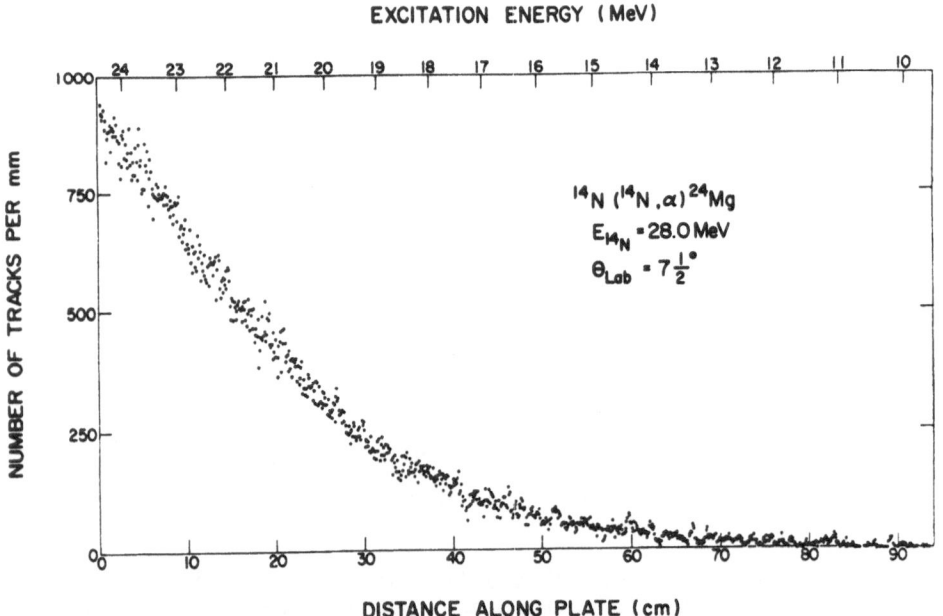

Fig. 5. Alpha particle spectra from the $^{12}C(^{16}O,\alpha)^{24}Mg$ and $^{14}N(^{14}N,\alpha)^{24}Mg$ reaction

fer reactions induced by heavy ions on ^{54}Fe (cf [9], [10]) excite no
more than $\sim$ 10 levels (or closely grouped levels in the same region
(fig. 4)); these structures have been connected with the new ideas on
quartet excitations [15], [11]. Another example of remarkable channel
selectivity is shown on fig. 5. At the same excitation energy of ^{24}Mg
the (^{14}N,α) reaction [19] shows a smooth spectrum while the (^{16}O,α)
reaction leads to sharp isolated peaks [12].

3. Simple Structures in Exit Channels

Let us now turn to the main subject of this talk, i.e. the role
of simple structures in exit channels observed by resonances in the
compound system. (I will not discuss the giant dipole resonance).

The first resonances in an exit channel were found for analogues
of Sn^{119} [1] on excitation functions of inelastic proton scattering
to the first 3$^-$ state of Sn^{118} and interpreted as due to weak coupling
of single quasi-particles to the 3$^-$ excited core. Later inelastic scattering excitation functions have been extensively used for analogue
resonances in order to look for such coupling to the excited cores.
The nicest story in this field was, I think, the discovery of 2p-1h
structures in ^{209}Pb. For a time, with the FN tandems, the study of analogues of ^{209}Pb was limited to the first excited states. Then with
their MP tandem getting into operation the Yale group immediately started the investigation of ^{209}Pb analogues at higher energies. Nothing
however appeared on the elastic scattering excitation function. Poor
justification for the Emperor! Then they repeated the experiment with
counting rates for inelastic scattering and the 2p-1h structures were
found in the decay to particular excited states of ^{208}Pb.

How resonances in an exit channel can occur?

Let us consider a nuclear reaction schematically described by
successive steps of interaction from the single particle configuration
(direct reaction) to the many particle-many hole compound nucleus (fig.
6). The wave function of the compound system can be expanded in terms
of quasi-bound states of different configurations [16]. Now, is the
wave function of the compound nucleus always very complex? The different configurations are connected by the two-body residual interactions. If all matrix elements M_{jj} (fig. 7) are equivalent no structures
will appear and the system will always have the full complexity. One
feels immediately that strong selection rules on M_{jj} may give rise to
structures.

We know how the vanishing of matrix elements due to nuclear forces
between $T_>$ and $T_<$ configurations gives rise to analogue resonances.

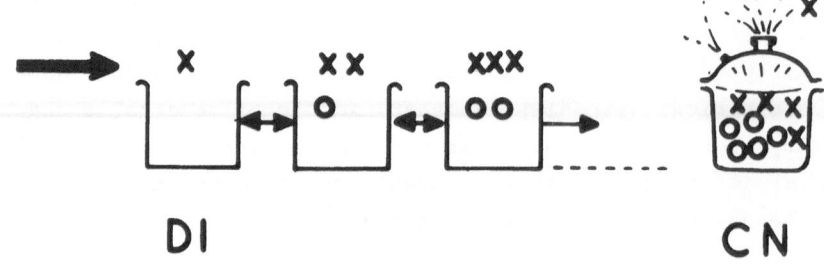

DI

CN

$$\psi_E = \alpha|\psi_0\rangle + \beta|\psi_1\rangle + \gamma|\psi_2\rangle + \dots \quad \xi|\psi_\phi\rangle$$

Fig. 6. See text

ψ_{CN} always very complex ?

$$\langle\psi_i| \vee |\psi_j\rangle$$
$$|M_{ij}|$$

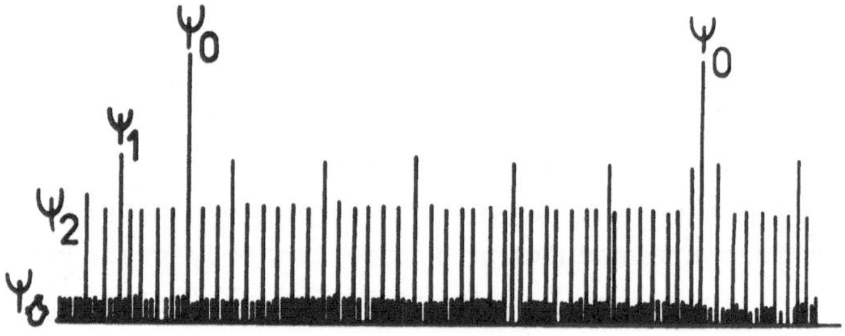

No structures if all
M_{ij} equivalent

Fig. 7. See text

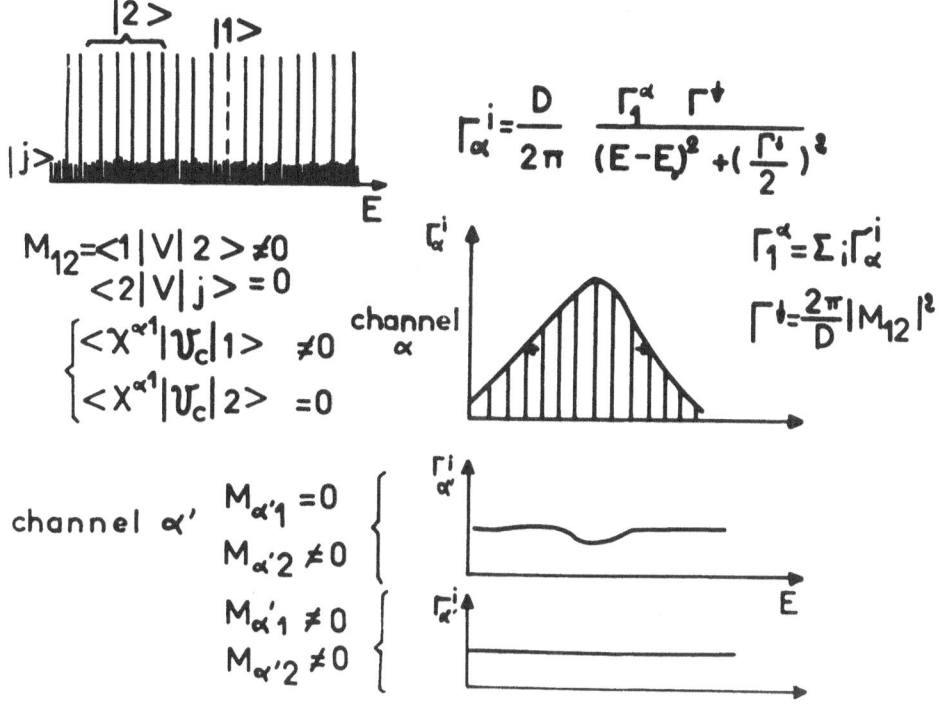

Fig. 8. See text

For a time the absence of clearly identified intermediate resonances
others than the analogue states lead to the idea that only the strong
isospin selection rules could produce intermediate structure. In the
following I will concentrate only on selectivities which do not come
from isospin.

If all M_{ij} are not equivalent, quasi bound states can exist with
few components in the wave function and even corresponding to one single configuration. But these quasi-bound states are embedded in the
continuum and we must consider also the couplings to the different
open channels.

Let us suppose a simple state $|1>$ among a class of states 12 and
states $|j>>$ (fig. 8). Γ_α^i is fine structure partial width, $\Gamma^\downarrow$ the spre-
ading width.

Consider a channel α:

If the states $|2>$ are coupled only to state $|1>$ and if only this
state is coupled to the continuum of $|1>$ in channel α, the distributi-
on of Γ_α^i has a Lorentzian shape [16]. The spreading width obtained from
the half-width of the distribution is related to the matrix element
M_{12}; the sum of the partial widths Γ_1^α compared to the simple-state

model-width will give the spectroscopic factor for the state |1>.

Now consider a <u>second channel α'</u>: different situations are to be considered
i) state |1> is common doorway for channels α and α' i.e. only the state |1> is coupled to the continuum of |1> in channel α'
the partial widths in channel α' will display the same intermediate structure;
the two channels are correlated.
ii) only states |2> are coupled to channel α'; the partial widths in channel α will present a dip at perturbed energy of |1>

This can introduce an anti-correlation between channels α and α'
iii) both the state |1> and states |2> are coupled to channel α' (for example by a common doorway). The partial width Γ_α^i, does not display the structure observed in channel α,
the two channels are uncorrelated.

If the channel α is the entrance channel the structure can be observed (more or less washed out) in the other channels and we get the well known correlation between cross-sections.

But it is obvious that the point of interest for us is the case where <u>α is an exit channel</u>, since we expect intermediate structures to appear for channels leading to excited states. This means the following:
- the elastic scattering excitation function will not display the structure
- it is possible that the structure will be observed only for one partial cross-section
- the different partial cross-sections can be completely uncorrelated.

Here I will open a parenthesis: I think it is evident from the preceeding results that <u>the absence of correlations is not a valid test</u> to decide whether a "bump" in the cross section is due to the presence of a simple state or to statistical fluctuations.

Now, do the conditions summarized in fig. 8 exist? Or more precisely let us consider the situation in fig. 9, where α_o is the entrance channel, Ψ_o a common doorway states |1> and |2>; is there experimental evidence for such a case? The answer is yes!

The first well identified, non-analogue, intermediate structure has been found in subthreshold fission on ^{237}Np [18],[21] . The top of fig. 10 shows the resonances as observed in the total cross section: the reduced neutron widths are distributed with fluctuating strengths of about equal mean intensity. In contrast the fission cross sections shows resonances bunched together around 40 eV exhibiting a strong intermediate structure effect. There are approximately 40 clusters below

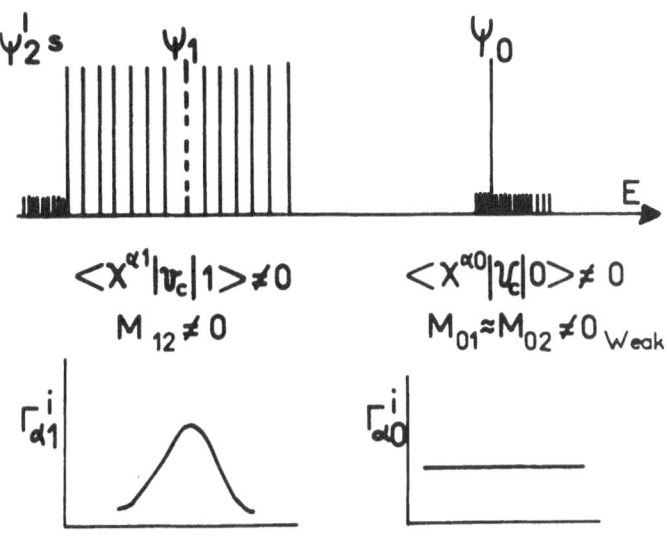

Fig. 9. See text

2 keV with an average spacing of about 50 eV whereas the average level spacing of the resonances is about 0.5 eV; between the clusters the fission cross-section is so small that it can hardly be measured.

The second example lies on the opposite side of the periodic table. The excitation functions of the reaction $^{40}Ca(pp')^{40}Sc$ have been measured between 4.8 and 8.2 MeV proton energies for several excited states of ^{40}Ca [20]. There is a very strong enhancement of the cross section for the 3^- state (3.73 MeV) at a proton energy of about 6.2 MeV (fig. 11) and nearly all states which strongly resonate on this excitation function have been assigned a spin $5/2^+$. The distribution of the partial widths (fig. 12) is well reproduced by a Lorentzian shape as predicted by the theory of intermediate structure [16]. The parameters (see fig. 9) $\Gamma^{\downarrow}$, M_{12}, Γ_1^{α} and the spectroscopic factor for the simple state have been obtained.

The ratio $\Gamma_p/\Gamma_{p'}(3^-)$ is not constant, which means that the simple state responsible for the observed structure is not a common doorway for the entrance and exit channels. The elastic partial widths are rather constant for all the fine structure states; that indicates that all fine structure states are weakly coupled to a common doorway (see fig. 9).

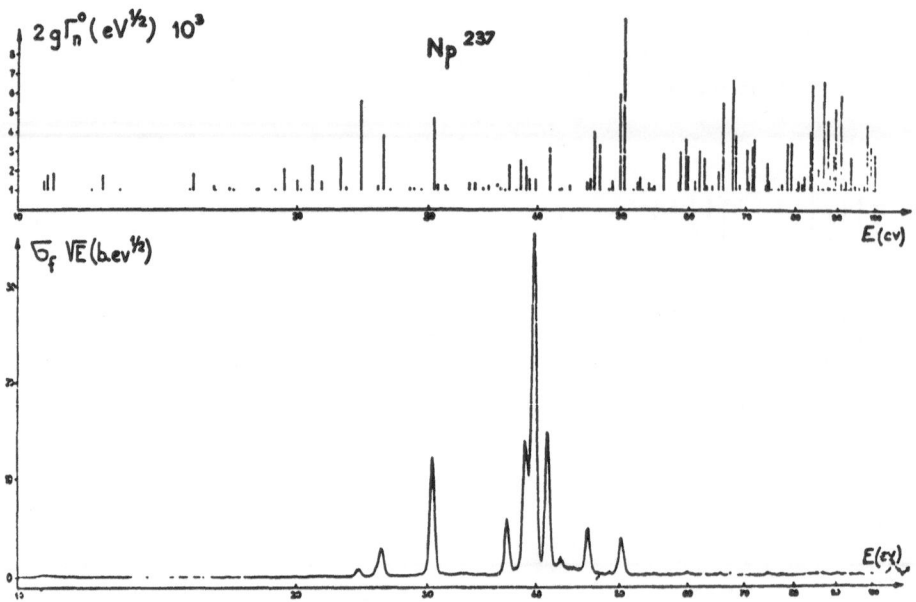

Fig. 10 The observed fission cross section of ^{237}Np between 10 eV and 100 eV. The upper part of the figure shows the positions of the resonances as observed in the total cross section. For each resonance the height of the bar is proportional to the reduced neutron width $2g$ Γ_n^o

A priori it might be surprising to <u>compare the two</u> just mentioned <u>reactions</u>; what can be common between fission of a heavy deformed nucleus and inelastic scattering on the closed-shell nucleus ^{40}Ca? But one can discover that there are <u>common features which enter exactly the picture I described for the occurence of intermediate structure in exit channel. The difference lies</u> only in <u>the nature of the simple state</u> and the physics of its decay.

Let us consider, in this respect, the display on fig. 12 - for the two reactions the partial widths in the entrance channel shows no structure; there is a common doorway ψ_o

in ^{41}Se ψ_o is the far away $2d5/2$ single particle state

in ^{238}Np I suppose it is some neutron scattering state.

- the partial widths of a given channel show the structure that indicates for both reactions the existence of a simple state $|1\rangle$ sharing its strength on neighbouring states of same spin and parity $|2\rangle$.

In ^{42}Sc the simple state is due to weak coupling of the 3^- core of ^{40}Ca to a single particle in the $2p1/2$ orbit.

In ^{238}Np the simple state is a quasi-bound state in the second well of a double-humped fission barrier.

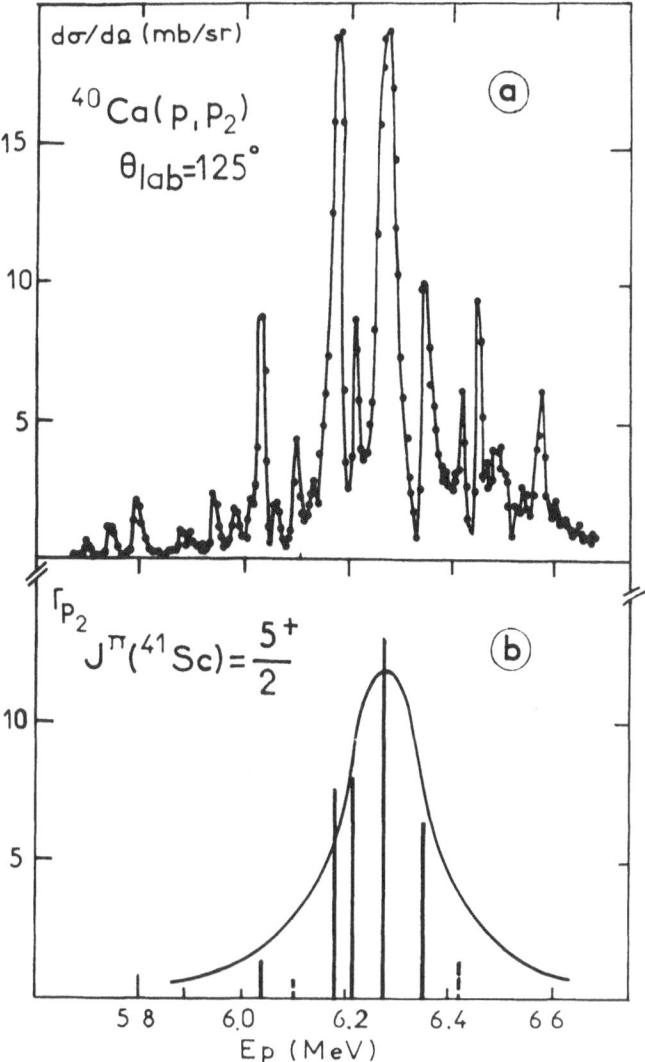

Fig. 11. a) Excitation function of the reaction $^{40}Ca(pp_2)$
^{41}Sc leaving ^{40}Ca in the first 3^- state.
b) Partial widths for resonances with assigned spin
5/2+; the solid curve is a fit with the predicted
Lorentzian shape for the following values of para-
meters: $D=100$ keV, $E_0=7.2$ MeV, $\Gamma_1^\alpha=38$ keV, $\Gamma^\downarrow=200$ keV

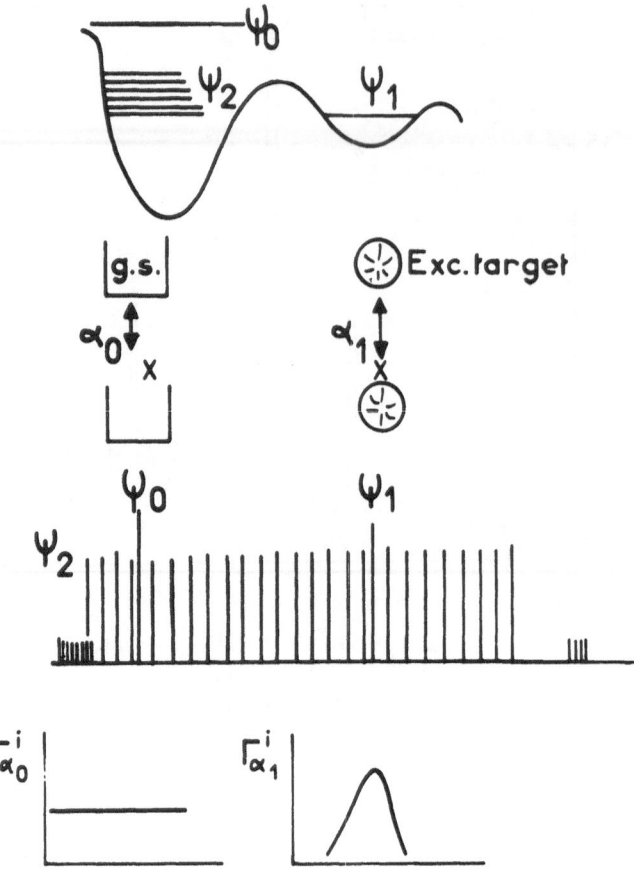

Fig. 12. See text

4. Conclusion - Future in Heavy-ion Reactions

The common scheme of the above reactions has shown that it is ju-stified to expect that intermediate structure will appear in a specific channel if selective overlapping conditions exist.

Interesting information on new structure aspects of the nucleus can be obtained when one is able to find the nature of the underlying "simple" state. I think that the main interest really begins with this research on the nature of the simple state.

If I could give an advise I would say:don't search systematically for intermediate structure for itself; but if you find a large bump it is worthwhile to make an effort to search out if the nucleus is telling you something.

This is particularly true for heavy-ion reactions. From the very beginning heavy-ion excitation functions have shown structures.

Twelve years ago, three marked resonances were found in the excitation function of α, p, n and γ yields of the reaction $^{12}C + ^{12}C$ just below the Coulomb barrier [2]. The first interpretation of these resonances was based on the formation of a quasi-bound $^{12}C - ^{12}C$ molecule. Recently these resonances have been re-examined for selective exit channels of the $^{12}C(^{12}C,\alpha)^{20}Ne$ reaction [22] . The results support the α-particle doorway state hypothesis [18] and a large overlap with quartet states in ^{20}Ne.

The survey of structures in heavy-ion reactions is the subject of another talk of this conference. I show only two examples of correlations observed in the study of $^{12}C + ^{16}O$ excitation functions [13] on fig. 13 and fig. 14.

I confess I do not understand why in the presence of strong structures in heavy-ion excitation functions so many people want absolutely to demonstrate that these are statistical fluctuations.

Statistical fluctuations in my mind have never been anything else than a correct parametrization of our ignorance of all details of nuclear structure.

Confronted with new theoretical concepts as quartet excitations for example, I think it is the role of the experimentalists to search for new experimental criteria to explain the observed structures.

I hope you are convinced that simple structures can appear in selected exit channels and be a useful tool for nuclear spectroscopy. But don't forget that "simple" is a relative concept and that we are searching with heavy ions many-particle correlations and collective properties of high excited nuclei.

This means that we search for simple relations between complex structures.

I tried to illustrate this on fig. 15. Both the elephant and the butterfly are complex structures. But the butterfly landing on the elephant's trunk represents a simple relation between them.

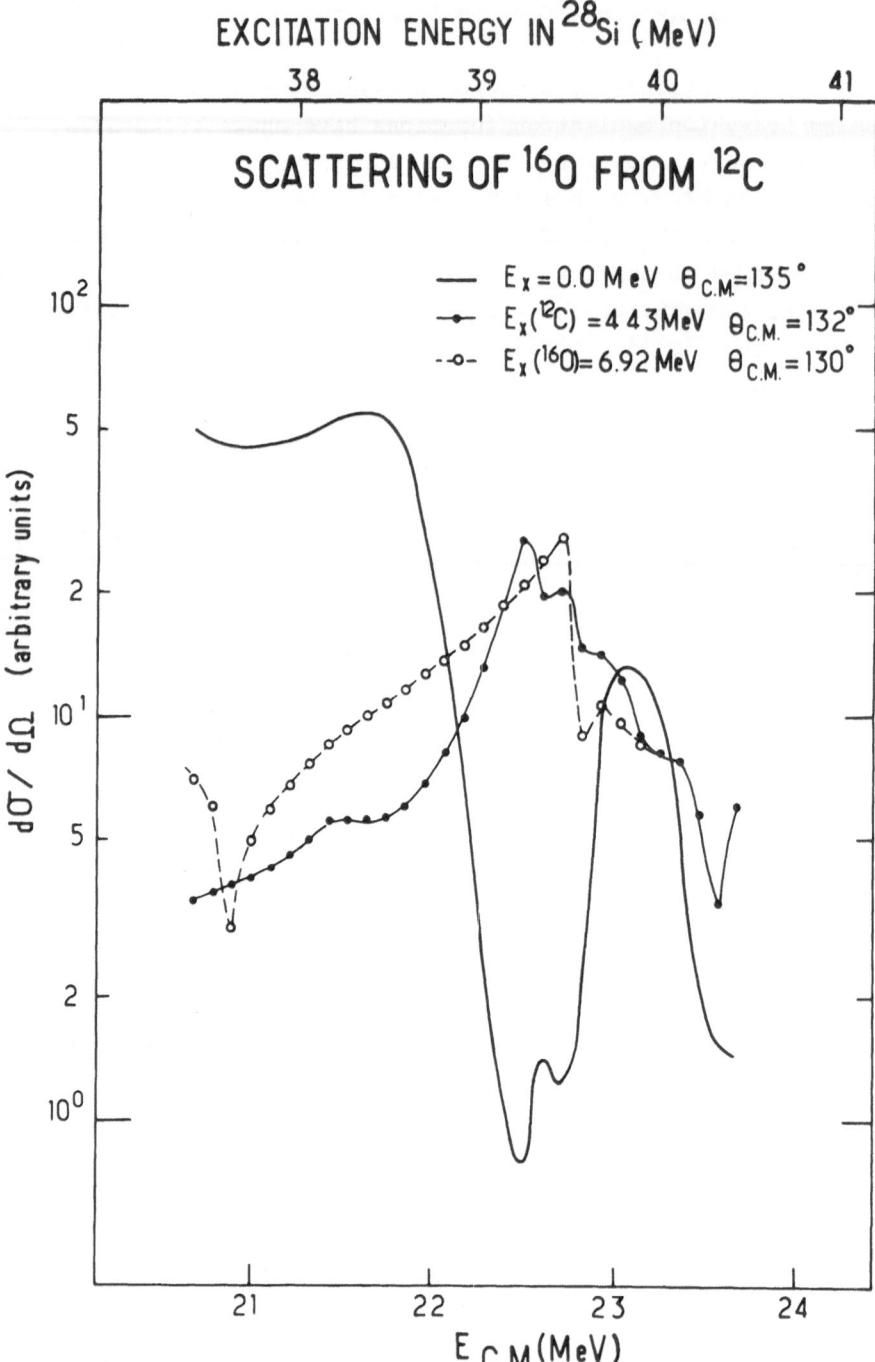

Fig. 13. Excitation function of elastic and inelastic scattering for ^{16}O from ^{12}C. Notice the strong anticor-
relation, or opposition in phase between the elastic
and inelastic scattering

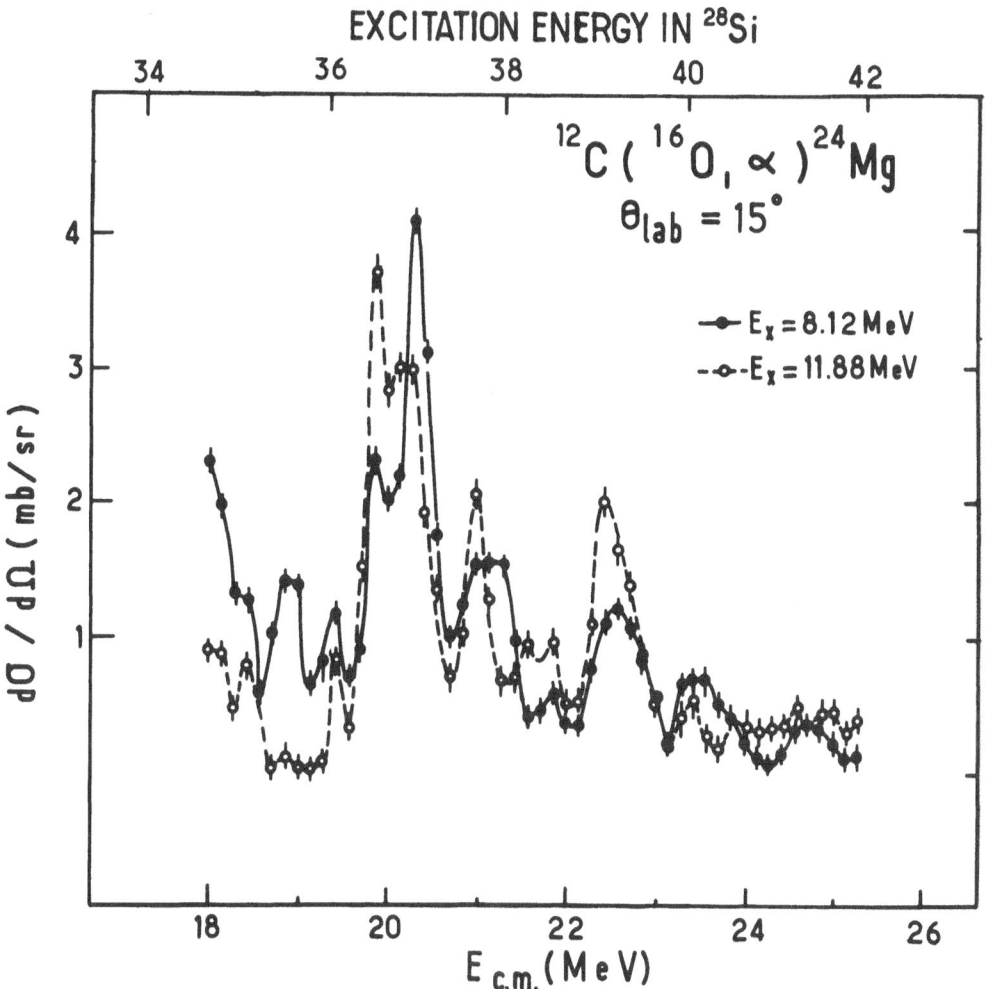

Fig. 14. Excitation function of the reaction $^{12}C(^{16}O,\alpha)^{24}Mg$.
Example of two strong correlated channels

314

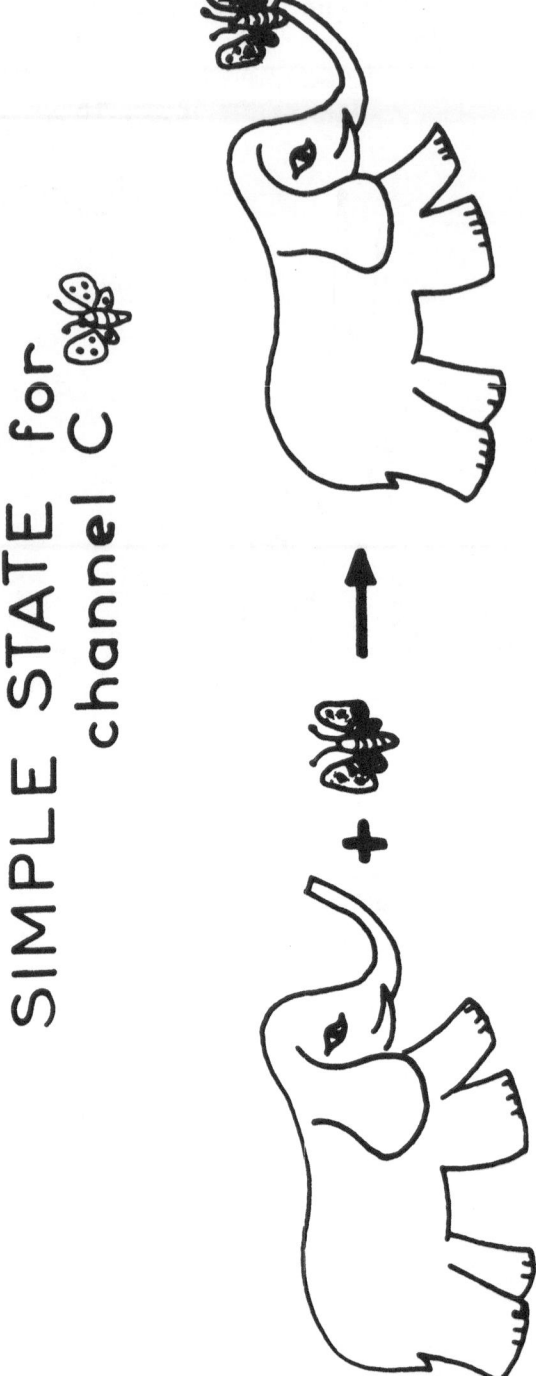

Fig. 15. See text

References

1. ALLAN, D.L., JONES, G.A., MORRISON, G.D., TAYLOR, R.B., WEINBERG, R.B., Phys. Lett. $\underline{17}$, 1 (1965).

2. ALMQUIST, E., BROMLEY, D.A., KUEHNER, J.A., Phys. Rev. Lett. $\underline{4}$, 515 (1960).

3. BEUZIT, R., DELAUNAY, J., FOUAN, J.P., RONSIN, H., Nucl. Phys. $\underline{A137}$ 97 (1969).

4. BLOCH, D., in Proceedings of the Symposium on Recent Progress in Nuclear Physics with Tandems, Heildelberg 1966.

5. BLOCH, C., CINDRO, N., HARAR, S., Progress in Nuclear Physics Vol. $\underline{10}$, p. 77 (1968).

6. BOHR, A. and MOTTELSON, B.R., Nuclear Structure Vol. $\underline{1}$, p. 434.

7. BRUGE, G., CHAUMEAUX, A., DeVRIES, R., MORRISON, G.C., Phys. Rev. Lett. $\underline{29}$ 295 (1972).

8. CONJEAUD, M., HARAR, S., PICARD, J., Phys. Lett. $\underline{23}$, p. 104 (1966).

9. COTTON, E., BIANCHI, L., PAPINEAU, A., Aix-en-Provence, Europ. Conf. on Nucl. Phys. vol. II, p. 82 (1972).

10. CUNSOLO, A., LEMAIRE, M.C., MERMAZ, M.C., QUEBERT, J.L., SZTARK, H. Europh.Conf. on Nucl. Phys. Aix-en-Provence, Vol. II, p.66 -(1972).

11. FARAGGI, H. et al., Ann. Phys. Vol. 66, p. 905 (1971 - de Shalit Memorial).

12. GASTEBOIS, J., Journal de Physique, Colloque $\underline{C6}$, vol. 32, p. 57 (1971).

13. GASTEBOIS, J., BALLINI, R., CHARLES, P., FERNANDEZ, B., FOUAN, J., Lett. Nuovo Cim. Vol. 2, n^0 3, p. 90 (1971).

14. GILLET, V., unpublished (1967).

15. GILLET, V., Prodeeding International Conference on properties of Nuclear States, Montréal (1969) p. 483 and Phys. Lett. $\underline{34B}$, I, p. 24 (1971).

16. MAHAUX, C., WEIDENMULLER, H.A., Shell Model Approach to Nuclear Reactions, North-Holl, Publ. Comp. 1969.

17. MICHAUD, G.J., VOGT, E.W., Phys. Rev. $\underline{C5}$ 350 (1972).

18. MICHAUDON, A., Neutron Cross-section, Conf. NBS spec. publ. n^0 229, Vol.1, p. 427, (1968).

19. MIDDLETON, R., GARRETT, J.D., FORTUNE, H.T., Phys. Rev. $\underline{C4}$, n^0 6, pl 1987 (1971).

20. MITTIG, W., CASSAGNOU, Y., CINDRO, N., PAPINEAU, L., SETH, K.K., Topical Conference on the Structure of 1 f7/2 nuclei Editrice Compositori International Physics Series p. 339 (1971) and MITTIG, W., Thesis, Orsay 1971.

21. PAYA, D., BONS, J., DERRIEN, H., FUBINI, A., MICHAUDON, A., RIBON, P., Colloque de la Soc. Franc. de Physique, Bordeaux (1967), Journ. Phys. T29 Cl. p. 159 (1968).
22. VOIT, H., ISCHENKO,G., SILLER, F., to be published.
23. WEISSKOPF, V.F., Physics Today, 14, n° 7, 18 (1961).

SUMMARY AND CONCLUSIONS

HERMAN FESHBACH

Massachusetts Institute of Technology

Cambridge, Massachusetts

The summary will be restricted to the two main themes which dominated this conference. The first involved the observation of intermediate structure and the understanding of its physical significance in terms of doorway states. The second involves the concept of a hierarchal classification of states which are ordered according to their complexity, that is from the simple to the more complicated. This classification is exploited in the "pre-equilibrium" theory of nuclear reactions which attempts to calculate the probability for each of these states and their probability for decay into a given channel. The two themes are closely connected by the fact that the doorway states are members of the hierarchy, the more complex states being designated as second and higher order doorway states. We shall come back to this connection in the second section of this report which is concerned with the pre-equilibrium phenomena.

The theory of intermediate structure was discussed by Mahaux. I shall comment on a few of the points that he raised and at the same time attempt to indicate the significance of the discussions and papers presented at this meeting with respect to fundamental issues connected with this field.

There are a number of very clear and well-known examples of intermediate structure: isobar analog resonance, the giant dipole and isomeric fission. Many cases, at least initially, are not as clear, with the consequence that it is often suggested that the observed structure is a statistical fluctuation. It is probably necessary to use statistical tests to determine the statistical significance of a structure. However the results obtained are not particularly definitive since a small statistical significance does not necessarily mean that the structure is not intermediate structure nor does a large value guarantee that it does. Much more significant and much more useful are the physical tests which can be made. Let us outline some of these tests.

Bear in mind that a doorway state resonance can be treated like any resonance if one realizes that the total width contains an additive term the down or spreading width, for decay into more complex states. Decay into various channels with varying branching ratios is also possible. This has the consequence that the partial cross-section to

individual channels will be correlated. An example of such a correlation described at this conference is the $^{12}C(^{16}O,\alpha)^{24}Mg$ reaction at
$E_{c.m.}$ of 19.7 MeV of Malmin et al. [8] described by Stokstad. Of course
if the branching ratio is small the correlation may not be observable
in the background of non-resonating processes. Thus the absence of observable correlation with a particular channel does not prove the absence of intermediate structure. Examples of this phenomenon are to be
seen in the scattering of protons by ^{58}Ni and by ^{208}Pb. The elastic
scattering in both cases shows very little structure while the inelastic scattering to particular states shows a great deal of structure
and cross-correlations so that much of it is intermediate in character.

Another test can be made if it is possible to observe the decay
of an excited state itself formed by decay from a doorway state resonance. The angular correlation in, for example, a (p,p'γ) between the
emergent proton and gamma ray would determine directly if there is a
sufficiently long lived doorway state.

Another test, which was applied by Mittig et al [9], in the inelastic proton scattering by ^{40}Ca to its 3^- state, is that the fine
structure resonances which are observed when high energy resolution experiments are performed have the same quantum numbers as that of the
doorway state. This is of course a property which has been demonstrated many times for isobar analog resonances in, for example, the experiments of the Duke group.

These expamples of tests emphasize what can be done in experiments
with a given target. But an equally important criterion emerges from
the question: is the physics of the phenomenon interesting? If the phenomenon is an isolated one occuring only in one nuclear system and at
one energy, it is not interesting. It is no a "nuclear property" common to many nuclei and whether or not it does or does not exhibit intermediate structure is not worth investigating. But if it has a systematic dependence on the nature of the target or residual nucleus and/
or on the nature of the entrance or exit channels it presents an interesting property of nuclear structure. It is often possible to construct models which illuminate the nature of the process and indicate
under what circumstances it might be observed. Once such a systematic
behavior is demonstrated discussions as to its statistical significance bocome unnecessary.

Mahaux and many others, myself included, have emphasized the model
depedence of the doorway state. However it would be more accurate to
say that the doorway state is representation dependent as we shall now
indicate. Fundamental to the concept of the doorway state is the possi-

bility of analyzing the wave function for a nuclear system into components of successively increasing complexity. The notion of complexity
is however representation dependent. A vibrational state which would
be considered as a simple excitation in terms of the photon model is
quite complicated when expressed in terms of the states of the spherical shell model. The question naturally arises as to how we are to determine the complexity of a given excitation: that is what representation should be used in a particular case? The answer is fairly obvious.
We adopt the convention that the asymptotic states of the particles in
each channel be designated as the simplest states of the system. These
particles will interact without changing their internal state, i.e.
without leaving the channel. Such a picture implies a particular Hamiltonian acting in the channel and a residual interaction which can
only lead to excitations of the particles involved. Increasing complexity is now defined in terms of this residual interaction. A single
application of it on the simplest state will lead to a state of a first
order of complexity. The doorway states associated with this channel
are in this class of wave functions. For purposes of this discussion
we shall refer to these as doorway states of the first order. A state
of the next order of complexity is obtained by applying the residual
interaction for a second time. The resultant wave function will contain some components which are simple or first order in complexity.
These must be projected out so that the remainder which is now of second order in complexity containing doorway states of second order can
be isolated. The process can obviously be continued.

This set of definitions is not model dependent. However in practice it is often necessary as well as convenient to describe the ground
and excited states in a given channel in terms of the representation
provided by a given model. As a consequence the meaning of complexity
will be phrased in terms of the concepts of the model; increasing complexity often corresponding to the presence of an increasing number of
quasi-particles. The model chosen will of course depend upon what we
know of the physics of the nuclei involved. And if this is not too well
known it can be that investigation of the intermediate structure will
help resolve this uncertainty. However it should be clear that the use
of a model is a matter of convenience and not one of principle. The
existence of a hierarchy of states is well known in the case of the dipole resonance. In this conference the observation of doorway states
of second order for an isobar analog resonance were reported by Petrascu.

These qualitative reamarks can be formalized. The results for the circumstances under which the simplest state and the doorway state of first order are considered explicitly while the more complex state wave functions are energy averaged lead to a pair of coupled Schroedinger equations for the appropriate amplitudes

$$(E - H_{oo}) \psi_o = H_{od} \psi_d$$

$$(E - H_{dd}) \psi_d = H_{do} \psi_o$$

(1)

The coupling potentials H_{od} and H_{do} are proportional to the residual interaction of which we spoke earlier. A single doorway state is assumed. If the higher order doorway states are included explicitly and/or if more than one doorway state is considered a corresponding additional number of coupled equations are required. Because of the energy averaging the Hamiltonian matrix in Eq. (1) is generally complex. If the imaginary values particularly of H_{dd} are not large, the spreading width will be corresponding small, a doorway state will be observable leading to observable intermediate structure. The assymetry of the transmission coefficients is also obtained. But if this width is large, that is if Im H_{dd} is large, intermediate structure will not be seen.

An example of this discussion is contained in the reports of Stokstad and Scheid to this conference. Coupled equations like those of Eq. (1) were employed by Asciutto and also by Scheid and Greiner. They however obtained quite different results because these two authors assumed differing values for Im H_{dd}. Stated more positively the nature of intermediate structure can be used to determine properties of the coupling Hamiltonian matrix, the observability being particularly sensitive to the ratio of Im H_{dd} to H_{od}.

The single channel optical model is obtained if an additional energy average is performed, the width of the averaging function being larger than the width of the intermediate structure. This optical model can then predict the average energy and therefore dependence of the elastic cross-section and by a simple suitable generalization the average energy dependence of reaction cross-sections. It in fact should be possible to determine the parameters of the optical potential from cross-sections obtained in the neighborhood bordering on the region with intermediate structure.

Under what circumstances will intermediate structure and thus the related doorway states be most readily observable? Since the doorway state has good quantum numbers such as angular momentum and parity the corresponding structure will be present only in those partial waves which have these quantum numbers. Since generally many channels and partial waves contribute to the cross-section the doorway state contribution is often a small component of the total amplitude. It will thus be more easily observable if the experiment is designed to select a particular channel and even further a particular partial wave.

The canonical examples, the giant dipole resonance and isomeric fission certainly have this selectivity property. The observability of the giant dipole resonance follows from the fact that the incident gamma ray will selectively excite states which are related to the ground state of the target nucleus according to the angular momentum, parity and isospin of the multipole involved in the gamma absorption. In the electric dipole case the gamma-ray carries a unit angular momentum, odd parity and unit isospin.

In the case of isomeric fission the selectivity occurs simply because the fission channel is so easily identified. In the neutron channel the effect of the intermediate structure is not noticeable.

The condition of specificity is not satisfied by the analog resonance case. Here it is the very narrow width of the resonance which permits its observation if sufficiently good energy resolution prevails. The narrowness of the resonance originates in the approximate validity of a symmetry principle - isospin conservation. Observation of such narrow intermediate structure will usually signify the operation of a symmetry principle and may be one way to discover such symmetry principles.

From reports presented at this conference some general experimental situations in which selectivity can be exploited to improve the observability of intermediate structure can be abstracted. As emphasized by Mme. Papineau the principle of selectivity tells us that intermediate structure can be more easily observed in an exit rather than an entrance channel, a point which was originally made by Bloch, Cindro and Harar [2]. Examples discussed at this conference include the isomeric fission case referred to above and included by Mittig et al. and Sperber in their reports. Inelastic scattering has furnished several expamples in the past; the elastic channel exhibiting little structure but the inelastic channels having a great deal. In this conference the cross-section for inelastic scattering of protons by ^{40}Ca reported by Mittig et al. [9] showed intermediate structure. It was moreover possible to resolve the fine structure and thus obtain an estimate of the spreading width.

A single channel can also be isolated even in the case of elastic scattering if a phase shift analysis can be made or if for some reason

only a single partial wave is involved in the reaction. A particularly
important case of the latter is discussed by Stokstad, namely the heavy ion reaction

$$^{16}O + ^{16}O \rightarrow ^{12}C + ^{20}Ne \qquad (2)$$

going to the ground state and first excited of ^{20}Ne and to the unresolved sum of the 4^+ state in ^{20}Ne and the 2^+ in ^{12}C. The energy range
explored was between 17.5 and 30 MeV in the center of mass reference
frame. Strong angular and cross-channel correlations were found. The
angular distribution was found to be almost pure P_L^2 where L is 16 for
one set and 18 for another. These particular L waves could be traced as
a function of the energy and were found to be present within energy
bands of 350 keV width. (Interestingly these bands do not coincide with
the gross-structure peaks in the elastic scattering channel.) Structure
within the peaks of reaction cross-section for process (2) was found.
This was interpreted as intermediate structure since this structure
was present when the final state in ^{20}Ne was the ground, the 2^+ or 4^+
excited states. Moreover the peaks for these partial cross-sections
coincided.

Heavy ion transfer reactions as indicated by these results seem
to be a fruitful area to search for intermediate structure. The selectivity principle which is effective here is a consequence of angular
momentum matching: when the angular momentum of the system before and
after the collision agree the cross-section will be a maximum. Such a
selectivity principle is important at high excitation energies for then
the width of most of the possible candidates for doorway states are
greatly broadened by the large density of open channels into which the
state can decay. Stokstad points to this large density as a possible
mechanism responsible for the absence of intermediate structure below
the Coulomb barrier in a large variety of heavy ion systems in spite
of its appearance in the $^{12}C + ^{12}C$ and $^{16}O + ^{16}O$ reactions. This mechanism becomes more effective as the excitation energy increase. In the
case of heavy ion reactions this is countered by the angular momentum
selectivity which in turn permits the selection of the high angular
states for which the density of open channels is much reduced compared
with the density of low angular momentum states. Heavy ion reactions
provide thus a particularly unique tool for the observation of intermediate structure energies.

The method of phase shift analysis is most easily applied when
each of the scattering particles have zero spin. An example of this me-

thod is given in the paper by Singh et al but was not used in the work
described at this conference. One of the difficulties of the method is
the possible lack of uniqueness of the phase shift analysis.

Intermediate structure may also become visible in kinematic regions (that is angle and energy) in which the non-resonating component
of the cross-section is reduced, for example because of destructive interference. Such cancellations are most often present for large scatte-
ting or reaction angles as can be seen from the following amplitude for
elastic scattering of spin zero particles:

$$\sum_{\ell} (2\ell + 1) \, t_\ell \, P_\ell \, (\cos \theta)$$

Since $P_\ell (-1) = (-)^\ell$ considerable cancellation among the amplitudes for
different partial waves can occur at some large angle depending upon
the ℓ dependence of t_ℓ. Resonant effects which are restricted to a particular ℓ have thus an improved visibility. The paper of Strzalkowski
on back angle alpha particle scattering offers a number of possible
examples of this phenomenon. Back angle enhancement is seen in a number
of nuclei and resonances at alpha particle energies between 20-30 MeV
with widths of the order of several hundred keV. Detailed examples were
discussed by Mayer-Böricke and by Drentje in contributions to this conference. Rinat (Reiner) has proposed that the structure is a consequence of rotational quasi-molecular doorway states [10]. Agassi and Wall
[1] present an alternative mechanisms, namely the exchange scattering
with alpha particle clusters on the nuclear surface. Note that in
either explanation some sort of collective four particle groupings are
indicated.

The non-resonant amplitudes obviously interfere destructively in
a diffraction minimum. It was pointed out in a discussion with Brentano
that examining the energy dependence of the cross-section at such a minimum might reveal an anomaly which appears in a given partial wave.

Another kind of mechanism which Gerumb and Strobel claim explains
a number of results in the inelastic scattering of protons by ^{16}O involves the doorway state as an intermediate state in a two step process. Direct reactions are often thought of as "single step" processes
since the DWBA often gives such an excellent description of the cross-
section and with a substantial value for the spectroscopic factor. However in some cases, particularly for large angles or for particular
kinds of residual nuclear states, two step processes must be included
before an accurate prediction of the observed cross-section can be obtained. It is reasonable to presume that the first step of the two step

process may in many cases go to an intermediate state of the doorway
type. Thus by observing the energy dependence of this two step component of the cross-section one may be able to observe intermediate structure in a direct process.

As a final remark in this section on the observation of intermediate structure, I would like to return to a remark I made at the 1967
Gordon Conference on nuclear reactions. At that time there was at least
one case of an isobar analog resonance in a nucleus (Z,N) whose parent
state in the nucleus (Z-1, N+1) was in the continuum. As a consequence
neutron scattering from the target (Z-1, N) should show an anomaly at
the appropriate energy. Observation of the expected anomaly should provide insight into the sort of effects one can expect for intermediate
structure and thus perhaps generate new criteria for its observation.
A similar remark applies to the isobar analogs of the giant dipole state.

The above discussion has concentrated mostly upon intermediate
structure in the continuum. Intermediate structure for bound states
particularly in the more excited domain where they are dense can be expected to be present in for example (d,p) and (p,d) reaction cross-sections.

Our discussion now turns to the structure information which can
be obtained from the analysis of intermediate structure. What is the
nature of the corresponding doorway state and what does it reveal with
regard to the nuclear system involved? A number of examples which may
provide such structure information and which were discussed at this
conference are described below.

Intermediate structure seems to be more visible when the target·
(or residual) nucleus has a collective state. The doorway state formed
by the coupling of the incident (or exit) particle and this collective
state appear to be weakly coupled to the continuum states. (See for example the ^{40}Ca(p,p′) experiment referred to above). The underlying reason for this seems to me to be of very great interest. Apparently most
of the strength of a particular spin and parity and possibly other properties are concentrated in the collective state and thus is only weakly present in other states.

A second example suggests that long lifetimes of a system composed
of two heavy ions can be achieved in a state such that the ions rarely
interpenetrate. If as a consequence there is relatively weak coupling
of this state to other channels, and if the angular momentum of the state is large the ions may move in stable orbits and form a nuclear molecule. The coupling to other channels will then produce the fine struc-

ture which is observed with improved resolution.

Another suggestion (see paper by Scheid and reference to Imanishi [7] by Stokstad) depends upon the deformability of the ions involved. When the ions are deformed, states of some stability compared to the spherical configurations may become accessible. These can then serve as doorways to a fusion process or more generally to the formation of the compound system.

In still another hypothesis which has been proposed for reactions involving light nuclei, the doorway state is supposed to arise from the state formed when one of the ions goes to an intermediate state in which it is disassociated into alpha particles. The fine structure is generated by the residual interaction between the nucleons. The residual interaction consists of those components of the nuclear force which have not been included in the description of the interaction among alpha particles, between the alpha particles and the ions, and between the ions.

So far in this review we have concentrated on the isolated doorway state. However, historically the situation considered when this terminology was first introduced in the paper with B. Block [3] was that of overlapping doorway state resonances. The Block paper was concerned with the strength function for s-wave neutrons. In it the estimate was made that the strength function is proportional to the density of doorway states near zero neutron energy. This problem was discussed at this conference by Newstead who paid particular attention to the minima in the s and p wave neutron strength functions. In Newstead's calculations the density of doorway states was estimated by counting the number of appropriate three quasi-particle states. Comparison with experiment indicates that Block's suggestion is valid in the neighborhood of the closed shell nuclei.

This brings us to the second theme of this conference - pre-equlibrium decay which is as we shall shortly note very much connected to the statistical theory of doorway states, one application of which we have just described in discussing neutron strength functions. The properties of the low and high energy ends of the spectrum of particles emitted in a reaction, for example the neutron spectrum in a (p,n) reaction, have now been understood for some time. The low energy end can be understood on the basis of the evaporation model while the high energy end involves the direct reaction mechanism and can be understood using the DWBA. Pre-equilibrium decay is concerned with the energy region in-between, where the spectrum cannot be described in terms of either the evaporation or direct reaction models. The procedure which was developed primarily by Griffin [5] has been discussed in the presentations of Blann, Gadioli, Obložinski and Chevarier. With this method an enormous number of reactions have been understood including (p,p'), (α,p), as well as (α, xn) on a variety of targets. I shall make no attempt to describe the details of the method here but refer the reader to these papers for a most adequate review. Suffice it to say that it is based upon the hierarchy of states obtained by successive applications of the residual interaction as described earlier in this report. The probability for forming each of these states from a less complex one is taken to depend upon the density of these states and the square of the magnitude matrix element of the residual interaction between these states. Each of the states can also emit a particle and by adding up their inidividual contributions the total particle spectrum can be obtained. It is clear that the Bohr independence hypothesis is violated, particularly in the early stages of this process, so that the nuclear temperature for the process will depend upon not only the excitation energy in the residual nucleus as is the case in the evaporation model, but also upon the incident kinetic energy. Moreover the angular distribution will no longer be isotropic as is the case for the evaporation model.

In the generalized Griffin method as described by Blann and Gadioli at this meeting the calculation of the relative probabilities of the various states is made along semi-classical kinetic theory lines. The parameters entering are essentially the number of degrees of freedom n_o involved in the first excitation (the doorway state), the density of states, ρ, taken for each excitation type and the average matrix element M connecting excitation types of differing complexity. The density of states is taken from Ericson's review article. As we have said earlier the results obtained with this theory using only two parameters are remarkably good considering its crudity.

Let me criticize the theory first at a relatively superficial level. The models which have been used are for the most part not able to predict the angular distribution of the particles. There is one model which can,but then the predictions are poor. Second, no account is taken of the influence of special resonances like the isobar analog resonance. These are for example clearly visible in the (p,n) data of the Livermore group [6]. Third, in the "cascade" model the kinematics of the nucleon-nucleon collision, the excitation mechanism in that model, are assumed to be that of free paritcles. This assumption is approximately valid for high energy incident particles but is difficult to justify at the relatively low energies at which the theory has been appli-

ed in the papers presented at this conference. At a more trivial level
the mean free path which has been used to describe the motion of a nucleon through the nucleus has been based on Goldberger's kinetic theory model. It would seem more correct to employ the empirical values
provided by the optical model.

Within the framework of these theories it would also be useful to
determine the sensitivity of the results to various possible assumptions regarding ρ. The ad-hoc quality of the assumptions made in some
cases regarding n_o and also the probability of emission of composite
particles clearly suggest the need for further theoretical investigations.

These considerations are of course well known to the practitioners
at this conference. At a more fundamental level, the interesting problem of incorporating the pre-equilibrium theory into the quantum mechanical reaction theory remains. One is reminded of a similar problem
with respect to the evaporation model. As first formulated by Weisskopf
the model was based on kinetic theory, much like those discussed at
this conference. The modern approach (see for example the discussion
by Hauser) is able to obtain the main results of the evaporation model
by applying the assumptions of the statistical modes to the quantum reaction cross-sections. It would appear that a generalization of the method would provide the means for deriving the pre-equilibrium model. A
hint that this is in fact the case is furnished by the analysis of Grimes et al [6] of their experiments. They indicate that their experimental results can be understood by presuming the presence of overlapping
doorway states and applying the statistical assumption, the results
for this case having been obtained by this author together with Kerman
and Lemmer [4]. By assuming a power law state density for the doorway
states the correct spectrum was obtained as well as the prediction that
the angular distribution would be symmetric about 90° in agreement with
experiment.

This concludes the summary. It was not possible to describe many
important contributions, even with the restriction to two major themes
as stated at the beginning of the summary. However, if I have been able
to transmit the excitement which prevailed throughout this meeting my
major goal for the summary would have been achieved.

References

1. AGASSI, D. and WALL, N.S., to be published.
2. BLOCH, C., CINDRO, N. and HARAR, S., Progr. Nucl. Phys., $\underline{10}$ 77 (1968).
3. BLOCK, B. and FESHBACH, H., Ann. Phys. (N.Y) $\underline{23}$ 47 (1963).
4. FESHBACH, H., KERMAN, A.K. and LEMMER, R.H., Ann. Phys. $\underline{41}$ 230 (1967).
5. GRIFFIN, J.J., Phys. Rev. Lett. $\underline{17}$ 478 (1966).
6. GRIMES, S.M., ANDERSON, J.D., Mc CLURE, J.W., POHL, B.A. and WONG, C., Phys. Rev. $\underline{C3}$ 645 (1971); $\underline{C4}$ 607 (1971) and $\underline{C6}$ 236 (1972).
7. IMANISHI, B., Nucl. Phys. $\underline{A125}$ 33 (1969).
8. MALMIN, R.E., SIEMSSEN, R.H., SINK, D.A. and SINGH, P.P., Phys. Rev. Lett. $\underline{28}$ 1590 (1972).
9. MITTIG, W., CASSAGNOU, Y., CINDRO, N., PAPINEAU, L. and SETH, K.K., to be published (1972).
10. RINAT, A.S. (Reiner), Phys. Lett. $\underline{38B}$ 281 (1972).

Seminars Held at the Conference

1. BLEULER, K., New Trends in Nuclear Physics
2. BOHLEN, H.G., Molecular Scattering States of Valence Nucleons in the Elastic Scattering of Heavy Ions
3. BRZOSKO, J., Compound Nucleus Exit States
4. CHEVARIER, N., Pre-Compound Emission for α and Proton Induced Reactions at Medium Energies
5. ČAPLAR, R., The Analysis of (n,α) and (p,α) Spectra by the Pre-Equilibrium Model
6. DERRIEN, H., Statistical Tests for the Existence of Intermediate Structure in ^{239}Pu Neutron Cross-Sections
7. GADIOLI, E., Pre-Equilibrium Emission in Proton Induced Reactions at Intermediate Energy
8. HATEGAN. C., The Structure of the Anomaly Observed at the Threshold of the Analogous Channel
9. LOVAS, I., A Three-Body Model for the Intermediate Processes in Nuclear Reactions
10. MITTIG, W., Intermediate Structure in the Reaction ^{40}Ca(p,p')
11. NEWSTEAD, C., The Influence of Intermediate Reactions on the Neutron Strength Function
12. OBLOŽINSKY, P., Pre-Equilibrium Processes in (α,n) Reactions
13. STRZALKOWSKI, A., Resonance in Backward Scattering of Alpha Particles

Lecture Notes in Physics

Selected Issues from

Lecture Notes in Mathematics

Vol. 7: Ph. Tondeur, Introduction to Lie Groups and Transformation Groups. Second edition. VIII, 176 pages. 1969. DM 16,–

Vol. 40: J. Tits, Tabellen zu den einfachen Lie Gruppen und ihren Darstellungen. VI, 53 Seiten. 1967. DM 16,–

Vol. 52: D. J. Simms, Lie Groups and Quantum Mechanics. IV, 90 pages. 1968. DM 16,–

Vol. 55: D. Gromoll, W. Klingenberg und W. Meyer, Riemannsche Geometrie im Großen. VI, 287 Seiten. 1968. DM 20,–

Vol. 56: K. Floret und J. Wloka, Einführung in die Theorie der lokalkonvexen Räume. VIII, 194 Seiten. 1968. DM 16,–

Vol. 81: J.-P. Eckmann et M. Guenin, Méthodes Algébriques en Mécanique Statistique. VI, 131 pages. 1969. DM 16,–

Vol. 82: J. Wloka, Grundräume und verallgemeinerte Funktionen. VIII, 131 Seiten. 1969. DM 16,–

Vol. 89: Probability and Information Theory. Edited by M. Behara, K. Krickeberg, and J. Wolfowitz. IV, 256 pages. 1969. DM 18,–

Vol. 91: N. N. Janenko, Die Zwischenschrittmethode zur Lösung mehrdimensionaler Probleme der mathematischen Physik. VIII, 194 Seiten. 1969. DM 16,80

Vol. 103: Lectures in Modern Analysis and Applications I. Edited by C. T. Taam. VII, 162 pages. 1969. DM 16,–

Vol. 128: M. Takesaki, Tomita's Theory of Modular Hilbert Algebras and its Applications. II, 123 pages. 1970. DM 16,–

Vol. 140: Lectures in Modern Analysis and Applications II. Edited by C. T. Taam. VI, 119 pages. 1970. DM 16,–

Vol. 144: Seminar on Differential Equations and Dynamical Systems II. Edited by J. A. Yorke. VIII, 268 pages. 1970. DM 20,–

Vol. 167: Lavrentiev, Romanov and Vasiliev, Multidimensional Inverse Problems for Differential Equations. V, 59 pages. 1970. DM 16,–

Vol. 170: Lectures in Modern Analysis and Applications III. Edited by C. T. Taam. VI, 213 pages. 1970. DM 18,–

Vol. 183: Analytic Theory Differential Equations. Edited by P. F. Hsieh and A. W. J. Stoddart. VI, 225 pages. 1971. DM 20,–

Vol. 193: Symposium on the Theory of Numerical Analysis. Edited by J. Ll. Morris. VI, 152 pages. 1971. DM 16,–

Vol. 198: M. Hervé, Analytic and Plurisubharmonic Functions in Finite and Infinite Dimensional Spaces. VI, 90 pages. 1971. DM 16,–

Vol. 206: Symposium on Differential Equations and Dynamical Systems. Edited by D. Chillingworth. XI, 173 pages. 1971. DM 16,–

Vol. 214: M. Smorodinsky, Ergodic Theory, Entropy. V, 64 pages. 1971. DM 16,–

Vol. 228: Conference on Applications of Numerical Analysis. Edited by J. Ll. Morris. X, 358 pages. 1971. DM 26,–

Vol. 230: L. Waelbroeck, Topological Vector Spaces and Algebras. VII, 158 pages. 1971. DM 16,–

Vol. 233: C. P. Tsokos and W. J. Padgett, Random Integral Equations with Applications to Stochastic Systems. VII, 174 pages. 1971. DM 18,–

Vol. 235: Global Differentiable Dynamics. Edited by O. Hájek, A. J. Lohwater, and R. McCann. X, 140 pages. 1971. DM 16,–

Vol. 240: A. Kerber, Representations of Permutation Groups I. VII, 192 pages. 1971. DM 18,–

Vol. 243: Japan-United States Seminar on Ordinary Differential and Functional Equations. Edited by M. Urabe. VIII, 332 pages. 1971. DM 26,–

Vol. 247: Lectures on Operator Algebras. Tulane University Ring and Operator Theory Year, 1970–1971. Volume II. XI, 786 pages. 1972. DM 40,–

Vol. 257: R. B. Holmes, A Course on Optimization and Best Approximation. VIII, 233 pages. 1972. DM 20,–

Vol. 261: A. Guichardet, Symmetric Hilbert Spaces and Related Topics. V, 197 pages. 1972. DM 18,–

Vol. 267: Numerische Lösung nichtlinearer partieller Differential- und Integro-Differentialgleichungen. Herausgegeben von R. Ansorge und W. Törnig, VI, 339 Seiten. 1972. DM 26,–

Vol. 275: Séminaire Pierre Lelong (Analyse) Année 1970–1971. VI, 181 pages. 1972. DM 18,–

Vol. 276: A. Borel, Représentations de Groupes Localement Compacts. V, 98 pages. 1972. DM 16,–

Vol. 277: Séminaire Banach. Edité par C. Houzel. VII, 229 pages. 1972. DM 20,–

Vol. 280: Conference on the Theory of Ordinary and Partial Differential Equations. Edited by W. N. Everitt and B. D. Sleeman. XV, 367 pages. 1972. DM 26,–

Vol. 282: W. Klingenberg und P. Flaschel, Riemannsche Hilbertmannigfaltigkeiten. Periodische Geodätische. VII, 211 Seiten. 1972. DM 20,–

Vol. 284: P.-A. Meyer, Martingales and Stochastic Integrals I. VI, 89 pages. 1972. DM 16,–

Vol. 285: P. de la Harpe, Classical Banach-Lie Algebras and Banach-Lie Groups of Operators in Hilbert Space. III, 160 pages. 1972. DM 16,–

Vol. 293: R. A. DeVore, The Approximation of Continuous Functions by Positive Linear Operators. VIII, 289 pages. 1972. DM 24,–

Vol. 294: Stability of Stochastic Dynamical Systems. Edited by R. F. Curtain. IX, 332 pages. 1972. DM 26,–

Vol. 300: P. Eymard, Moyennes Invariantes et Représentations Unitaires. II, 113 pages. 1972. DM 16,–

Vol. 301: F. Pittnauer, Vorlesungen über asymptotische Reihen. VI, 186 Seiten. 1972. DM 18,–

Vol. 307: J. L. Bretagnolle, S. D. Chatterji et P.-A. Meyer, Ecole d'été de Probabilités: Processus Stochastiques. VI, 198 pages. 1973. DM 20,–

Selected Issues from

Springer Tracts in Modern Physics

Vol. 45: P. D. B. Collins and E. J. Squires, Regge Poles in Particle Physics. VII, 292 pages. 1968. DM 78,–

Vol. 50: Current Algebra and Phenomenological Lagrange Functions. International Summer School, Karlsruhe 1968. VII, 156 pages. 1969. DM 44,–

Vol. 52/53: Weak Interactions, International Summer School, Karlsruhe 1969.

Vol. 52: VII, 214 pages. 1970. DM 58,–

Vol. 53: V, 106 pages. 1970. DM 38,–

Vol. 55: Low Energy Hadron Interactions, Compilation of Coupling Constants, International Meeting, Ruhestein 1970. VII, 290 pages. 1970. DM 78,–

Vol. 57/60: Strong Interaction Physics, Heidelberg-Karlsruhe International Summer Institute 1970.

Vol. 57: VII, 270 pages. 1971. DM 78,–

Vol. 60: V, 233 pages. 1971. DM 78,–

Vol. 59: Symposium on Meson-, Photo-, and Electroproduction at Low and Intermediate Energies, Bonn 1970. VI, 222 pages. 1971. DM 78,–

Vol. 62/63: Photon-Hadron Interactions, International Summer Institute, Desy 1971.

Vol. 62: VII, 147 pages. 1972. DM 58,–

Vol. 63: VII, 189 pages. 1972. DM 78,–